International Workshop: Meson Production at Intermediate and High Energies 2011

Journal of Physics: Conference Series Volume 349

Messina, Italy
10-11 November 2011

Editors:

Giorgio Giardina
Fabio Bossi
Paolo Levi Sandri
Paolo Pedroni
Hartmut Schmieden

ISBN: 978-1-62276-695-6
ISSN: 1742-6588

Printed from e-media with permission by:

Curran Associates, Inc.
57 Morehouse Lane
Red Hook, NY 12571

Some format issues inherent in the e-media version may also appear in this print version.

Printed by Curran Associates, Inc. (2013)

For permission requests, please contact the Institute of Physics
at the address below.

Institute of Physics
Dirac House, Temple Back
Bristol BS1 6BE UK

Phone: 44 1 17 929 7481
Fax: 44 1 17 920 0979

techtracking@iop.org

Additional copies of this publication are available from:

Curran Associates, Inc.
57 Morehouse Lane
Red Hook, NY 12571 USA
Phone: 845-758-0400
Fax: 845-758-2634
Email: curran@proceedings.com
Web: www.proceedings.com

TABLE OF CONTENTS

INVITED TALKS

ORAL CONTRIBUTIONS

Preface

The International Workshop 'Meson Production at Intermediate and High Energies' was held in the 'Capo Peloro Resort' Hotel in Messina, Italy on November 10–11, 2011. The workshop was organized by the University of Messina and 'Fondazione Bonino-Pulejo', in the wonderful setting of the confluence between the Ionian and Tyrrhenian seas, the center of the ancient historical and mythological civilizations of the Mediterranean countries.

The main purpose of this workshop was to deal with aspects of electromagnetic and strong forces by meson photoproduction and the electron–positron collider, and to search for dark energy. The subjects covered at the workshop in Messina involved the main activities of the laboratories of Europe and countries overseas.

The topics included:

- Baryon spectroscopy and 'missing resonances'
- Polarization observables
- Pseudoscalar and vector meson production through e.m. and hadronic reactions
- Hadron cross section measurements
- Measurements with polarized target or/and beam

Editors:
Giorgio Giardina (University of Messina)
Fabio Bossi (INFN – Laboratori Nazionali di Frascati)
Paolo Levi Sandri (INFN – Laboratori Nazionali di Frascati)
Paolo Pedroni (INFN – Sezione di Pavia)
Hartmut Schmieden (University of Bonn)

Organizing Committee:
Chairman: G Giardina (Messina - Italy)
Co-Chairman: F Bossi (Frascati - Italy)
Co-Chairman: P Levi Sandri (Frascati - Italy)
Co-Chairman: P Pedroni (Pavia - Italy)
Co-Chairman: H Schmieden (Bonn - Germany)

Scientific Secretary: G Mandaglio (University of Messina – Italy)

Journal of Physics: Conference Series **349** (2012) 011001 doi:10.1088/1742-6596/349/1/011001

Local Organizing Committee:
F Curciarello, V De Leo, G Fazio, G Giardina, G Mandaglio and M Romaniuk

Organizing Institutions:

University of Messina | Fondazione Bonino-Pulejo (Messina)

Sponsored by:
University of Messina, Fondazione Bonino-Pulejo (Messina) and INFN Sezione di Catania

Web-Site: http://newcleo.unime.it/workshop2011

Journal of Physics: Conference Series **349** (2012) 011001 doi:10.1088/1742-6596/349/1/011001

All papers published in this volume of *Journal of Physics: Conference Series* have been peer reviewed through processes administered by the proceedings Editors. Reviews were conducted by expert referees to the professional and scientific standards expected of a proceedings journal published by IOP Publishing.

International Workshop: Meson Production at Intermediate and High Energies IOP Publishing
Journal of Physics: Conference Series **349** (2012) 012001 doi:10.1088/1742-6596/349/1/012001

Search for Dark Forces at KLOE and KLOE-2

Fabio Bossi for the KLOE-2 Collaboration
Laboratori Nazionali di Frascati INFN, via Enrico Fermi 40, 00044 Frascati, Italy
E-mail: fabio.bossi@lnf.infn.it

Abstract. Following recent puzzling astrophysical results, searches for a relatively low mass ($\sim$ 1 GeV) new vector boson (the U), weakly coupled with SM particles, are being pursued in several different laboratories in the world. In particular the KLOE-2 Collaboration has searched for this new particle in the decays of the ϕ meson into an η and an e^+e^- pair, with null result. An upper limit on the existence of the U has been set, in the mass range $50 < M_U < 420$ MeV.

1. Introduction

The lagrangian of the Standard Model (SM) of particle physics is built to obey the $SU(2)_L \otimes U(1)_Y \otimes SU(3)_C$ gauge symmetry. Massless matter fermions are charged under at least one of the above symmetries, which are mediated by the proper set of massless vector bosons. Spontaneous symmetry breaking induced by the Higgs potential provides then mass to both the matter fields and to the vector bosons. This theory perfectly describes all of the known reactions of elementary particles observed to date.

On the other hand, in the context of newtownian gravity, there is compelling astrophysical evidence of the existence of Dark Matter (DM), whose nature is still undiscovered. The only thing we know about DM is that it is *not* composed of SM particles, and that is sensitive to gravitation. Actually, there is no fundamental reason not to assume that matter fields different from the SM ones are sensitive to different types of gauge symmetries, and thus coupled to different types of vector bosons. If (at least some of) these new matter fields are also coupled to SM gauge bosons (for instance they have electric charge), higher order loop diagrams can allow SM particles to couple the new symmetries, even if they are not charged under them.

The above ideas were already developed in the early 80's in the framework of supersymmetric extensions of the SM [1] [2], but have in recent years become particularly popular since they might allow explaining some intriguing and thus far unexplained experimental astrophysical results. A further reason for these models to be particularly attractive is that they can induce observable effects in present day collider experiments, also at energies of order 0.1 - 10 GeV.

In this paper I will firstly briefly present some of the most interesting astrophysical puzzles of the recent years, and their possible explanation in terms of the existence of a new hidden gauge symmetry. I will then discuss some of the possible collider signatures of this new symmetry, finally concentrating on the recent KLOE-2 limit on the existence of a "dark photon" using ϕ meson Dalitz decays.

2. Experimental results and their possible explanation

Several recent experiments, as for instance PAMELA [3], FERMI [4], and ATIC [5] have observed in cosmic ray data a large excess of electrons and positrons with energies between approximately 10 and 100 GeV, with respect to what can be accounted by supernova shocks and interactions of cosmic ray protons with the interstellar medium.

The INTEGRAL satellite [6] observes a 511 keV signal from the galactic core, which suggsts the existence of an abundant positron annhilation source, far exceeding what expected from supernovae only.

Another long standing puzzle is the annual modulation signal reported by the DAMA/LIBRA experiment in the Gran Sasso laboratory [7], that is consistent with what expected by nuclear scattering of WIMP dark matter. The CoGent Collaboration observes an excess of events in their lowest energy electrons bins, with a possible annual modulation in the signal [8]. However, other experiments [9] using different detection techniques contradict both DAMA and CoGent.

All in all, these observations do not have easy interpretations in terms of standard astrophysical and/or particle physics processes, which makes their study an intringuing task per se. More interestingly, however, there have appeared in literature papers [10], [11], [12] [13], [14] arguing that they can all be interpreted by some common physical process, i.e. by the same model of DM production and annihilation. Although these papers differ between each other by some specific detail, all of them in general postulate the existence of relatively heavy ($\sim$ 1 TeV) WIMP DM states together with at least one relatively light ($\sim$ 1 GeV) vector boson, mediator of a new hidden gauge symmetry. SM particles are *not* charged under this new symmetry, however they can still couple with the "dark photon" (baptized under different names in the literature: A', U, ϕ...) through the kinetic mixing mechanism with ordinary SM bosons, and specifically with the photon. Typically, the mixing strength is parametrized by a single parameter ϵ, whose value has to be determined experimentally. In order to better accommodate the above mentioned experimental results, however, preferred values of ϵ are in the ball-park of 10^{-3}. As a consequence of that, the U can be produced and observed at present day colliders depending on its mass and on the value of ϵ, as discussed in the following section.

Before discussing the collider signatures of the U boson, it is important to underline that another intriguing aspect of the models discussed above, is the role that they can play in explaining the observed $\sim 3\sigma$ discrepancy between the measured and calculated values of the muon g-2 (see for instance [15]). Actually, the presence of diagrams of U boson exchange can increase the theoretical prediction towards the measured value, for properly choosen M_U and ϵ.

3. U boson production mechanisms

The U boson can be produced in e^+e^- collisions via the radiative reaction $e^+e^- \to U\gamma$, with subsequent decay of the U into a lepton pair. If the two leptons are charged, it can be observed as a resonant peak of the lepton pair invariant mass distribution over the standard continuous QED background. An estimate based on a *BaBar* measurement of Υ(2S,3S) decays into $\mu^+\mu^-\gamma$, gives null result. The analysis was actually motivated by the search for a light scalar particle [16], which should have in fact a different acceptance with respect to a vector one. With some caveat one can still translate the *BaBar* null result into a limit of $\sim 10^{-3}$ on the mixing parameter in the range $2m_\mu$ - 9.3 GeV.

U bosons can be produced in electron collisions on a fixed target in a process analogous to ordinary bremsstrahlung. In this case, production cross sections are much higher with respect to e^+e^- processes. However backgrounds, both from ordinary QED reactions and from possible beam related sources are also higher. A comprehensive discussion of the possible detection strategies for this kind of experiments can be found in [17]. Schematically, beam dump experiments are useful to probe the region of low masses and very low couplings, a condition which favours relatively long U lifetimes, while for higher masses and couplings forward spectrometers with the best possible vertex resolution are required.

Some old beam dump experiments is actually being data-mined, while new experimental activities have started in several laboratories all over the world. In particular, during the last year two papers have been published by the MAMI A1 experiment in Mainz (Germany) [18], and by the APEX experiment at JLAB (USA) [19], setting limits on the existence of the U boson with mass between approximately 200 and 300 MeV, for ϵ down to $\sim 10^{-3}$.

A further line of reasearch also available at e^+e^- colliders is the study of the decays of a vector meson into a pseudoscalar and a U, as suggested by Reece and Wang [20]. This decays should occur at a rate suppressed by a factor ϵ with respect to the standard radiative ones, which have typical branching ratios of $\sim$1%. In particular Reece and Wang have focussed their attention on the channel $\phi(1020) \to \eta U$. With the statistics acquired so far by the KLOE experiment at the DAΦNE facility in Frascati, they have argued that one could probe mixing parameters down to $\sim 10^{-3}$, for U masses below $m_\phi - m_\eta \sim 470$ MeV. This search has actually been performed by the KLOE-2 Collaboration, as described in the rest of the paper.

4. The KLOE experiment

The KLOE experiment operated from 2000 to 2006 at DAΦNE, the Frascati ϕ-factory. DAΦNE is an e^+e^- collider running at a center-of-mass energy of $\sim$ 1020 MeV, the mass of the ϕ meson. Equal energy positron and electron beams collide at an angle of π-25 mrad, producing ϕ mesons nearly at rest. The detector consists of a large cylindrical Drift Chamber (DC), surrounded by a lead-scintillating fiber electromagnetic calorimeter (EMC). A superconducting coil around the EMC provides a 0.52 T field. The beam pipe at the interaction region is spherical in shape with 10 cm radius, it is made of a Beryllium-Aluminum alloy of 0.5 mm thickness. Low beta quadrupoles are located at about $\pm$50 cm distance from the interaction region. The drift chamber [21], 4 m in diameter and 3.3 m long, has 12,582 all-stereo tungsten sense wires and 37,746 aluminum field wires. The chamber shell is made of carbon fiber-epoxy composite with an internal wall of $\sim$ 1 mm thickness, the gas used is a 90% helium, 10% isobutane mixture. The spatial resolutions are $\sigma_{xy} \sim 150$ μm and $\sigma_z \sim 2$ mm. The momentum resolution is $\sigma(p_\perp)/p_\perp \approx 0.4\%$. Vertexes are reconstructed with a spatial resolution of $\sim$ 3 mm. The calorimeter [22] is divided into a barrel and two endcaps, for a total of 88 modules, and covers 98% of the solid angle. The modules are read out at both ends by photomultipliers, both in amplitude and time. The readout granularity is $\sim(4.4 \times 4.4)$ cm^2, for a total of 2440 cells arranged in five layers. The energy deposits are obtained from the signal amplitude while the arrival times and the particles positions are obtained from the time differences. Cells close in time and space are grouped into energy clusters. The cluster energy E is the sum of the cell energies. The cluster time T and position $\vec{R}$ are energy-weighted averages. Energy and time resolutions are $\sigma_E/E = 5.7\%/\sqrt{E\ (\mathrm{GeV})}$ and $\sigma_t = 57\ \mathrm{ps}/\sqrt{E\ (\mathrm{GeV})} \oplus 100$ ps, respectively.

The physics program of KLOE covers a wide range of topics, including tests of discrete symmetries conservation/violation, precision measurement of SM parameters, studies on low energy QCD. A good review can be found in [23].

5. U boson production in ϕ decays

We have studied the process $\phi \to \eta\, U$, using a sample of 1.5 fb^{-1} of data collected in 2004-2005. The U boson is searched for looking at its decay into an electron-positron pair, since $e^\pm$ are easily identified in KLOE using time-of-flight (ToF), while the η meson is tagged by its $\pi^+\pi^-\pi^0$ decay channel, which provides a clean final state with four charged particles and two photons.

Radiative decays of the ϕ into $\eta\gamma$ have a branching ratio of about 1.2%, which translates in a cross section of approximately 40 nb. The photon can convert into an electron-positron pair while traversing some detector material with probability of $\sim$1%, resulting in a final state that mimicks our signal. An irreducible background due to the Dalitz decay of the ϕ meson, is also present, with branching ratio of order 10^{-4}. Differently from our signal, however, it is not resonant. In order to study the effect of these events on our analysis, they have been simulated by the Monte Carlo, using a Vector Meson Dominance model [24], using the form factor parametrization from the SND experiment [25]. Signal production and decay has been simulated according to [20].

Events are preselected requiring the presence of four charged tracks from the interaction point (IP) and two photons. The best match to the η mass of the two photons and two oppositely charged tracks (assumed to be pions) is found, the remaining two tracks are assumed to be an $e^{\pm}$ pair. The recoil mass to the $e^{\pm}$ pair is then computed and only events with $535 < M_{\rm recoil}(ee) < 560$ MeV are retained.

Although a large part of the backgrounds are already rejected at this level, some contamination from photon conversions and from misreconstructed ϕ decay channels still remains. The former are rejected thanks to a specific photon-conversion recognition alghoritm, the latter by identifying fake $e^{\pm}$ by time-of-flight to the calorimeter. In fact this last cathegory of backgrounds is dominated by multi-pionic events.

The analysis efficiency, estimated by MC, ranges between 10 and 20%, depending on the invariant mass value of the $e^{\pm}$ pair.

About 14,000 $\phi \rightarrow \eta e^+e^-$ events survive the cuts, with a negligible background contamination. No evident peak is seen in the invariant mass distribution of the lepton pair, as shown in fig. 1. In order to extract the correct upper limit on the U boson production, an accurate description of the Dalitz decay background is needed. For the purpose a fit is performed on the M_{ee} distribution, with a function taken from [24]. The binning of the fit is of 5 MeV. When estimating the background of a given bin, the fit is performed removing the five bins centered around it. Details of the procedure are found in [26].

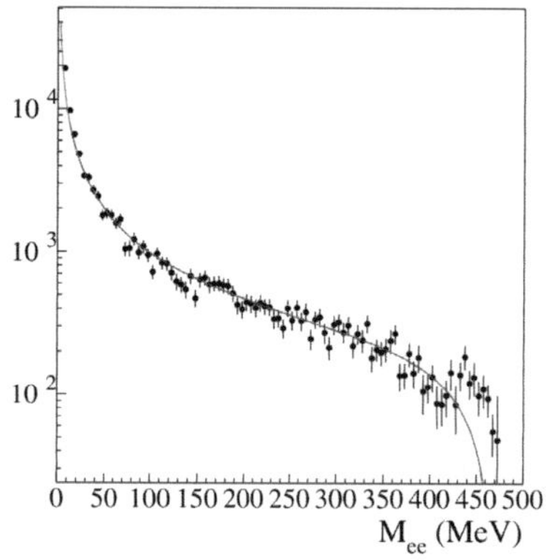

Figure 1. Fit to the corrected M_{ee} spectrum for the Dalitz decays $\phi \rightarrow \eta e^+e^-$.

In fig. 2 the exclusion plot at 90% C.L. on the number of events for the decay chain $\phi \rightarrow \eta U$, $\eta \rightarrow \pi^+\pi^-\pi^0$, $U \rightarrow e^+e^-$, is shown. Taking into account the analysis efficiency this result is then reported in terms of the parameter $\alpha'/\alpha = \epsilon^2$, where α' is the coupling of the U boson to electrons and α is the fine structure constant. The opening of the $U \rightarrow \mu^+\mu^-$ threshold, in the hypothesis that the U boson decays only to lepton pairs and assuming equal coupling to e^+e^- and $\mu^+\mu^-$, has been included. In fig. 3 the smoothed exclusion plot at 90% C.L. on α'/α is compared with existing limits from the muon anomalous magnetic moment a_μ and from recent measurements of the MAMI/A1 [18] and APEX [19] experiments. The gray line is where the

U boson parameters should lay to account for the observed discrepancy between measured and calculated a_μ values.

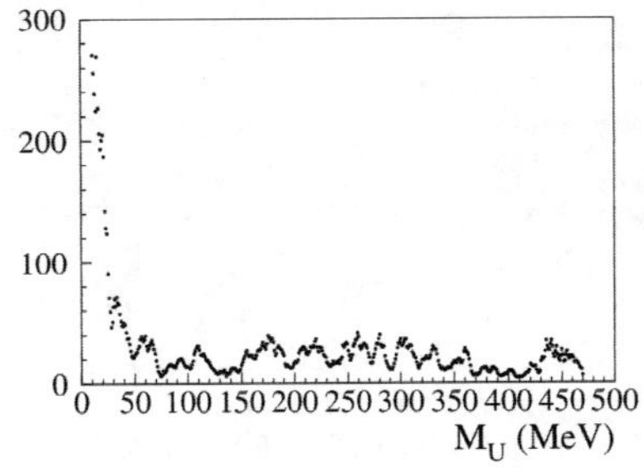

Figure 2. Upper limit at 90% C.L. on the number of events for the decay chain $\phi \to \eta U$, $\eta \to \pi^+\pi^-\pi^0$, $U \to e^+e^-$.

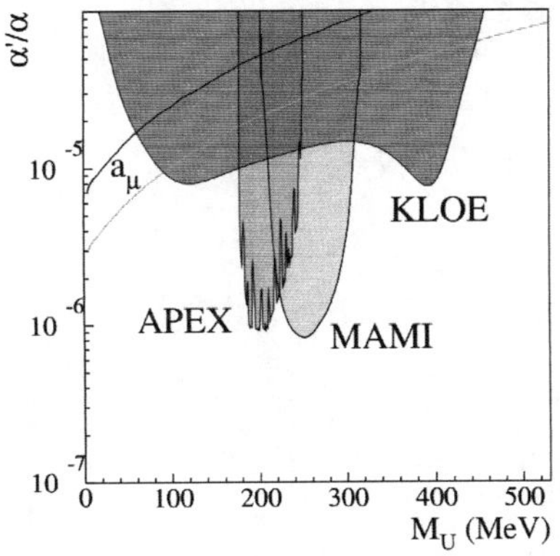

Figure 3. Exclusion plot at 90% C.L. for the parameter $\alpha'/\alpha = \epsilon^2$, compared with existing limits in our region of interest.

Our result greatly improves existing limits in a wide mass range, resulting in an upper limit on the α'/α parameter of $\leq 2 \times 10^{-5}$ @ 90% C.L. for $50 < M_U < 420$ MeV, thus covering part of the expected ϵ range. We exclude that the existing a_μ discrepancy is due to a U boson with mass ranging between 90 and 450 MeV.

6. Future plans

In the described analysis, only the $\pi^+\pi^-\pi^0$ decay channel has been used to tag the presence of the η meson. We are presently studying the possibility to use also other η dominant decay channels, as the 2γ and the $3\pi^0$ ones. Moreover, a new data taking campaing with a slightly modified detector has started with the aim of increasing the acquired luminosity by about an order of magnitude [27]. This will allow us increasing our sensitivity to lower ϵ values in the same mass range.

We are also studying the possibility of using the $e^+e^- \to e^+e^-\gamma$ reaction, which will allow us to test the very low mass region.

References

[1] P. Fayet, 1980 *Phys. Lett.* B **95** 285
[2] P. Fayet, 1981 *Phys. Lett.* B **187** 184
[3] O. Adriani *et al* 2009 *Nature* **458** 607
[4] A.A. Abdo *et al* 2009 *Phys. Rev. Lett.* **102** 181101
[5] J. Chang *et al* 2008 *Nature* **456** 362
[6] P. Jean *et al* 2003 *Astron. Astrophys.* **407** L55
[7] R. Bernabei *et al* 2008 *Eur. Phys. J.* C **56** 333
[8] C.E. Aalseth *et al* 2011 *Phys. Rev. Lett* **107** 141301
[9] E. Aprile *et al* 2011 *Phys. Rev. Lett* **107** 131302
[10] M. Pospelov, A. Ritz, M.B. Voloshin 2008 *Phys. Lett.* B **662** 53
[11] N. Arkani-Hamed, D.P. Finkbeiner, T.R. Slatyer, N. Weiner, 2009 *Phys. Rev.* D **79** 015014
[12] D.S.M. Alves, S.R. Behbahani, P. Schuster, J.G. Wacker, 2010 *Phys. Lett.* B **692** 323
[13] R. Essig, P. Schuster, N. Toro 2009 *Phys. Rev.* D **80** 015003
[14] J.M. Cline and A.R. Frey 2011 *Phys. Rev.* D **84** 075003
[15] F. Jegerlehner and A. Nyffeler 2009 *Phys. Rept.* **477** 1
[16] B Aubert *et al* 2009 *Phys. Rev. Lett.* **103** 081803
[17] J.D. Bjorken, R. Essig, P. Schuster, N. Toro, 2009 *Phys. Rev.* D **80** 075018
[18] H. Merkel *et al* , 2011 *Phys. Rev. Lett.* **106** 251802
[19] S. Abrahamyan *et al* , 2011 *Phys. Rev. Lett* **107** 191804
[20] M. Reece, L.T. Wang, 2009 *JHEP* **07** 051
[21] M. Adinolfi *et al* 2002 *Nucl Inst.and Meth.* A **488** 51
[22] M. Adinolfi *et al* 2002 *Nucl.Inst.and Meth.* A **482** 364
[23] F. Bossi *et al* 2008 *Riv. Nuovo Cim.* **31** 531
[24] L.G. Landsberg 1985 *Phys. Rep.* **128** 301
[25] M. N. Achasov *et al* 2001 *Phys. Lett.* B **504** 275
[26] F Archilli *et al* . 2012 *Phys. Lett.* B **706** 251
[27] G. Amelino Camelia *et al* 2010 *Eur. Phys. J.* C **68** 619

KLOE results on light meson spectroscopy and prospects for KLOE-2

Paolo Gauzzi on behalf of the KLOE-2 Collaboration
Sapienza Università di Roma e INFN Sezione di Roma, P.le A.Moro, 5 I-00185 Rome, Italy
E-mail: paolo.gauzzi@roma1.infn.it

Abstract. The results obtained by the KLOE Collaboration on light meson spectroscopy are presented. The radiative decay $\phi \to \eta\gamma$ have been used to study several η decay channels. The Dalitz plot distributions of the $\eta \to 3\pi$ decays, both in charged and neutral final states have been measured. The box anomaly contribution in $\eta \to \pi^+\pi^-\gamma$ has been investigated in, and the rare decays $\eta \to \pi^0\gamma\gamma$, $\eta \to \pi^+\pi^-e^+e^-$ and $\eta \to e^+e^-e^+e^-$ have been measured. Also the strategy for the measurement of the transition form factor of $\phi \to \eta e^+e^-$ is described. The radiateve process $\phi \to \eta'\gamma$ has been used to study the $\eta' \to \eta\pi\pi$ decay channels, obtaining a measurement of the pseudoscalar mixing angle, and finding an evidence for a gluonium content of η'. The decays $\phi \to PP\gamma$, where P means a pseudoscalar meson, have been exploited to investigate the light scalar mesons, $f_0(980)$, $a_0(980)$, and $\sigma(600)$. The couplings of the scalar mesons to $K\bar{K}$, $\pi\pi$ or $\eta\pi^0$, and to the ϕ resonance have been measured. The prospects for the new KLOE-2 data-taking just started at the upgraded DAΦNE with an upgraded detector are described.

1. Introduction

The KLOE Collaboration took data from 2001 to 2006 at the Frascati ϕ-factory DAΦNE, collecting about 2.5 fb^{-1} at the peak of the $\phi(1020)$ resonance, and 250 pb^{-1} off-peak, mainly at $\sqrt{s} = 1$ GeV. KLOE has covered several aspects of the light meson physics by exploiting all the main decay channels of the ϕ resonance:

(i) kaon physics: test of CKM unitarity and lepton universality, rare decays, tests of CPT violation and of quantum coherence by exploiting the kaon interferometry;

(ii) η, η' decays: tests of Chiral Perturbation Theory (ChPT) and of Vector Meson Dominance (VMD) models, measurement of the pseudoscalar mixing angle;

(iii) study of the low mass scalar mesons, $f_0(980)$, $a_0(980)$ and $\sigma(600)$.

(iv) precision measurement of the hadronic cross-section using the process $e^+e^- \to \pi^+\pi^-\gamma$ when the photon is emitted by an initial lepton (Initial State Radiation), relevant to settle the hadronic contribution to the anomalous magnetic moment of the muon;

(v) hadron production in $\gamma\gamma$ collisions;

The results obtained in items (ii) and (iii) are reported in this paper, while items (iv,v) are covered by other talks in this Workshop[1][2].
In 2008 the Accelerator Division of the LNF tested a new interaction scheme for DAΦNE, aiming to reach an increase of a factor of three in luminosity with respect to the KLOE data taking.

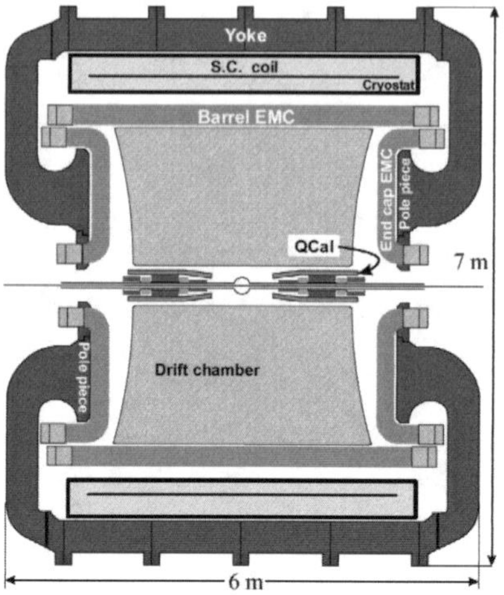

Figure 1. Vertical cross-section of the KLOE detector.

Following the success of this test, a new data-taking campaign of the KLOE experiment (KLOE-2 in the following) with an upgraded detector has been proposed[3], with the goal to collect 20 fb^{-1} of data in the next three years. The detector upgrade is organized as a two step process:

(i) two tagger devices for scattered leptons in $\gamma\gamma$ processes have already been installed in view of the restarting of the data-taking in December 2011;

(ii) a major upgrade consisting in the insertion of an Inner Tracker (IT) and two small angle calorimeters is planned for Summer 2012.

2. The KLOE detector and the KLOE-2 upgrades

The KLOE detector is shown in fig.1. It consists of a large cylindrical drift chamber (DCH) surrounded by an electromagnetic calorimeter (EMC), both immersed in an axial magnetic field of 0.52 T produced by a superconducting solenoid.

The DCH[4], of internal radius 25 cm and external one of 2 m, has about 52000 wires, 12500 of wich are sense wires, with all stereo geometry, and it is filled with a 90% He - 10% isobutane gas mixture. The momentum resolution is $\sigma_p/p \simeq 0.4\%$, and the space resolution is 150 μm in the transverse plane (x,y) and 2 mm along the DCH axis (z).

The EMC[5] has been designed to detect photons in the range 20 - 500 GeV with high efficiency, and with good energy and time resolutions. A fast response of the calorimeter is also an essential requirement since it is used as the main trigger of the experiment. The chosen solution has been a sampling calorimeter made of scintillating fibers, of 1 mm diameter, embedded in groved lead foils 0.5 mm thick, forming a composite with fiber:lead:glue ratio of 48:42:10. The total thickness of 23 cm corresponds to about 15 X_0. The energy and time resolutions are $\sigma_E/E = 5.7\%/\sqrt{E(\mathrm{GeV})}$ and $\sigma_t = 55\ \mathrm{ps}/\sqrt{\mathrm{E(GeV)}} \oplus 140$ ps, respectively.

2.1. Present upgrade

For the forthcoming data-taking a tagger system to detect the scattered leptons from the $\gamma\gamma$ processes $e^+e^- \to e^+e^-\gamma^\star\gamma^\star$ has been installed. It consists of two different devices:

(i) a Low Energy Tagger (LET), made of two small calorimeters of LYSO crystals, readout by SiPM, to detect leptons with energy between 130 and 300 MeV; they are placed near the

Journal of Physics: Conference Series **349** (2012) 012002 doi:10.1088/1742-6596/349/1/012002

Interaction Point (IP), inside the main KLOE detector and the required energy resolution is $\sigma_E/E < 10\%$ for $E > 150$ MeV;

(ii) a High Energy Tagger (HET) to detect leptons with $E > 400$ MeV, consisting in two scintillator hodoscopes readout by PMT's, placed after the first bending dipoles of DAΦNE, about 11 m far from the IP; given the energy-position correlation the 5 mm pitch corresponds to an energy resolution of 2.5 MeV.

2.2. *Future upgrade*

A major upgrade is planned for Summer 2012:

(i) an Inner Tracker (IT) made of four layers of cylindrical triple GEM will be installed between the beam-pipe and the DCH, to improve the acceptance for low momentum tracks;

(ii) two Crystal Calorimeters (CCALT) will cover the low polar angle regions to increase the acceptance for photons and $e^{\pm}$ down to 10°;

(iii) the focusing quadrupoles will be instrumented with sampling calorimeters (QCALT), made of tungsten and scintillator tiles.

3. η, η' decays

The dominant decay channels of the ϕ meson are $K\bar{K}$ pairs, however other interesting final states are produced with high rates. The radiative decays $\phi \to$ meson$+\gamma$ produce large samples of pseudoscalar and scalar mesons. Since $Br(\phi \to \eta\gamma) = 1.3\%$ we collected about 100 millions of η mesons, that can be easily identified by their recoil against the photon of $E = 363$ MeV. In the KLOE data sample we have also about half a million η' mesons ($Br(\phi \to \eta')\gamma = 6.25 \times 10^{-5}$), identified through their decays into $\eta\pi^+\pi^-$ and $\eta\pi^0\pi^0$ and the subsequent η decays.

3.1. $\eta \to 3\pi$

The $\eta \to 3\pi$ decay violates isospin symmetry, and it is mainly induced by a QCD Lagrangian term proportional to the d and u quark mass difference. Once the amplitude for this decay is known, the measurement of the decay rate is suitable for a precise determination of the ratio of the light quark masses

$$Q^2 = \frac{m_s^2 - \hat{m}^2}{m_d^2 - m_u^2}$$

where $\hat{m} = \frac{1}{2}(m_d + m_u)$. At leading order in ChPT, $\Gamma_{LO}(\eta \to \pi^+\pi^-\pi^0) = 66$ eV[6], to be compared with the experimental value $\Gamma_{exp} = (295 \pm 16)$ eV; a one loop calculation gives $\Gamma_{NLO(\eta\to\pi^+\pi^-\pi^0)} = (176 \pm 50)$ eV[7]. Higher order corrections improve the calculation[8], but there is yet no agreement with the experimental value.
We measured the Dalitz plot of both the charged, $\eta \to \pi^+\pi^-\pi^0$, and the neutral decay, $\eta \to 3\pi^0$.

$\eta \to \pi^+\pi^-\pi^0$. In a sample of 450 pb^{-1} of data we selected about 1.3 millions of $\eta \to \pi^+\pi^-\pi^0$ events. The Dalitz plot(fig.2) is described in terms of the variables X and Y that are linear combinations of the pion energies:

$$X = \sqrt{3}\frac{E_+ - E_-}{\Delta}$$
$$Y = 3\frac{E_0 - m_0}{\Delta} - 1 \qquad (1)$$

where $\Delta = m_\eta - 2m_\pm - m_0$.

Journal of Physics: Conference Series **349** (2012) 012002 doi:10.1088/1742-6596/349/1/012002

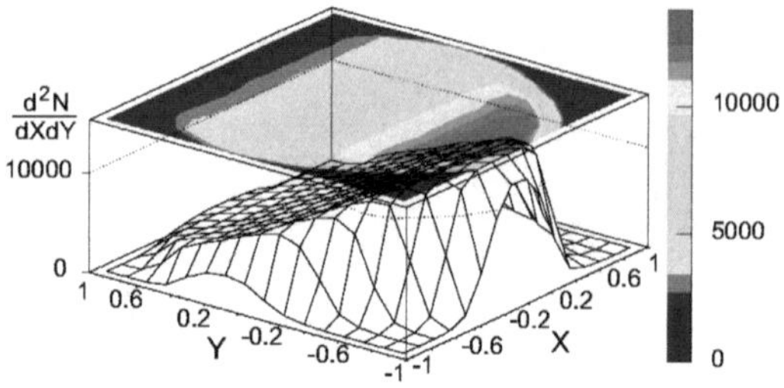

Figure 2. Dalitz plot density for $\eta \to \pi^+\pi^-\pi^0$.

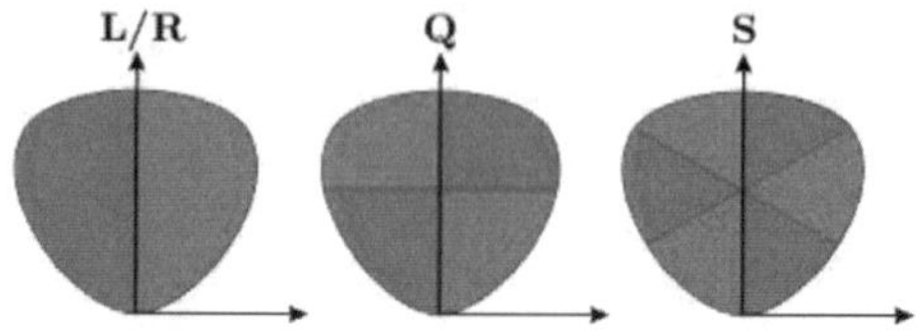

Figure 3. Definition of the η Dalitz plot integrated asymmetries.

The Dalitz plot density is usually parametrized as a power expansion around the center ($X = Y = 0$):

$$|A(X,Y)|^2 = 1 + aY + bY^2 + cX + dX^2 + eXY + fY^3 + ... \quad (2)$$

The parameters of the expansion are obtained from a fit to eq.(2) and are listed in tab.1.

Table 1. Results of the Dalitz plot fit.

a	$-1.090 \pm 0.005^{+0.008}_{-0.019}$
b	$0.124 \pm 0.006 \pm 0.010$
c	$0.002 \pm 0.003 \pm 0.001$
d	$0.057 \pm 0.006^{+0.007}_{-0.016}$
e	$-0.006 \pm 0.007^{+0.005}_{-0.003}$
f	$0.14 \pm 0.01 \pm 0.02$
$P(\chi^2)$	73%

From the fit results the following points can be emphasized:

- the c and e parameters, that multiply the odd powers of X, are compatible with zero as expected since their presence would be a signal of charge conjugation violation;
- the quadratic slope in Y, b, does not agree with the Current Algebra prediction ($b = a^2/4$);
- a large cubic term in Y is needed to fit the Dalitz plot; if the fit is repeated by forcing $f = 0$, the fit quality becomes very poor (χ^2 probability of the order of 10^{-6}).

The integrated asymmetries(see fig.3 for definitions) of the Dalitz plot are sensitive to charge conjugation violation. In particular the left-right asymmetry is a test the overall C conservation, like the c and e parameters of the fit, the quadrant asymmetry is sensitive to C violation in $\Delta I = 2$ amplitudes, while the sextant asymmetry tests C violation with $\Delta I = 1$.

All the measured values of the three asymmetries, reported in tab.2, turn out to be compatible with zero, and are the most stringent tests at present.

Table 2. Integrated asymmetries of the $\eta \to \pi^+\pi^-\pi^0$ Dalitz plot.

	(units 10^{-4})
A_{RL}	$9 \pm 10^{+9}_{-14}$
A_Q	$-5 \pm 10^{+3}_{-5}$
A_S	$8 \pm 10^{+8}_{-13}$

$\eta \to 3\pi^0$ The Dalitz plot of the neutral decay is symmetric, and its density can be parametrized as:

$$|A|^2 \propto 1 + 2\alpha Z + ...$$

where Z is the distance from the Dalitz plot center normalized to its maximum.

$$Z = \frac{2}{3}\sum_{i=1}^{3}\left(\frac{3E_i - m_\eta}{m_\eta - 3m_\pi}\right)^2$$

In the same data-set analyzed for the charged channel we selected a clean sample of about 6.5×10^5 fully neutral events. The fit result is shown in fig.4, and from this fit we get the slope:

$$\alpha = -0.0301 \pm 0.0035^{+0.0022}_{-0.0036}$$

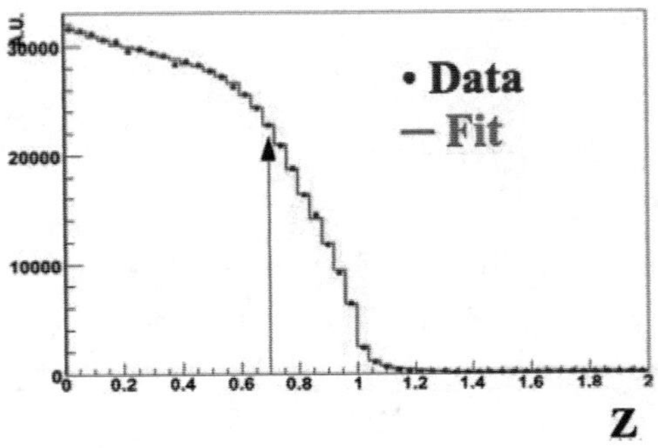

Figure 4. Fit of the Z distribution for $\eta \to 3\pi^0$, the arrow indicates the fit range $(0 \div 0.7)$.

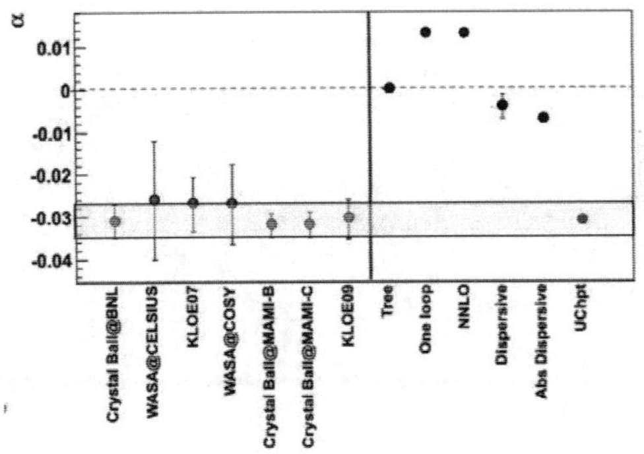

Figure 5. Slope of the $\eta \to 3\pi^0$ Dalitz plot: comparison of the experimental results and some theoretical predictions.

In fig.5 the KLOE results is compared to the other measurements and to some theoretical prediction. All the calculations based on ChPT predict positive values for the slope α. A better agreement is found by the authors of ref.[9] in the framework of the U(3)ChPT in combination with a coupled-channels method, including final state interactions through the Bethe-Salpeter equation. Also the calculation performed whitin the framework of the modified Non-Relativistic Effective Field Theory (NREFT)[10] gives a value $\alpha = -0.025 \pm 0.005$ in good agreement with the world average, however in this calculation there are large discrepancies for what concerns the parameters of $\eta \to \pi^+\pi^-\pi^0$. Recently, dispersive approaches to the calculation of the $\eta \to 3\pi$ amplitudes have been proposed[11][12] that, using the experimental information of the charged decay, are able to reproduce the negative slope of the neutral one. In both cases the authors derive the Q ratio, 21.31 ± 0.59 and 23.1 ± 0.7, respectively, different from the value expected in the Dashen limit, $Q_{Dashen} = 24.2$

A new analysis of the charged channel on the full KLOE data-set is in progress with the aim to reduce the systematic uncertainties, by using a different analysis scheme and an improved MC simulation.
Also the decay $\eta' \to \pi^+\pi^-\pi^0$ can be related to the d and u quark mass difference; at present only the measurement by CLEO exists[13], based on 24 events detected; at KLOE-2 with 5 fb^{-1} of integrated luminosity we expect to collect about 8000 of such events.

3.2. η, $\eta' \to \pi^+\pi^-\gamma$

The decays η, $\eta' \to \pi^+\pi^-\gamma$ in principle proceed through two different contributions: a resonant one dominated by the exchange of the ρ meson and a non-resonant one described by a contact term, the box anomaly which is a higher term of the Wess-Zumino-Witten Lagrangian. Both the total decay rate and the shape of the invariant mass distribution of the two pions in the final state are sensitive to the presence of such a contact term.
Previous measurements of $\eta \to \pi^+\pi^-\gamma$ date back to the 70's[14][15] and the analysis of these two data-sets show some contradictions[16]. Moreover there is a 2 - 3 standard deviations discrepancy between these old results and the more recent one by the CLEO Collaboration[17], as shown in tab.3.

Table 3. Past measurements of $\frac{\Gamma(\eta\to\pi^+\pi^-\gamma)}{\Gamma(\eta\to\pi^+\pi^-\pi^0)}$.

author	year	events	value
CLEO Coll.[17]	2007	859	$0.175 \pm 0.007 \pm 0.006$
Thaler et al. [15]	1973	18000	0.209 ± 0.004
Gormley et al.[14]	1970	7250	0.201 ± 0.006
PDG average			0.203 ± 0.008

Using a sample of about 560 pb^{-1} of data, we measured the branching fraction of $\eta \to \pi^+\pi^-\gamma$ normalized to $\eta \to \pi^+\pi^-\pi^0$[18]. We extracted 2×10^5 signal events from the fit of the distribution of $E - p$ (fig.6), where E and p are the energy and momentum of the photon from the η decay, reconstructed from all the other particles in the event ($p^\mu = p^\mu_\phi - p^\mu_{\pi+} - p^\mu_{\pi-} - p^\mu_{\gamma_\phi}$). The main background comes from $\phi \to \pi^+\pi^-\pi^0$.

Concerning the normalization sample(fig.7), we selected about 1.2×10^6 events, corresponding to a $Br(\eta \to \pi^+\pi^-\pi^0) = (22.41 \pm 0.03 \pm 0.35)\%$, in very good agreement with the PDG value, $(22.74 \pm 0.28)\%$[19].
The resulting ratio of branching ratios is in agreement with the recent CLEO measurement, favouring a sizeable contribution of the contact term[16]:

$$\frac{\Gamma(\eta \to \pi^+\pi^-\gamma)}{\Gamma(\eta \to \pi^+\pi^-\pi^0)} = 0.1838 \pm 0.0005 \pm 0.0030$$

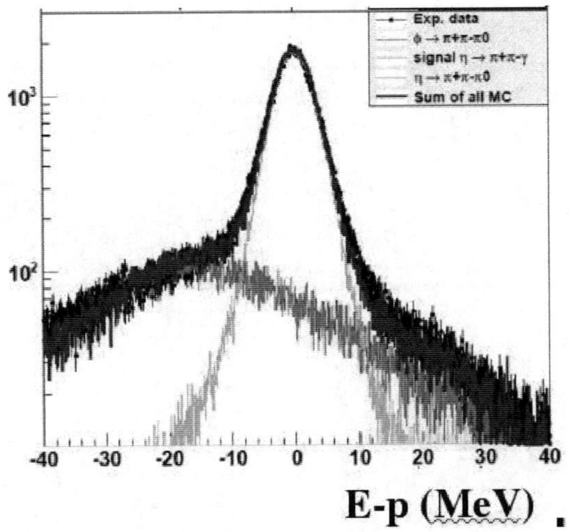

Figure 6. $E-p$ distribution; being E and p the energy and momentum of the photon from the η decay, the signal is peaked at zero.

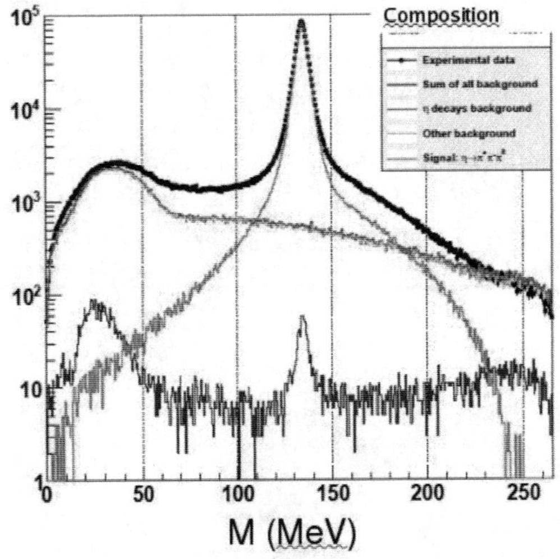

Figure 7. Recoiling mass against $\pi^+\pi^-\gamma$: the normalization sample ($\eta \to \pi^+\pi^-\pi^0$) is peaked at the π^0 mass.

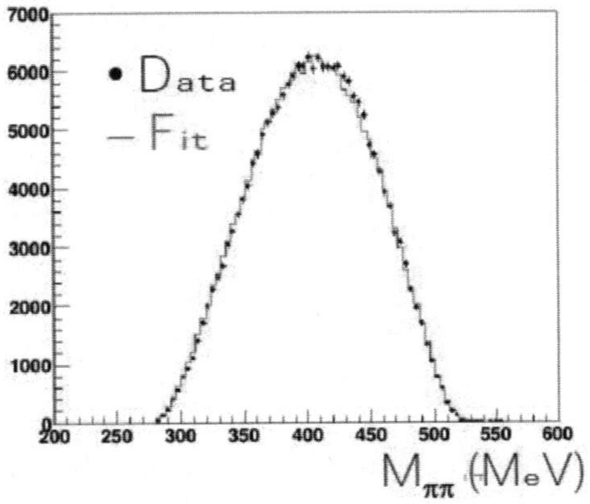

Figure 8. Dipion invariant mass distribution after backgrund subtraction, the red histogram is the fit result.

In fig.8 is shown the $\pi^+\pi^-$ invariant mass distribution after background subtraction, and with superimposed a fit according to the parametrization of ref.[20].

Also the η' meson data show some contradictions: in 1997 the Crystal Barrel Collabration claimed the evidence of the box anomaly in the two pion invariant mass distribution[21], while in 1998 the data by L3 were well described in terms of resonant contribution only[22]. Then it is important to measure with high statistics this decay in order to shed light on the contribution of the box anomaly. At KLOE-2, with 5 fb^{-1}, about 10^5 $\eta' \to \pi^+\pi^-\gamma$ events are expected, including an overall analysis efficiency of 20%, evaluated on a subsample of KLOE data.

3.3. $\eta \to \pi^+\pi^-e^+e^-$

The predictions for the $Br(\eta \to \pi^+\pi^-e^+e^-)$ based on ChPT and on VMD are in the range $26 \div 36 \times 10^{-5}$. Before KLOE only two low statistics measurements were avalaible, based on 4 events by CMD-2[23] and on 16 events by WASA@CELSIUS[24]. This process can be used as a probe of physics beyond the SM since an asymmetry in the angle ϕ between the decay planes of the two pions and of the two leptons(see fig.9), would be signal of CP violation. Such an asymmetry is originated by the interference of the electric transition amplitude, CP violating, with the CP conserving magnetic one, as observed for the $K_L \to \pi^+\pi^-e^+e^-$ decay[25][26].

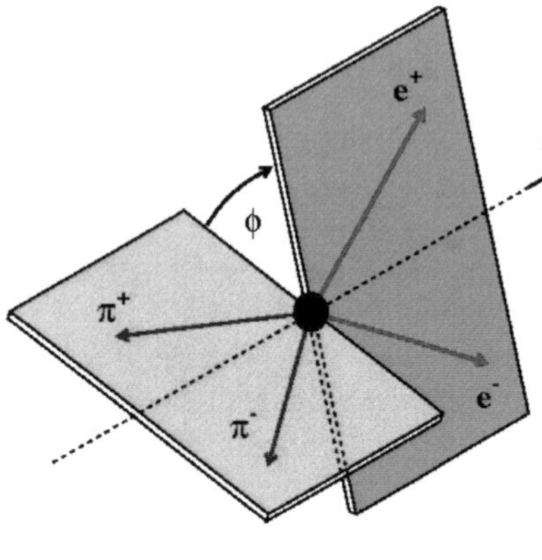

Figure 9. ϕ is the angle between the decay planes of the two pions and of the two leptons.

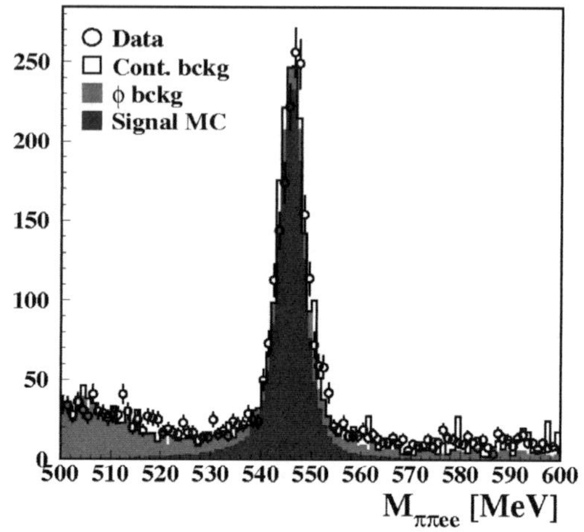

Figure 10. Invariant mass of the four charged tracks for $\eta \to \pi^+\pi^-e^+e^-$ candidate events.

$$A_{CP} = \frac{N(\sin\phi\cos\phi > 0) - N(\sin\phi\cos\phi < 0)}{N(\sin\phi\cos\phi > 0) + N(\sin\phi\cos\phi < 0)}$$

This asymmetry is constrained in the SM by the $\eta \to \pi^+\pi^-$ decay: from the experimental limit on the branching fraction by KLOE[27], one can expect $A_{CP} < 10^{-4}$, while according to theory $A_{CP} < 10^{-15}$. However a non-CKM mechanism has been proposed[28] that could raise the asymmetry up to 2% level.
We analyzed 1.7 fb^{-1} of data, and we used the four track invariant mass distribution to extract the signal(fig.10).

From a fit with signal and background shapes evaluated by MC we found 1555 ± 52 signal events with 368 background ones. This corresponds to

$$Br(\eta \to \pi^+\pi^-e^+e^-(\gamma)) = (26.8 \pm 0.9 \pm 0.7) \times 10^{-5}$$

inclusive of radiative effects. From the distribution of $\sin\phi\cos\phi$ (fig.11) we could produce the first measurement of the asymmetry, consistent with zero at 3% level:

$$A_{CP} = (-0.6 \pm 2.5 \pm 1.8) \times 10^{-2}$$

The insertion of the IT in the second phase of KLOE-2, will allow to increase the efficiency and to reduce the background. With a sample of 20 fb^{-1} we could reduce the total uncertainty in the branching fraction to 1.4%, and we could also improve the sensitivity to the asymmetry by a factor of at least two.

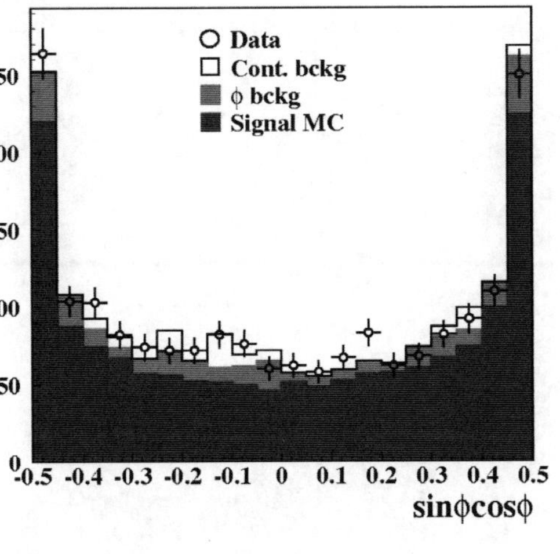

Figure 11. Distribution of $\sin\phi\cos\phi$ for $\eta \to \pi^+\pi^-e^+e^-$ events.

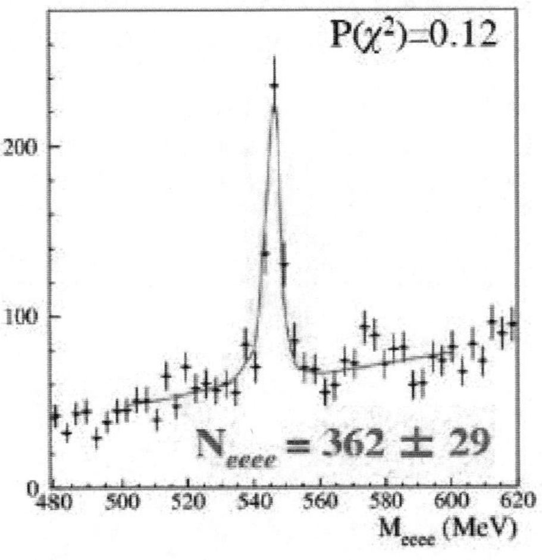

Figure 12. Invariant mass of the four leptons for $\eta \to e^+e^-e^+e^-$ candidate events.

3.4. $\eta \to e^+e^-e^+e^-$

The double Dalitz decays in principle can be used to study the internal structure of the η meson, by measuring the 4-momenta of the virtual photons via the invariant mass of the lepton pairs. The theoretical predictions for the branching fraction lay in the range $2.4 \div 2.6 \times 10^{-5}$ and only two experimental upper limits at 90% C.L. existed before the KLOE analysis, $Br(\eta \to e^+e^-e^+e^-) < 6.9 \times 10^{-5}$ (CMD-2)[23] and $Br(\eta \to e^+e^-e^+e^-) < 9.7 \times 10^{-5}$ (WASA)[24].
In the same data sample analyzed for $\eta \to \pi^+\pi^-e^+e^-$, we observed for the first time this decay as shown in fig.12, obtaining 362 ± 29 $\eta \to e^+e^-e^+e^-$ events, that correspond to

$$Br(\eta \to e^+e^-e^+e^-(\gamma)) = (2.4 \pm 0.2 \pm 0.1) \times 10^{-5}$$

fully inclusive of final state radiation.

3.5. $\eta \to \pi^0\gamma\gamma$

The decay $\eta \to \pi^0\gamma\gamma$ is a unique test of ChPT, since the lowest order amplitude, $O(p^2)$, is vanishing for neutral particles as well as the tree level amplitude at $O(p^4)$; the loop contributions at $O(p^4)$ are suppressed by G-parity and by the large kaon mass, thus the dominant terms are $O(p^6)$.
From the experimental point of view, several measurements have been published since '70s, with branching fraction decreasing with time by some order of magnitude. The most recent measurements are from GAMS-2000[29], $Br(\eta \to \pi^0\gamma\gamma) = (7.2 \pm 1.4) \times 10^{-4}$, Crystal Ball@AGS, $Br(\eta \to \pi^0\gamma\gamma) = (2.21 \pm 0.24 \pm 0.47) \times 10^{-4}$[30], and Crystal Ball@MAMI, $Br(\eta \to \pi^0\gamma\gamma) = (2.25 \pm 0.46 \pm 0.17) \times 10^{-4}$[31].
We analyzed the fully neutral decay with five prompt photons in the final state, $\phi \to \eta\gamma$ with $\eta \to \pi^0\gamma\gamma$. There is a large irreducible background coming from the other, more abundant, five prompt photons processes (see Sect.4), and from $\phi \to \eta\gamma$ with $\eta \to 3\pi^0$, with lost photons or merged clusters in the EMC. The four photon invariant mass distribution is reported in fig.13; after the subtraction of the residual background our preliminary result is[32]

$$Br(\eta \to \pi^0\gamma\gamma) = (8.4 \pm 2.7 \pm 1.4) \times 10^{-5}$$

Journal of Physics: Conference Series **349** (2012) 012002 doi:10.1088/1742-6596/349/1/012002

about two standard deviations lower than the Crystal Ball measurements and the ChPT predictions. From the first period of KLOE-2 data taking, with about 10 fb^{-1} of data, we

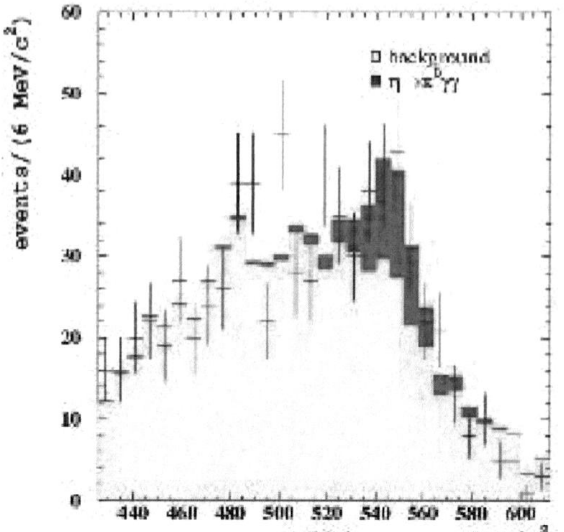

Figure 13. Distribution of the four photon invariant mass.

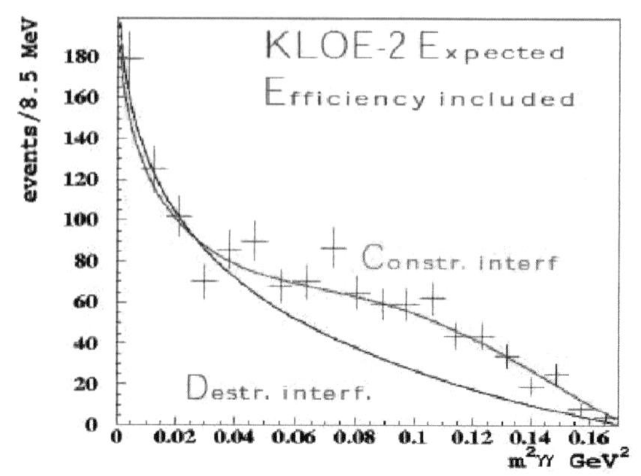

Figure 14. Simulation of the $m^2_{\gamma\gamma}$ expected at KLOE-2 in case of constructive and destructive interference between the VMD and $a_0(980)$ contributions.

should be able to measure the branching fraction at 3% level, and to provide the $\gamma\gamma$ invariant mass spectrum with sufficient accuracy to solve the ambiguity connected to the sign of the interference term between the scalar amplitude, dominated by the exchange of the $a_0(980)$, and the VMD one(fig.13).

3.6. Transition form factor of $\phi \rightarrow \eta e^+e^-$

The Dalitz decays, like $A \rightarrow B\ell^+\ell^-$, can be used to study the internal electromagnetic structure of the decaying meson. For non point-like mesons the vertex $A \rightarrow B$ is described in terms of a transition form factor (FF), which is a function of the four momentum squared q^2 of the virtual photon (invariant mass of the $\ell^+\ell^-$ pair). Usually a pole parametrization is used for the FFs:

$$F(q^2) = \frac{1}{1 - \frac{q^2}{\Lambda^2}}$$

The Λ parameter is a characteristic mass, that according to VMD can be roughly identified with the mass of the nearest vector meson. The dilepton invariant mass distributions of $\eta \rightarrow e^+e^-\gamma$ and $\eta \rightarrow \mu^+\mu^-\gamma$, recently measured by NA60 [33] and Crystall Ball@MAMI[34] are described by $\Lambda_\eta^{-2} = 1.92 \div 1.95$ GeV^{-2} in agreement with the VMD predictions $\Lambda_\eta^{-2} = 1.88$ GeV^{-2}. On the other hand there is a discrepancy between the dilepton spectrum of $\omega \rightarrow \pi^0\mu^+\mu^-$, with $\Lambda_\omega^{-2} = 2.24$ GeV^{-2} also measured by NA60, and the VMD expectations 1.68 GeV^{-2}.
Concerning $\phi \rightarrow \eta e^+e^-$, there is only one low statistics measurement of the FF, by SND, $\Lambda_\phi^{-2} = (3.8 \pm 1.8)$ GeV^{-2}[?], to be compared with the VMD expectation, $\Lambda_\phi^{-2} \simeq m_\phi^{-2} \simeq 1$ GeV^{-2}.
We analyzed 1.5 fb^{-1} of data looking for $\phi \rightarrow \eta e^+e^-$ with $\eta \rightarrow \pi^+\pi^-\pi^0$. In fig.15 is shown the recoiling mass against e^+e^-; by applying a tight cut around the η peak, we selected abut 14000 candidate events, whose e^+e^- invariant mass spectrum is shown in fig.16.

A preliminary evaluation of the branching fraction has been performed

$$Br(\phi \rightarrow \eta e^+e^-) = (1.070 \pm 0.012) \times 10^{-4}$$

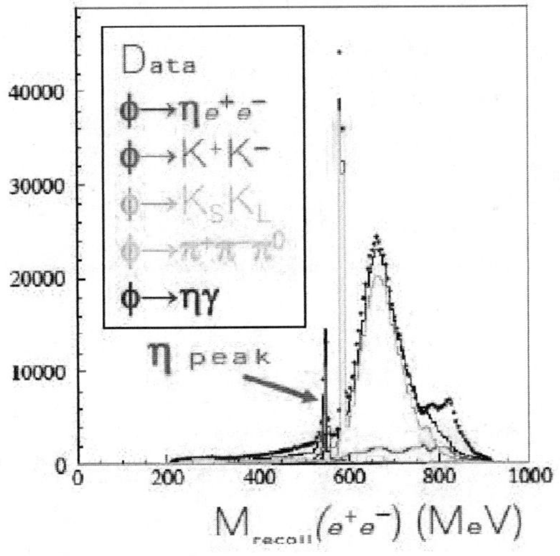

Figure 15. Recoil mass against the e^+e^- pair; the $\phi \to \eta e^+e^-$ is clearly visible in the peak corresponding to the η mass, the peak at 590 MeV is due to $K_S \to \pi^+\pi^-$ events with wrong mass assignment.

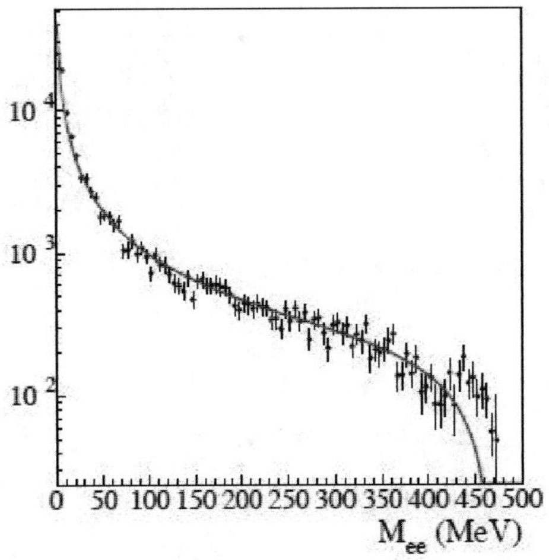

Figure 16. e^+e^- invariant mass distribution with fit superimposed.

in good agreement with the PDG average $(1.15 \pm 0.10) \times 10^{-4}$[19]. The quoted uncertainty is only statistics, the evaluation of the systematics is in progress. The extraction of the FF parameter is still in progress, however the collected statistics will allow to reach an accuracy of about 3% on Λ_ϕ.
This process is also suitable for the search of a possible low mass gauge boson, mediator of dark forces, as discussed by F.Bossi in this Workshop[35][36].

3.7. Precise η mass measurement

KLOE has performed the most precise determination to date of the η mass. Before that measurement there was an eight standard deviation discrepancy between the value measured in $p + d \to {}^3He + \eta$ at threshold by GEM@COSY[37], and the values obtained by looking at η decays by NA48[38] and CLEO[39].
We used the $\phi \to \eta\gamma$ decay, with $\eta \to \gamma\gamma$, by looking for three prompt photon events. In the Dalitz plot of fig.17 three bands are clearly visible: a low mass band given by the $\phi \to \pi^0\gamma$ decay, process used to check of the whole procedure, and the other two bands from $\phi \to \eta\gamma$. In order to select pure η and $\pi^0 \to \gamma\gamma$ samples we apply a cut by retaining the events below the black line of fig.17. The projection (fig.18) is fitted to extract the mass value. We found:

$$\begin{aligned} m_\eta &= (574.874 \pm 0.007 \pm 0.031)\text{MeV} \\ m_{\pi^0} &= (134.906 \pm 0.012 \pm 0.048)\text{MeV} \end{aligned} \qquad (3)$$

The π^0 mass is in agreement within 1.4 standard deviations with the world average[19]. The η mass agrees with the measurements by NA48 and CLEO, and is the most accurate one up to now.

Journal of Physics: Conference Series **349** (2012) 012002 doi:10.1088/1742-6596/349/1/012002

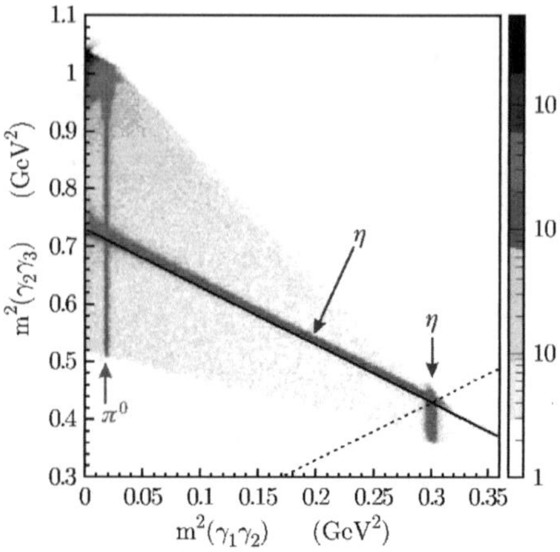

Figure 17. Dalitz plot of the selected 3γ events ($E_{\gamma_1} < E_{\gamma_2} < E_{\gamma_3}$).

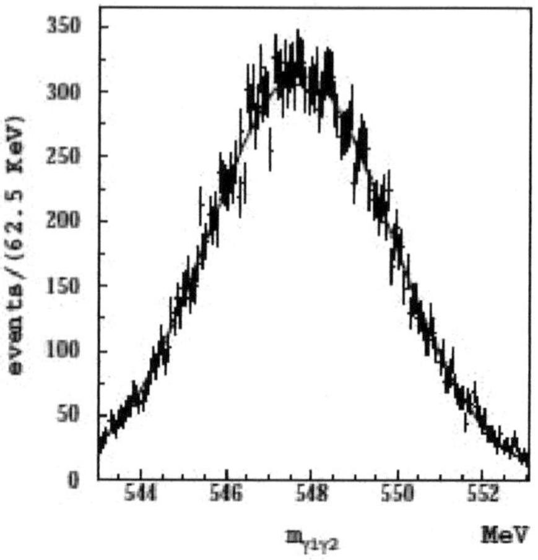

Figure 18. Invariant mass of the $\gamma_1 - \gamma_2$ pair after the cut indicated in fig.17.

Recently a preliminary result has been presented by the ANKE Collaboration at COSY, based on the kinematics of the reaction $p + d \rightarrow^3 He + \eta$, in agreement with our value, $m_\eta = (574.869 \pm 0.007 \pm 0.040)$ MeV[40].

3.8. $\eta - \eta'$ mixing

The $\eta - \eta'$ system can be parametrized in terms of only one mixing angle in the quark-flavour basis.

$$|\eta> = \cos\varphi_P|q\bar{q}> - \sin\varphi_P|s\bar{s}>$$
$$|\eta'> = \sin\varphi_P|q\bar{q}> + \cos\varphi_P|s\bar{s}> \tag{4}$$

where $|q\bar{q}> = \frac{1}{\sqrt{2}}|u\bar{u} + d\bar{d}>$. The ratio of branching fractions $R = Br(\phi \rightarrow \eta'\gamma)/Br(\phi \rightarrow \eta\gamma)$, can be related to the pseudoscalar mixing angle φ_P[41]:

$$R = \cot^2\varphi_P \left(1 - \frac{m_S}{\hat{m}}\frac{C_{NS}}{C_S}\frac{\tan\varphi_V}{\sin 2\varphi_P}\right)^2 \left(\frac{p_{\eta'}}{p_\eta}\right)^3 \tag{5}$$

where $\varphi_V = 3.2°$ is the vector mixing angle, $p_{\eta(\eta')}$ is the $\eta(\eta')$ momentum in the ϕ center of mass, and $C_{NS} = 0.91 \pm 0.05$ and $C_S = 0.89 \pm 0.07$ represent the effect of the OZI-rule which reduces the vector and pseudoscalar wave-function overlap, and finally $m_S/\hat{m} = 1.24 \pm 0.07$[41]. We selected $\phi \rightarrow \eta'\gamma$ with either $\eta' \rightarrow \eta\pi^+\pi^-$, $\eta \rightarrow 3\pi^0$, and $\eta' \rightarrow \eta\pi^0\pi^0$, $\eta \rightarrow \pi^+\pi^-\pi^0$, both corresponding to the $\pi^+\pi^- + 7$ prompt photon final state. We used the $\phi \rightarrow \eta\gamma$ decay where $\eta \rightarrow \pi^+\pi^-\pi^0$, which is background free, as the normalization process. In a sample of 427 pb^{-1} we selected 3407 $\eta'\gamma$(fig.19) and about 1.7×10^7 $\eta\gamma$ events, and we obtained

$$R = (4.77 \pm 0.09 \pm 0.19) \times 10^{-3}$$

from which we derive

$$Br(\phi \rightarrow \eta'\gamma) = (6.20 \pm 0.11 \pm 0.25) \times 10^{-5}$$

where the systematic uncertainty is dominated by the knowledge of the $\eta' \rightarrow \eta\pi\pi$ branching fractions.

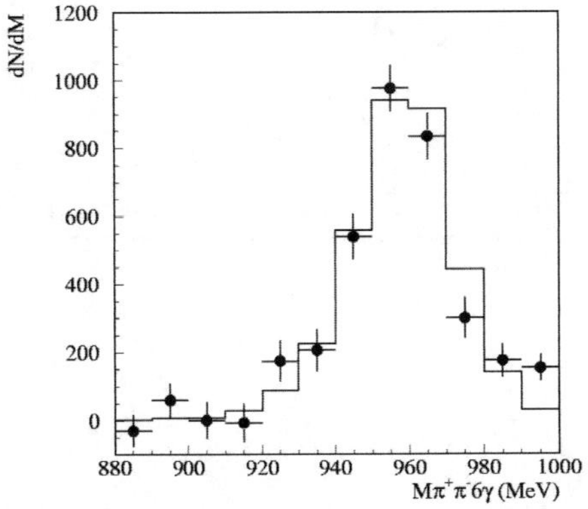

Figure 19. Invariant mass of 6 out of 7 photons, after the combinatorial background subtraction.

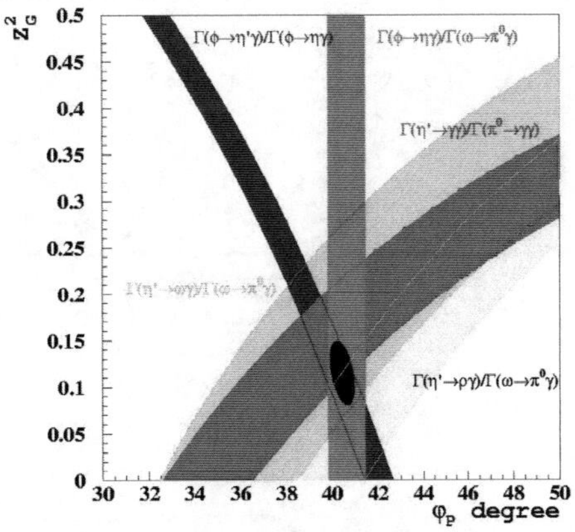

Figure 20. φ_P, $Z^2_{\eta'}$ plane; the black ellipse is the 68% C.L. region of the fit.

From eq.5, we extracted the mixing angle

$$\varphi_P = (41.4 \pm 0.3 \pm 0.9)^\circ$$

that in the singlet-octet basis becomes: $\vartheta_P = (-13.3 \pm 0.3 \pm 0.9)^\circ$.
The η' meson, being an almost pure $SU(3)_F$ singlet, is considered a good candidate to host a gluon condensate. Allowing for a η' gluonium content ($|G>$), eq.(4) should be modified[42]

$$|\eta'>= X_{\eta'}|q\bar{q}> + Y_{\eta'}|s\bar{s}> + Z_{\eta'}|G>$$

where a non-zero value of $Z^2_{\eta'}$ is a signal of gluonium content of η',

$$X_{\eta'} = \sin\varphi_p \cos\varphi_G$$
$$Y_{\eta'} = \cos\varphi_p \cos\varphi_G$$
$$Z_{\eta'} = \sin\varphi_G$$

and eq.(5) becomes

$$R = \cot^2\varphi_P \cos^2\varphi_G \left(1 - \frac{m_S}{\hat{m}}\frac{C_{NS}}{C_S}\frac{\tan\varphi_V}{\sin 2\varphi_P}\right)^2 \left(\frac{p_{\eta'}}{p_\eta}\right)^3$$

We performed a fit by combining our R measurement with other experimental constraints[43]; the fit parameters are listed in tab.4, and the result is shown graphically in fig.20. We then estimate a gluonium content of η', different from zero at three standard deviation level.

The fit result is driven by the measurement of $\Gamma(\eta' \to \gamma\gamma)/\Gamma(\pi^0 \to \gamma\gamma)$. Since the theoretical frameworks in which the $\eta' \to \gamma\gamma$ and the $P \to V\gamma$ decays are described are slightly different[44], it would be very important to obtain a sensitivity to gluonium fraction independent from

Table 4. Result of the fit to the gluonium content of η'.

$Z^2_{\eta'}$	0.12 ± 0.04
φ_P	$(40.4 \pm 0.6)^\circ$
C_{NS}	0.94 ± 0.03
C_S	0.83 ± 0.05
φ_V	$(3.32 \pm 0.10)^\circ$
$m_S/\hat{m}$	1.24 ± 0.07
χ^2/ndf	4.6/3
$P(\chi^2)$	20%

$\eta' \to \gamma\gamma$. It can be shown[43] that this could be achieved at KLOE-2 if we will be able to measure the main η' branching fractions at 1% level, in that case the sensitivity to $Z^2_{\eta'}$ would increase to four σ level. A further improvement in the gluonium determination could be obtained by measuring the also the η' decay width at 1% through the measurement of $\sigma(e^+e^- \to e^+e^-\gamma^\star\gamma^\star \to e^+e^-\eta')$; however this would require to run DAΦNE at least at $\sqrt{s} \simeq 1.2$ GeV.

4. Light scalar mesons

The problem of the classification of the scalar mesons of mass below 1 GeV is still open. It is controversial whether they are ordinary $q\bar{q}$ mesons[45], compact $qq\bar{q}\bar{q}$ states[46], or bound states of a K and a $\bar{K}$[47].

In KLOE we studied the radiative decays $\phi(1020) \to PP\gamma$, where P means a pseudoscalar meson, as they are dominated by the exchange of a scalar meson S in the intermediate state ($\phi \to S\gamma$, and $S \to PP$), and then they are suitable processes to study the light scalars, since the branching ratios and also the PP invariant mass shapes are expected to depend on their structure.

The $f_0(980)$ has been searched for in the dipion final states: $f_0(980) \to \pi^+\pi^-$ has been observed as a deviation from the expected Initial State Radiation plus Final State Radiation contributions in $e^+e^- \to \pi^+\pi^-\gamma$, while the Dalitz plot of the reaction $e^+e^- \to \pi^0\pi^0\gamma$ has been used to study the $f_0(980) \to \pi^0\pi^0$ decay, looking also for a possible signal of the $\sigma(600)$. The $a_0(980)$ has been studied by exploiting the decay chains $\phi \to \eta\pi^0\gamma$ with $\eta \to \gamma\gamma$ and $\eta \to \pi^+\pi^-\pi^0$.

In order to discriminate between models, data have been fit to different theoretical approaches. For each of them the $a_0(980)/f_0(980)$ masses and the couplings to $K\bar{K}$, $\pi\pi$ or $\eta\pi^0$ and to the ϕ have been extracted. The decay $\phi \to (a_0/f_0) \to K^0\bar{K}^0\gamma$ has also been searched for.

4.1. $e^+e^- \to \pi^0\pi^0\gamma$

The analyzed data sample consists of 450 pb^{-1}. This sample has been analyzed by grouping all runs in center of mass energy bins of 100 keV; only those runs belonging to the bin with the highest statistics have been used for the fit of the Dalitz plot. This subsample corresponds to 145 pb^{-1} collected at $\sqrt{s} = 1019.7 - 1019.8$ MeV.

The data analysis consists of four steps: *(i)* events with five prompt photons have been selected; *(ii)* a kinematic fit with the constraint of the total 4-momentum conservation has been performed; *(iii)* a pairing algorithm of photons to π^0's is applied; *(iv)* a second kinematic fit has been performed with the further constraints of the π^0 masses, according to the chosen photon pairing. The residual background mainly comes from $\phi \to \eta\pi^0\gamma$, same final state of the signal, and from $\phi \to \eta\gamma$ with $\eta \to \pi^0\pi^0\pi^0$, that can mimic the signal when there are lost or merged photons. It

has been evaluated by MC and then subtracted.
The Dalitz plot after the background subtraction is shown in fig.21: there are two entries per event, corresponding to the two possible $\pi^0\gamma$ combinations. The two bands in the region $M_{\pi\pi} < 700$ MeV are due to the non resonant process $e^+e^- \to \omega\pi^0$, with $\omega \to \pi^0\pi^0$, while the region of high $M_{\pi\pi}$ is dominated by the f_0.
The Dalitz plot has been fitted using the parametrization of the scalar ampitude according to the Kaon Loop (KL) model in which the ϕ is coupled to the scalar meson through a loop of charged kaons[48]; all the formulae can be found in ref.[49]. The fit function includes also the contribution of the $\sigma(600)$; the free parameters or the f_0 are its mass and the couplings $g_{f_0K^+K^-}$ and $g_{f_0\pi^+\pi^-}(=\sqrt{2}g_{f_0\pi^0\pi^0})$. The vector part ($e^+e^- \to \omega\pi^0$ and $\phi \to \rho^0\pi^0$, with $\rho^0 \to \pi^0\gamma$) plus the interference one are described by seven free parameters. The $\sigma(600)$ parameters have been fixed to the values of the best fit of ref.[48].
The relevant parameters of the fit are reported in tab.5. If we perform a fit by excluding the $\sigma(600)$ contribution, the fit quality becomes very poor.

Table 5. Parameters of the $f_0(980)$.

	$f_0 \to \pi^0\pi^0$	$f_0 \to \pi^+\pi^-$
M_{f_0} (MeV)	984.7	983.7
$g_{f_0K^+K^-}$ (GeV)	3.97	4.74
$g_{f_0\pi^+\pi^-}$ (GeV)	-1.82	-2.22
$R_{f_0} = (g_{f_0K^+K^-}/g_{f_0\pi^+\pi^-})^2$	4.8	4.6

We tried also an alternative parametrization of the amplitude in which a point-like coupling of the scalar to the ϕ meson is assumed, then this model is called "No Structure" (NS)[50]. The NS model does not include explicitly the $\sigma(600)$ contribution. In this case the coupling of the ϕ to the f_0 can be determined directly from the fit, and we obtain $g_{\phi f_0\gamma} = 2.61 \pm 0.31$ GeV^{-1}. From the integral of the scalar contribution only, the following branching ratio can be extracted[49]:

$$Br(\phi \to S\gamma \to \pi^0\pi^0\gamma) = (1.07^{+0.01}_{-0.03(fit)}\,{}^{+0.04}_{-0.02(syst)}\,{}^{+0.05}_{-0.06(mod)}) \times 10^{-4} \tag{6}$$

where the third uncertainty is sistematic and it is due to the model formulation.

4.2. $e^+e^- \to \pi^+\pi^-\gamma$

Only a small fraction of $\pi^+\pi^-\gamma$ events originates from $\phi \to f_0\gamma$ with $f_0 \to \pi^+\pi^-$. The main contribution is given by $e^+e^- \to \pi^+\pi^-\gamma$ in which the photon is produced by either an initial state (ISR) or a final state (FSR) radiation. A small amount of events is expected also from the decay $\phi \to \rho^\pm\pi^\mp$ with $\rho^\pm \to \pi^\pm\gamma$.
Events with one vertex close to the interaction region and two tracks of opposite charge, emitted at polar angles $\vartheta_{\pi^\pm} > 45°$, have been selected. The ISR component, that is dominant for small photon polar angles, is suppressed by requiring the polar angle of the missing momentum $\vartheta_\gamma > 45°$. Furthermore the presence of a photon in the calorimeter matching the missing energy and momentum is required to reduce the residual $\pi^+\pi^-\pi^0$ background.
The $\pi^+\pi^-$ invariant mass distribution (fig.22) has been fitted to the differential cross-section ($m = M_{\pi^+\pi^-}$):

$$\frac{d\sigma}{dm} = \frac{d\sigma_{ISR}}{dm} + \frac{d\sigma_{FSR}}{dm} + \frac{d\sigma_{\rho\pi}}{dm} + \frac{d\sigma_{scal}}{dm} \pm \frac{d\sigma_{scal\otimes FSR}}{dm}$$

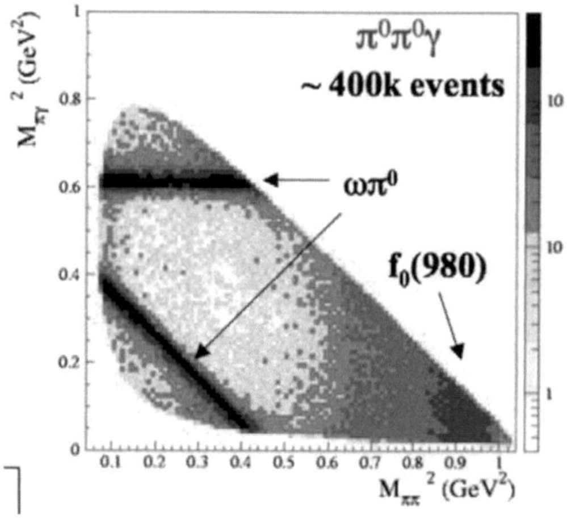

Figure 21. Dalitz plot of $e^+e^- \to \pi^0\pi^0\gamma$.

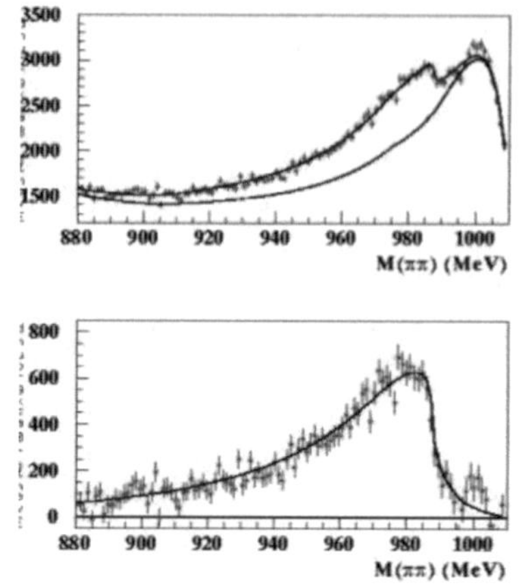

Figure 22. $M_{\pi^+\pi^-}$:upper plot - full spectrum; lower plot - only scalar.

The analytic expressions for the first two terms as well as for the $\rho\pi$ term are taken from ref.[51]. The scalar contribution is parametrized according to the KL model, and the last term $d\sigma_{scal\otimes FSR}/dm$ can be added (constructive interference between f_0 and FSR) or subtracted (destructive interference). The relevant parameters of the fit are listed in tab.5, and are in good agreement with the parameters from the fit of the Dalitz plot of the neutral channel.
A destructive interference between f_0 and FSR is preferred by the fit. Due to the huge irreducible background, the charged channel is not sensitive to the presence of the $\sigma(600)$. By integrating the f_0 term of the fit function, the following range for the branching ratio can be derived[52]:

$$Br(\phi \to f_0\gamma \to \pi^+\pi^-\gamma) = 2.1 \times 10^{-4} - 2.4 \times 10^{-4}$$

As expected this branching ratio is about twice the value of eq.(6).

4.3. $e^+e^- \to \eta\pi^0\gamma$

To study the $a_0(980)$, the decay $\phi \to \eta\pi^0\gamma$ has been exploited, by looking for the two final states corresponding to $\eta \to \gamma\gamma$ ($Br = 39.31\%$) and $\eta \to \pi^+\pi^-\pi^0$ ($Br = 22.74\%$).
The fully neutral final state is characterized by five photons originating from the DAΦNE interaction point, then the selection and the analysis scheme are very similar to those of $\pi^0\pi^0\gamma$. The main backgrounds come from the other five photon final states, $\phi \to f_0(980)\gamma \to \pi^0\pi^0\gamma$ and $e^+e^- \to \omega\pi^0 \to \pi^0\pi^0\gamma$, while the seven photon process, $\phi \to \eta\gamma$ with $\eta \to 3\pi^0$, can mimic five photon events due to either loss or merging of photons. Also the three photon final states, $\phi \to \eta(\pi^0)\gamma$ with $\eta(\pi^0) \to \gamma\gamma$, give a small contribution to the selected sample, due to either photon splittings or accidental coincidence with machine background. Other background processes are negligible. In the analyzed sample of 450 pb^{-1}, 29061 events have been selected. The expected irreducible background amounts to about 50% of the final sample, and has been evaluated by MC and checked on data with control samples largely dominated by background. After the background subtraction, the sample consists of 13269 ± 192 events. By normalizing to the number of ϕ mesons produced in the analyzed data set, the following branching ratio is obtained[53]:

$$Br(\phi \to \eta\pi^0\gamma) = (7.01 \pm 0.10 \pm 0.20) \times 10^{-5} \quad (7)$$

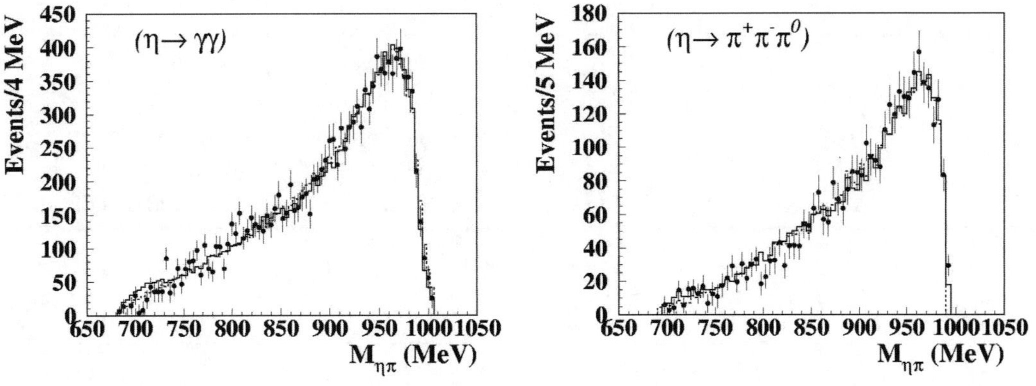

Figure 23. Fit to the $\eta\pi^0$ invariant mass distributions after background subtraction.

The decay chain with $\eta \to \pi^+\pi^-\pi^0$ is characterized by two opposite charge pions and five prompt photons. With respect to the fully neutral final state, it provides a lower statistics since the branching ratio of $\eta \to \pi^+\pi^-\pi^0$ is about half of the $\eta \to \gamma\gamma$ one. However in this case the background is in principle less relevant since no other final state with two tracks and five photons is expected to have a significant branching ratio from the ϕ. Thus the main sources of background are due to abundant final states with two tracks and either four or six photons.
At the end of the analysis chain 4181 events are selected with a total background of 15%. By normalizing to the produced ϕ's[53]:

$$Br(\phi \to \eta\pi^0\gamma) = (7.12 \pm 0.13 \pm 0.22 \times 10^{-5} \quad (8)$$

in good agreement with eq.(7) within the uncertainties. By combining the two results and taking into account the common normalization error, $Br(\phi \to \eta\pi^0\gamma) = (7.06 \pm 0.22) \times 10^{-5}$ is obtained, where the uncertainty is both statistic and systematic.
A combined fit, with the same set of free parameters, has been performed on the two $\eta\pi^0$ invariant mass distributions, to both KL and NS models. The free parameters of the fit are: M_{a_0}, $g_{a_0 K^+K^-}$, $g_{a_0\eta\pi^0}$, the branching ratio of the vector contribution, and, as a relative normalization between the two different final states, the ratio $R_\eta = Br(\eta \to \gamma\gamma)/Br(\eta \to \pi^+\pi^-\pi^0)$. For what concerns the NS fit also the $g_{\phi a_0 \gamma}$ coupling is left free. The fit results are shown in fig.23, and the parameters are listed in tab.6.

The good agreement of the ratio R_η with the PDG value, 1.73 $\pm$ 0.04[19], indicates that the two samples are consistent with each other, and that the background subtraction has been correctly performed.
From the small vector background found one can conclude that the $\phi \to \eta\pi^0\gamma$ process is completely dominated by $\phi \to a_0(980)\gamma$.

4.4. $\phi \to K^0\bar{K}^0\gamma$

The $K^0\bar{K}^0$ pair is produced in a $J^{PC} = 0^{++}$ state, then the signature of this decay is provided by the presence of either two K_S or two K_L and a low energy photon ($E_\gamma < 23.8$ MeV). This decay is expected to be dominated by $\phi \to (f_0 + a_0)\gamma \to K^0\bar{K}^0\gamma$, and its branching ratio is

Table 6. $a_0(980)$ parameters from the combined fit of $M_{\eta\pi^0}$.

	KL model	NS model
M_{a_0} (MeV)	982.5± 1.6± 1.1	982.5 (fixed)
$g_{a_0K^+K^-}$ (GeV)	2.15± 0.06± 0.06	2.01± 0.07± 0.28
$g_{a_0\eta\pi^0}$ (GeV)	2.82± 0.03± 0.04	2.46± 0.08± 0.11
$g_{\phi a_0\gamma}$ (GeV^{-1})	-	1.83± 0.03± 0.08
$Br_{vect} \times 10^6$	0.92± 0.40± 0.15	negligible
$R_\eta = Br(\eta \to \gamma\gamma)/Br(\eta \to \pi^+\pi^-\pi^0)$	1.70± 0.04± 0.03	1.70± 0.03± 0.01
$R_{a_0} = (g_{a_0K^+K^-}/g_{a_0\eta\pi^0})^2$	0.58± 0.03± 0.03	0.67± 0.06± 0.13
$P(\chi^2)$	10.4%	30.9%

expected to depend on the size of the couplings of the scalars to kaons; interference between f_0 and a_0 can also be present.
To study this decay the K_SK_S channel with both $K_S \to \pi^+\pi^-$ has been chosen: events with two vertices near the interaction point, each connected to two tracks of opposite charge have been selected. The photon detection is also required to reject the $\phi \to K_SK_L \to \pi^+\pi^-\pi^+\pi^-$ background.
The full KLOE data sample has been analyzed, 5 events have been found in the data, with 3.2 background events expected from MC. Then an upper limit on the branching ratio has been set: $Br(\phi \to K^0\bar{K}^0\gamma) < 1.89 \times 10^{-8}$ @ 90% C.L.
By using the couplings found in the $\phi \to \pi\pi\gamma$ and $\phi \to \eta\pi^0\gamma$ analyses an allowed range can be evaluated, $Br(\phi \to K^0\bar{K}^0\gamma) = 4 \times 10^{-9} - 6.8 \times 10^{-8}$ well compatible with the upper limit.

4.5. Summary on light scalar mesons

In order to summarize the results presented in this Section, the following statements can be made:

(i) the branching ratios of $\phi \to S\gamma$ are compatible, with much smaller uncertainties, with the previous measurements by the VEPP-2M experiments, SND and CMD-2 (fig.24);
(ii) the f_0 branching ratio and couplings are closer to the SU(3) predictions for the $qq\bar{q}\bar{q}$ structure than for the $q\bar{q}$ one (see fig.24 and tab.7);
(iii) the a_0 branching ratio and couplings are in between the SU(3) predictions for $qq\bar{q}\bar{q}$ and $q\bar{q}$ structures;
(iv) large $g_{\phi S\gamma}$ couplings have been found, pointing to a large strange quark contents of both f_0 and a_0, in fact both $g_{\phi f_0\gamma}, g_{\phi a_0\gamma} > g_{\phi\eta'\gamma} = 0.75$ GeV^{-1} (η' has about 66% content of $s\bar{s}$);
(v) another indication of large strange quark content in the a_0 is that $Br(\phi \to a_0\gamma) \simeq Br(\phi \to \eta'\gamma)$;
(vi) concerning the $\sigma(600)$, there are indications of the need of such a broad, low mass scalar state to fit the Dalitz plot of $e^+e^- \to \pi^0\pi^0\gamma$.

Concerning the KLOE-2 data-taking, no big improvement is expected for the $e^+e^- \to \pi^0\pi^0\gamma$ and $e^+e^- \to \eta\pi^0\gamma$ processes, since the results of the fits are dominated by the model systematics. For $\phi \to K^0\bar{K}^0\gamma$ the insertion of the IT will increase the vertex resolution and then will improve the background rejection; with 20 fb^{-1} the expected sensitivity to the branching

Table 7. Comparison of the couplings to the SU(3) predictions for $qq\bar{q}\bar{q}$ and $q\bar{q}$ structure of the scalars(n means a non-strange quark).

	Experimental values	SU(3) predictions	
		$qq\bar{q}\bar{q}$	$q\bar{q}$
$(g_{a_0K^+K^-}/g_{a_0\eta\pi^0})^2$	0.6 - 0.7	1.2 - 1.7	$\sim$0.4
$(g_{f_0K^+K^-}/g_{f_0\pi^+\pi^-})^2$	4.6 - 4.8	>> 1	>> 1 ($f_0 = s\bar{s}$) 0.25 ($f_0 = n\bar{n}$)
$(g_{f_0K^+K^-}/g_{a_0K^+K^-})^2$	4 - 5	1	2 ($f_0 = s\bar{s}$) 1 ($f_0 = n\bar{n}$)

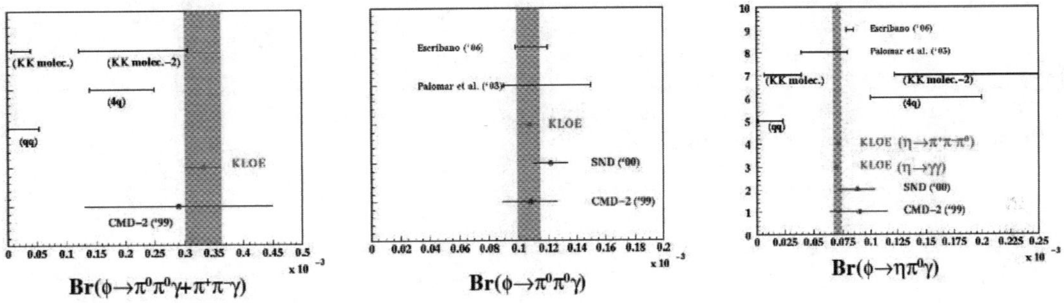

Figure 24. f_0 (left and center plots), and a_0 (right) branching ratios, compared with the previous measurements and with the theoretical predictions.

fraction is about 0.5×10^{-8}, of the same order of magnitude of the theoretical predictions, thus the first observation of this decay will be possible.
Another piece of information about the light scalars, will be provided by the study of the decays $\eta' \to \eta\pi\pi$: a scalar contribution mediated by $\sigma(600)$, $a_0(980)$ and $f_0(980)$ is expected[54] in the decay amplitude; the golden channel for KLOE-2 is $\eta' \to \eta\pi^+\pi^-$ with $\eta \to \gamma\gamma$, with 10 fb^{-1} of integrated luminosity we expect about 70000 fully reconstructed events (including a 23% efficiency evaluated from the analysis of this channel for other studies).
Moreover the $\sigma(600)$ will be studied with the $\gamma\gamma$ process $e^+e^- \to e^+e^-\pi^0\pi^0$[2].

5. Conclusions.

KLOE, with the 2.5 pb^{-1} collected during the 2001-2006 data-taking, has given a relevant contribution to the study of the light pseudoscalar and scalar mesons. In 2008 has been shown that with a new interaction scheme DAΦNE could increase the luminosity by a factor of about three. A new data-taking campaign is now starting with the KLOE detector upgraded, at first with taggers for $\gamma\gamma$ physics, and then with an Inner Tracker and new small angle calorimeters. We plan to collect about 20 pb^{-1} in the next three years, improving the precision on many of the performed measurements and studying new final states.

References

[1] Mandaglio G 2011 *these proceedings*
[2] Mascolo M 2011 *these proceedings*
[3] Amelino-Camelia G, Archilli F, Babusci D, Badoni D, Bencivenni G *et al.* 2010 *Eur.Phys.J.* **C68** 619–681 (*Preprint* 1003.3868)

[4] Adinolfi M, Ambrosino F, Andryakov A, Antonelli A, Antonelli M *et al.* 2002 *Nucl.Instrum.Meth.* **A488** 51–73
[5] Adinolfi M, Ambrosino F, Antonelli A, Antonelli M, Anulli F *et al.* 2002 *Nucl.Instrum.Meth.* **A482** 364–386
[6] Bijnens J and Gasser J 2002 *Phys.Scripta* **T99** 34–44 (*Preprint* hep-ph/0202242)
[7] Gasser J and Leutwyler H 1985 *Nucl.Phys.* **B250** 539
[8] Bijnens J and Ghorbani K 2007 *JHEP* **0711** 030 (*Preprint* 0709.0230)
[9] Borasoy B and Nissler R 2005 *Eur.Phys.J.* **A26** 383–398 (*Preprint* hep-ph/0510384)
[10] Schneider S P, Kubis B and Ditsche C 2011 *JHEP* **1102** 028 (*Preprint* 1010.3946)
[11] Colangelo G, Lanz S and Passemar E 2009 *PoS* **CD09** 047 (*Preprint* 0910.0765)
[12] Kampf K, Knecht M, Novotny J and Zdrahal M 2011 (*Preprint* 1103.0982)
[13] Naik P *et al.* (CLEO Collaboration) 2009 *Phys.Rev.Lett.* **102** 061801 (*Preprint* 0809.2587)
[14] Gormley M, Hyman E, Lee W Y, Nash T, Peoples J *et al.* 1970 *Phys.Rev.* **D2** 501–505
[15] Layter J, Appel J, Kotlewski A, Lee W Y, Stein S *et al.* 1973 *Phys.Rev.* **D7** 2565–2568
[16] Benayoun M, David P, DelBuono L, Leruste P and O'Connell H B 2003 *Eur.Phys.J.* **C31** 525–547 (*Preprint* nucl-th/0306078)
[17] Lopez A *et al.* (CLEO Collaboration) 2007 *Phys.Rev.Lett.* **99** 122001 (*Preprint* 0707.1601)
[18] Ambrosino F *et al.* (The KLOE-2 Collaboration) 2011 (*Preprint* 1107.5733)
[19] Nakamura K *et al.* (Particle Data Group) 2010 *J.Phys.G* **G37** 075021
[20] Picciotto C 1992 *Phys.Rev.* **D45** 1569–1574
[21] Abele A *et al.* (Crystal Barrel Collaboration) 1997 *Phys.Lett.* **B402** 195–206
[22] Acciarri M *et al.* (L3 Collaboration) 1998 *Phys.Lett.* **B418** 399–410
[23] Akhmetshin R *et al.* (CMD-2 Collaboration) 2001 *Phys.Lett.* **B501** 191–199 (*Preprint* hep-ex/0012039)
[24] Berlowski M, Bargholtz C, Bashkanov M, Bogoslawsky D, Bondar A *et al.* 2008 *Phys.Rev.* **D77** 032004
[25] Alavi-Harati A *et al.* (KTeV Collaboration) 2000 *Phys.Rev.Lett.* **84** 408–411 (*Preprint* hep-ex/9908020)
[26] Lai A *et al.* (NA48 Collaboration) 2000 *Phys.Lett.* **B496** 137–144
[27] Ambrosino F *et al.* (KLOE Collaboration) 2005 *Phys.Lett.* **B606** 276–280 (*Preprint* hep-ex/0411030)
[28] Gao D N 2002 *Mod.Phys.Lett.* **A17** 1583–1588 (*Preprint* hep-ph/0202002)
[29] Alde D *et al.* (Serpukhov-Brussels-Annecy(LAPP) Collaboration, Soviet-CERN Collaboration) 1984 *Z.Phys.* **C25** 225–229
[30] Prakhov S, Nefkens B, Allgower C, Bekrenev V, Briscoe W *et al.* 2008 *Phys.Rev.* **C78** 015206
[31] Unverzagt M (Crystal Ball at MAMI Collaboration) 2010 *Nucl.Phys.Proc.Suppl.* **198** 174–181 (*Preprint* 0910.1331)
[32] Di Micco B *et al.* (KLOE Collaboration) 2006 *Acta Phys.Slov.* **56** 403–409
[33] Uras A (NA60 Collaboration) 2011 *J.Phys.Conf.Ser.* **270** 012038 (*Preprint* 1108.0968)
[34] Berghauser H, Metag V, Starostin A, Aguar-Bartolome P, Akasoy L *et al.* 2011 *Phys.Lett.* **B701** 562–567
[35] Achasov M N *et al.* 2001 *Phys. Lett.* **B504** 275–281
[36] Bossi F 2011 *these proceedings*
[37] Archilli F *et al.* (KLOE-2 Collaboration) 2012 *Phys.Lett.* **B706** 251–255 (*Preprint* 1110.0411)
[38] Abdel-Bary M *et al.* (GEM Collaboration) 2005 *Phys.Lett.* **B619** 281–287 (*Preprint* hep-ex/0505006)
[39] Lai A *et al.* (NA48 Collaboration) 2002 *Phys.Lett.* **B533** 196–206 (*Preprint* hep-ex/0204008)
[40] Miller D H *et al.* (CLEO Collaboration) 2007 *Phys.Rev.Lett.* **99** 122002 (*Preprint* 0707.1810)
[41] Goslawski P 2011 *PrimeNet Workshop, Juelich*
[42] Bramon A, Escribano R and Scadron M 2001 *Phys.Lett.* **B503** 271–276 (*Preprint* hep-ph/0012049)
[43] Rosner J L 1983 *Phys.Rev.* **D27** 1101 revised version of Print-82-0634
[44] Ambrosino F, Antonelli A, Antonelli M, Archilli F, Beltrame P *et al.* 2009 *JHEP* **0907** 105 (*Preprint* 0906.3819)
[45] Escribano R and Nadal J 2007 *JHEP* **0705** 006 (*Preprint* hep-ph/0703187)
[46] Scadron M D, Rupp G, Kleefeld F and van Beveren E 2004 *Phys.Rev.* **D69** 014010 (*Preprint* hep-ph/0309109)
[47] Jaffe R L 1977 *Phys.Rev.* **D15** 267
[48] Weinstein J D and Isgur N 1982 *Phys.Rev.Lett.* **48** 659
[49] Achasov N and Kiselev A 2006 *Phys.Rev.* **D73** 054029 (*Preprint* hep-ph/0512047)
[50] Ambrosino F *et al.* (KLOE Collaboration) 2007 *Eur.Phys.J.* **C49** 473–488 (*Preprint* hep-ex/0609009)
[51] Isidori G, Maiani L, Nicolaci M and Pacetti S 2006 *JHEP* **0605** 049 (*Preprint* hep-ph/0603241)
[52] Achasov N and Gubin V 1998 *Phys.Rev.* **D57** 1987–1990 (*Preprint* hep-ph/9706363)
[53] Ambrosino F *et al.* (KLOE Collaboration) 2006 *Phys.Lett.* **B634** 148–154 (*Preprint* hep-ex/0511031)
[54] Ambrosino F *et al.* (KLOE Collaboration) 2009 *Phys.Lett.* **B681** 5–13 (*Preprint* 0904.2539)
[55] Fariborz A H and Schechter J 1999 *Phys.Rev.* **D60** 034002 (*Preprint* hep-ph/9902238)

Photoproduction of Mesons off Nuclei - The Photonuclear Programs at ELSA and MAMI

B. Krusche

Department of Physics, University of Basel, CH-4056 Basel, Switzerland
for the CBELSA/TAPS and Crystal Barrel/TAPS collaborations

E-mail: Bernd.Krusche@unibas.ch

Abstract. Recent results for the photoproduction of mesons from nuclei with tagged bremsstrahlung beams are summarized. The experiments have been done at the Mainz MAMI accelerator with the Crystal Ball/TAPS setup and at the Bonn ELSA accelerator with the Crystal Barrel/TAPS detector. Two main physics topics are covered. The electromagnetic excitation spectrum of the neutron has been studied with meson photoproduction reactions off quasi-free neutrons from light nuclei. Particularly interesting results have been obtained for η-photoproduction, where the excitation function of the neutron shows a pronounced, narrow structure which is not observed for the proton. The interaction of mesons with nuclear matter and the in-medium properties of hadrons under various aspects have been studied with meson photoproduction from nuclei, covering a large mass range (from the deuteron to lead). Questions like the possible formation of η-mesic nuclei and in-medium modifications of the σ-meson have been addressed.

1. Introduction

Photoproduction of mesons off nuclei is important for two different lines of research. The electromagnetic excitation spectrum of the neutron can only be studied in quasi-free kinematics with neutrons bound in light nuclei, in particular in the deuteron. Photoproduction of mesons from heavy nuclei is a very useful tool for the study of meson - nucleus interactions, or more generally the study of hadron-in-medium properties. Both lines of research form a significant part of the research programs at the Bonn ELSA and Mainz MAMI tagged photon facilities.

1.1. Quasi-free photoproduction off the neutron

In the same way in which the excitation spectra of atoms reflect the properties of the electromagnetic interaction, and the level schemes of nuclei reveal many details about the interaction between nucleons, are the excited states of the nucleon related to the fundamental properties of the strong interaction. Quantum Chromodynamics (QCD) at this energy scale is non-perturbative so that so far the experimental results could only be interpreted by phenomenological quark models. These models are based on different internal degrees of freedom (three equivalent constituent quarks, quark-diquark structures, additional collective modes) and different residual interactions of the quarks (gluon exchange, Goldstone boson exchange). Although it is evident, that such models can only serve as an approximation of the complicated structure of the nucleon, the idea was of course that a comparison of the observed excitation spectrum to the model results should at least give some guidance concerning the relevant properties of the interaction. However, so far this approach had only limited success. The confrontation of model results with data indicates similar problems for all models and no particular models can be clearly preferred. Already the positioning of some of the lowest lying, and therefore most basic, excited states is notoriously problematic. The N(1440)P_{11} ('Roper') and the Δ(1600)P_{33} states, which in the quark models both belong to the N=2 oscillator band, appear in nature well below the states assigned to the N=1 band. At higher excitation energies not only the ordering but already the match in simple number counting is poor; many more states are predicted than have been observed. Actually, with very few exceptions, for most combination of quantum numbers experimentally only the lowest lying state is known, while models predict a plethora of higher lying states. This problem became even more severe after the most recent partial-wave analysis of elastic pion scattering data by Arndt and collaborators [1] discarded almost half of the states listed in the Review of Particle Physics [2] as being due to statistical noise in earlier analyzes. However, here one must consider that also experimental bias could be at the center of the problem. Data analyses like the one cited above rely entirely on pion scattering and will therefore miss states that couple only weakly to πN. Since the higher the excitation energy the more decay channels are open and since there are quite some examples where nature prefers to invest in mass rather than in energy, it is not inconceivable that the higher lying states couple preferentially to decay channels involving heavier mesons or sequential decays via intermediate excited states.

Therefore, the still unanswered basic question is, whether this mismatch between experiment and model predictions is rooted in inappropriate internal degrees of freedom in the models or in bias in the experiments. Interestingly, first unquenched lattice results, which became available very recently [3], basically 're-discovered' the well-known SU(6)$\otimes$O(3) excitation structure of the nucleon and have a level counting consistent with the standard non-relativistic quark model, making simple arguments like a reduction of the effective number of degrees of freedom, e.g. by the formation of quark-diquark structures, less plausible. However, one should keep in mind, that these calculations are in a very early stage, still far away from the quality reached in the meantime for ground state properties of hadrons [4].

On the experimental side, during the last two decades photoproduction of mesons has developed into the prime tool for the experimental investigation of the excitation spectrum of the nucleon. It has now almost completely replaced meson induced reactions like elastic pion scattering. This development became possible due to the large progress in accelerator and detector technology, which nowadays allows to measure the photon induced reactions with at least comparable accuracy as hadron induced reactions, although the latter usually profit from production cross section which are larger by roughly three orders of magnitude. The choice of a different initial state and the investigation of many different final states avoid the bias against resonances that couple only weakly to $N\pi$ and the exploration of multiple meson production reactions like $\pi\pi$, $\pi\eta$... is supposed to give access to resonances that decay mainly via intermediate excited states. Furthermore, photon induced reactions have the additional advantage that the electromagnetic couplings are related to the spin-flavor correlations of the states and thus carry information about configuration mixing, which is sensitive to the details of the model wave functions.

The measurement of photoproduction reactions off the free proton has made tremendous progress during the last few years and now approaches its final phase with the measurement of single and double polarization observables However, the electromagnetic excitations are isospin dependent, so that also measurements of meson-production reactions off the neutron are required. Already the most simple observable, the total photoabsorption cross section, shows clearly different structures for the proton and neutron (see Fig. 1), demonstrating the expected differences in the electromagnetic excitation strengths. It is therefore possible, that resonances which couple (electromagnetically) only weakly to the proton may be more easily observed with

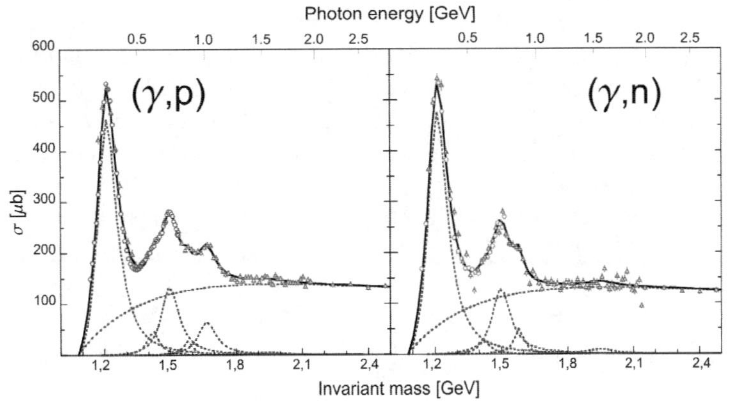

Figure 1. Cross section for total photoabsorption on the proton (left hand side) and the neutron (right hand side) [5]. Points: measured data, curves: fit of Breit-Wigner shapes of nucleon resonances ($P_{33}(1232)$, $P_{11}(1440)$, $D_{13}(1520)$, $S_{11}(1535)$, $F_{15}(1680)$ (only for proton), and $F_{37}(1950)$) and a smoothly varying background.

a neutron target. However, so far only few meson production reactions off the neutron have been measured, and most of them with much inferior quality than the corresponding reactions off the free proton. This is of course due to the complications related to the measurements off quasi-free neutrons bound in light nuclei. The detection of recoil neutrons, and even more the control of their detection efficiency, is non-trivial and the interpretation of the results is complicated by nuclear effects like Final State Interaction (FSI).

However, such programs have now been launched at CLAS, ELSA, and MAMI. These facilities are complementary because CLAS at Jlab is optimized for final states with charged particles, like e.g. from the $\gamma n \rightarrow p\pi^-$ reaction, while the almost 4π-covering electromagnetic calorimeters at ELSA and at MAMI can measure mixed charge and also complicated 'all-neutral' final states like $n\pi^0\pi^0$. We will discuss in this contribution recent results from ELSA and MAMI and pay special attention to the control of the systematic uncertainties in quasi-free production processes.

1.2. Meson photoproduction off heavier nuclei

Photoproduction of mesons from nuclei over a large range of mass numbers can be exploited for the investigation of meson-nucleus interactions, hadron-in-medium properties but also for less obvious topics like for example the study of nuclear mass form factors and nuclear matter transition form factors [6, 7, 8].

Elastic and inelastic reactions using secondary meson beams, in particular charged pions and kaons, have revealed many details of the nucleon - meson potentials. However, secondary meson beams are only available for long-lived, charged mesons. Much less is known for short-lived mesons like the η, η', and ω. Their interactions with nuclei can be studied only in indirect ways. The general idea is to produce them by some initial reaction in a nucleus and then study their interaction with the same nucleus. The absorption properties of mesons in nuclear matter and the related in-medium life times (respectively widths) have been studied at ELSA and MAMI during the last few years for a couple of pseudoscalar mesons (π^0 [9], η [10, 11], η' [12]) for the ω-meson [13, 14, 15, 16] meson, and by the LEPS collaboration at SPring-8 also for Φ-mesons [17], using the scaling of the cross sections with the atomic mass number A. The scaling is usually parameterized by

$$\frac{d\sigma}{dT}(T) \propto A^{\alpha(T)} \quad , \tag{1}$$

where T is the kinetic energy of the mesons. A value of α close to unity corresponds to a cross section scaling with the volume of the nucleus, i.e. with vanishing absorption, while a value of $\approx 2/3$ indicates surface proportionality, corresponding to strong absorption. These scaling coefficients can then be converted to absorption cross sections using for example Glauber-type approximation models. A basically equivalent concept are the so-called transparency ratios, used in some of the above cited work, which compare the nuclear cross sections scaled by the nucleon number to the free nucleon cross section (or for practical reasons to the cross section of a light reference nucleus scaled by the mass number of that nucleus). An important result of such analyses is an in-medium width of the ω meson on the order of 130 - 150 MeV in normally dense nuclear matter from which in a Glauber-type analysis, an absorption cross section of 70 mb was deduced [14]. This value is roughly a factor of three larger than the input previously used for $\sigma_{\omega N}$ in models. A similar result had been previously reported from the LEPS collaboration for the Φ meson [17]. In this case an absorption cross section of $\approx$ 30 mb was deduced, which has to be compared to the free nucleon absorption cross section of 7.7 - 8.7 mb. Both experiments give strong evidence for the much discussed in-medium modification of vector mesons.

An overview of the results for some mesons is summarized in Fig. 2. The absorption properties

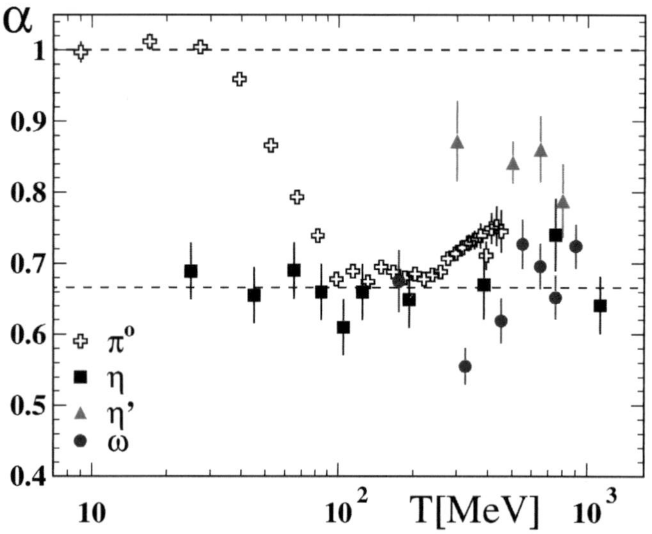

Figure 2. Scaling coefficients α as function of kinetic meson energy T for π^0, η, η', and ω mesons.

vary for the different meson types. We shortly comment on some features which are relevant for experiments discussed later in this contribution. The absorption coefficient for pions has a pronounced energy dependence. The reason is, that as soon as pions reach a kinetic energy of $\approx$100 MeV they can excite a nucleon into the Δ-resonance and are readily absorbed. Therefore, nuclei are basically 'black' for pions in this energy range. However, at low kinetic energies pions - the Goldstone bosons of chiral symmetry - interact only weakly with nucleons so that below kinetic energies of $\approx$40 MeV nuclei become transparent for pions. This behavior is important for the study of the in-medium invariant-mass distributions of pion pairs discussed below in view of in-medium modifications of the σ meson. Only pion pairs where both mesons have very small momenta are not significantly influenced by FSI effects. Another topic discussed below is the formation of mesic nuclei (quasi-bound systems of a nucleus and a meson). Due to the very weak low-energy interaction pions are obviously no candidates for such systems, but the situation is different for the heavier mesons, in particular the η-meson.

The low-momentum η-nucleon interaction is completely dominated by the s-wave excitation of the $S_{11}(1535)$ resonance, which overlaps with the $N\eta$ production threshold and has a $\approx$50 % decay branching ratio into $N\eta$ [18]. Therefore, nuclei are 'black' for η-mesons even for almost vanishing momenta of the meson. As a side remark, this strong coupling to the S_{11} resonance with almost equal branching ratios to $N\pi$ and $N\eta$ provides also a very efficient mechanism for the production of secondary η-mesons in nuclei via the $\gamma N \rightarrow N\pi$, $\pi N \rightarrow N\eta$ reaction chain at somewhat higher incident photon energies [11, 12], which must be properly accounted for in the extraction of the η-absorption probabilities.

Journal of Physics: Conference Series **349** (2012) 012003 doi:10.1088/1742-6596/349/1/012003

The special features of the $N\eta$ interaction have already 25 years ago lead to speculations about the existence of so-called η-mesic nuclei. Liu and Haider [19], based on the results from coupled channel analyses of η-production reactions, suggested the possible formation of quasi-bound η - nucleus states for $A > 10$ nuclei. Experimental evidence has been sought in pion induced reactions [20, 21], but those experiments did not produce conclusive evidence. More recently, Sokol and co-workers [22, 23] claimed evidence for the formation of η-mesic nuclei from bremsstrahlung induced reactions on ^{12}C

$$\gamma +^{12}\mathrm{C} \rightarrow p(n) +^{11}_{\eta}\mathrm{B}(^{11}_{\eta}\mathrm{C}) \rightarrow \pi^{+} + n + X \qquad (2)$$

where the $n\pi^{+}$ pairs were detected in the final state. The η-meson is produced in quasi-free kinematics on a nucleon (p,n) so that it is almost at rest in the residual $A = 11$ nucleus. When a quasi-bound state was produced, the η-meson can be re-captured by a nucleon into the S_{11} excitation, which may then decay into a pion-nucleon back-to-back pair. Sokol and Pavlyuchenko [23] claim an enhancement above background from quasi-free pion production for certain kinematic conditions.

Suggestions for the existence of very light η-mesic states, in particular involving helium or tritium nuclei, were later based on results from refined analyses of the η-nucleon, η-nucleus scattering length (see [24] for a summary), using precise data from threshold η-photoproduction off the proton [25], the deuteron [26, 27, 28, 29], and helium nuclei [30, 31]. Such states have been in particular sought in the threshold behavior of hadron and photon induced η-production reactions from light nuclei, following the conjecture that a resonant state close to production threshold will enhance the cross section with respect to phase-space expectations. Particularly large effects have been observed for the $dp \rightarrow \eta^{3}\mathrm{He}$ [32, 33] reaction and for the $\gamma^{3}\mathrm{He} \rightarrow \eta^{3}\mathrm{He}$ reactions. For the latter recently much more precise data have been measured at MAMI, which we will discuss below.

As a final remark to meson-nucleus interactions, we like to emphasize (see Fig. 2) that FSI effects seem to be considerably smaller for η' mesons than for the η and ω. This might be a first indication that the η'-nucleon interaction is not strong, which is important in view of the discussion about the possible existence of mesic states for this heavy meson.

Finally, we will discuss very recent results related to the in-medium properties of the σ-meson. A possible in-medium mass shift of the σ-meson has been been much discussed in connection with partial chiral symmetry restoration. The $J^{\pi} = 0^{+}$ σ-meson is the chiral partner of the $J^{\pi} = 0^{-}$ pion, but in vacuum their masses are very different, which is a well-known manifestation of chiral symmetry breaking. The naive assumption that the two masses should become degenerate in the chiral limit is supported by model calculations. A typical result is the density dependence of the mass calculated in the Nambu-Jona-Lasino model by Bernard, Meissner and Zahed [35], who predicted a considerable lowering of the σ-mass already at normal nuclear matter density ρ_0, where the pion mass is still stable. Due to the strong coupling of the σ to scalar, isoscalar pion pairs an in-medium shift of its mass should be reflected in the invariant mass distributions of $\pi^{0}\pi^{0}$ (and $\pi^{+}\pi^{-}$) pairs [36, 37, 38, 39]. Such effects have been sought in pion induced reactions at the CHAOS spectrometer (see e.g. [40, 41]) and with the Crystal Ball detector at BNL [42], as well as with photon induced reactions at MAMI [43, 44]. All experiments found some in-medium effects. However, they might also arise from final state interactions of the pions. Originally, the different behavior of possible decay channels of the σ ($\pi^{0}\pi^{0}$, $\pi^{+}\pi^{-}$) on one hand side, and isospin-channels that cannot couple to the σ (e.g. $\pi^{+}\pi^{+}$, $\pi^{0}\pi^{\pm}$) on the other hand, was taken as strong argument that FSI effects alone cannot explain the data. However, later [44] it was shown that intricate side-feeding effects from charge-exchange scattering of the pions can mimic σ in-medium modifications. It is therefore highly desirable to study such effects for pions with very low kinetic energies (that is double pion production in the vicinity of the production thresholds) in order to avoid the FSI effects.

2. Experiments

The experiments have been done at the tagged photon facilities of the Bonn ELSA accelerator and the Mainz MAMI accelerator. At both facilities almost 4π covering electromagnetic calorimeters combined with charged particle detectors have been used. The results discussed here have been obtained with unpolarized photon beams and unpolarized liquid or solid targets, but programs using linearly and circularly polarized beams and longitudinally and transversely polarized targets are also under way. A more comprehensive overview of experimental facilities for tagged photon experiments is given in [45].

2.1. The Crystal Barrel/TAPS setup at the Bonn ELSA accelerator

The electron stretcher accelerator facility ELSA [46, 47] delivers electron beams with energies up to 3.5 GeV and intensities of typically a few nA. Most of the experiments discussed here, have been done with the setup summarized in Fig. 3. The setup and the analysis procedures are discussed in detail in [48, 49]. The main detector was a 4π electromagnetic calorimeter combining the Crystal Barrel [50] (1290 CsI (Tl) crystals of 16 radiation lengths X_o all mounted in a target pointing geometry) with the TAPS detector [51, 52]. The barrel covered the full azimuthal angle for polar angles between 30° and 168°. The forward angular range was covered by TAPS mounted as a wall comprising 528 BaF_2 crystals of hexagonal shape with an inner diameter of 5.9 cm and a length of 25 cm corresponding to 12 X_o. The front face of the BaF_2 wall was located 1.18 m from the center of the targets, which were mounted inside the barrel, and each module of the detector was equipped with an individual plastic scintillator for charged particle identification. The TAPS wall was read out with photomultipliers while the barrel crystals were equipped with photodiodes, which did not deliver timing information. Therefore only hits in TAPS could be used in the first level trigger, which somewhat limited the detection efficiency of the device. This is not a disadvantage for measurements off the free proton, where in addition the proton can generate triggers in the 'Inner-detector' [53], a three-layer scintillating fiber detector surrounding the target at polar angles between 28° and 172°. It limits, however, the type of reactions that can be studied off quasi-free neutrons. Only reactions with large photon multiplicity have reasonable trigger efficiencies. For future experiments the trigger capability of the detector will be ungraded by a new readout system based on **A**valanche **P**hoto**D**iodes (APDs), which can deliver time information and trigger signals.

For most of the experiments discussed here, a liquid deuterium target was mounted inside the barrel. The more recent measurements with a polarized d-buthanol target were done with a slightly modified setup, using a smaller TAPS wall combined with a forward plug of the barrel of CsI crystals equipped with photomultipliers [45].

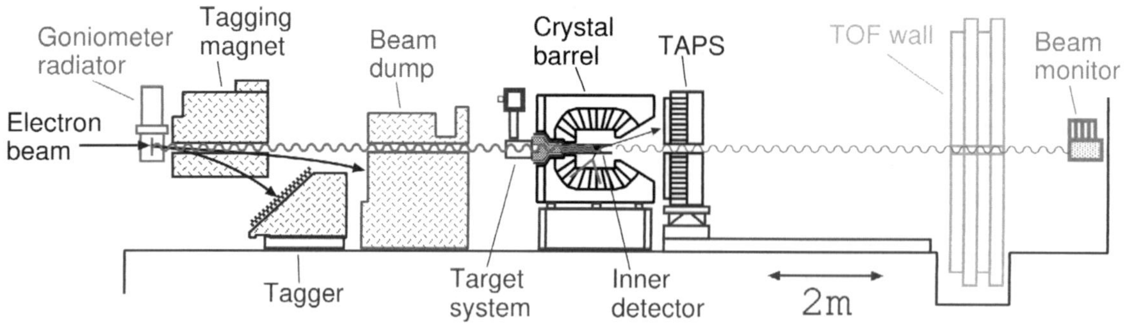

Figure 3. Experimental setup at the Bonn ELSA accelerator combining the Crystal Barrel detector, the TAPS detector as forward wall, inner detectors, and the photon tagging facility.

2.2. The Crystal Ball/TAPS setup at the Mainz MAMI accelerator

The MAMI accelerator in Mainz [54, 55] can now deliver electron beams with energies up to 1.5 GeV, however for some of the experiments lower beam energies have been used (883 MeV). Like at ELSA photon beams are produced by the bremsstrahlung process and are tagged with the Glasgow magnetic spectrometer [56]; at maximum electron energy with a typical resolution of 4 MeV (2 MeV for 883 MeV electron beam energy). Since the accelerator is designed also for electron scattering experiments, it can deliver electron beam intensities far larger than needed for tagged photon experiments ($\approx$ 100 μA, compared to some tens of nA used for tagging). For the experiments discussed here, Liquid hydrogen, liquid deuterium, liquid ^{3}He, and solid lithium, carbon, calcium, and lead targets have been used.

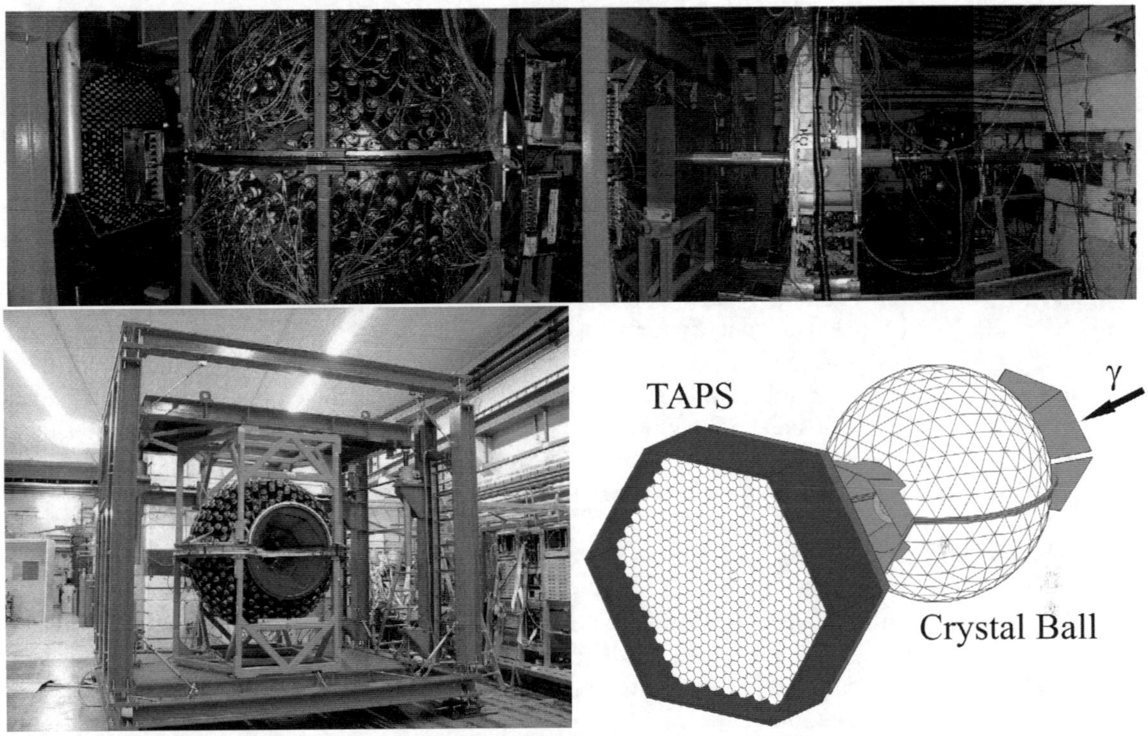

Figure 4. Overview of the MAMI setup. Upper part: beam coming from the right, Crystal Ball at the left of the picture, TAPS forward wall at the very left. Bottom part, left hand side: Crystal Ball in setup phase, right hand side: schematic drawing of the calorimeter.

The electromagnetic calorimeter (see Fig. 4) combined the Crystal Ball detector [57] and part of the TAPS detector. The Crystal Ball detector consists of 672 NaI crystals covering the full azimuthal angle for polar angles from 20° to 160°. For the TAPS forward wall, which covered angles from 21° to 2°, two different configurations have been used. The first was built from 510 BaF_2 crystals and placed at a distance of 1.75 m from the target center, the second comprised 384 elements and was placed at 1.457 m from the target. Both versions provided good resolution for time-of-flight and ToF-versus-energy measurements for particles detected in TAPS. The TAPS modules were again equipped with individual plastic scintillators for charged particle detection. The target inside the Ball was surrounded by a particle detection detector (PID) [58], which via $E - \Delta E$ measurements, comparing the energy loss in the PID and the deposited energy in the Ball, allowed also for separation of protons and charged pions. A detailed discussion of the setup and analysis procedures is given in [59].

3. Results and Perspectives

In this section we will summarize recent results, some of which are already published, but most are in preparation for publication and shortly indicate the direction of planned or already running follow-up experiments.

3.1. Quasi-free photoproduction of mesons off the neutron

Before we present the results for some selected recently studied meson production channels, we will shortly discuss some general issues, arising for the measurement of meson production reactions of quasi-free nucleons. The most important issues are:

- How well are the systematic uncertainties involved in the detection of mesons in coincidence with recoil neutrons and protons understood?
- How can we best deal with the effects of nuclear Fermi motion?
- What is the importance of other effects related to the nuclear environment like e.g. FSI?

These questions have been studied in detail for the measurements done at the Bonn facility (the data from MAMI are all still under analysis), using data from η- [49], η'- [48], $2\pi^0$-, and $\pi^0\eta$ production. We discuss here as an example the η-channel.

The electromagnetic calorimeters at both facilities are well capable of also detecting recoil protons and neutrons, although with very different detection efficiencies (up to 90 % for protons roughly 15 - 30 % for neutrons). Separation of photons, protons, and neutrons in the TAPS forward walls can be achieved with the help of the thin plastic scintillators in front of the BaF_2 modules, pulse-shape analysis, and time-of-flight, respectively ToF-versus-energy methods. Separation of protons from photons and neutrons can be done with the charged particle detectors surrounding the target, but no separation between photons and neutrons detected in the Barrel or Ball is possible (time-of-flight path too short). Therefore, the only possible strategy to analyze reactions in coincidence with such recoil neutrons is to accept events with (n+1)-neutral hits, where n is the number of expected photons. Then test the reaction hypothesis by looking at all possible invariant mass combinations of those hits (for example for $\eta \rightarrow 3\pi^0 \rightarrow 6\gamma$ three photon pairs must combine to invariant masses close to the π^0-mass and the invariant mass of these six photons should be close to the η-mass). In case of more than one possible combination the 'best' one is selected by a χ^2 test, and in the end the left-over neutral hit is tentatively accepted as recoil neutron. This procedure must be precisely reproduced by simulations for the reaction detection probability, which is not straight forward since already the simulation of the detector hardware response to neutrons is notoriously difficult. In case of recoil protons, detection efficiencies can be easily extracted using production reactions off free protons, but there are not many possible reactions one can use in the case of neutrons ($\gamma p \rightarrow n\pi^0\pi^+$ is one of the best suited). As far as the detector response to neutrons is concerned, simulations with the GCALOR program package [60] provide rather satisfactory results, but also they require very careful control of all parameters (in particular detector thresholds).

Therefore it is extremely helpful, that an intrinsic control exists. The cross sections for meson production off the deuteron (or other light nuclei) must obey the relation:

$$\sigma_{incl} = \sigma_p + \sigma_n + \sigma_{coh} \quad , \tag{3}$$

where σ_{incl} is the inclusive cross section without any condition for recoil baryons, σ_p, σ_n are the cross sections in coincidence with recoil protons and neutrons, and σ_{coh} is the coherent production cross section off the respective nucleus. In most cases, the latter one is completely negligible in comparison to the quasi-free reactions, which we will assume in the following discussion (if it is not, it can be also measured). This means, that the two exclusive cross sections with coincident protons and neutrons have to add up to the inclusive cross section. Only

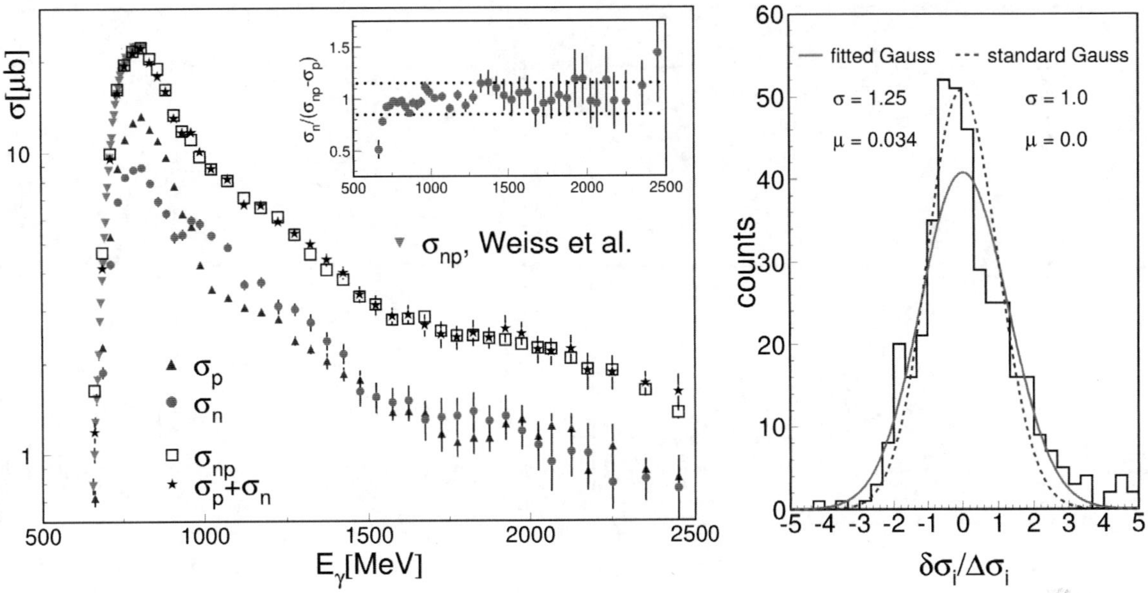

Figure 5. Comparison of cross sections for quasi-free η photoproduction [49]. Left hand side: total cross sections, (Blue) upward triangles: quasi-free proton cross section σ_p, (red) dots: quasi-free neutron cross section σ_n, (black) open squares: inclusive quasi-free cross section σ_{np}, (black) stars: $\sigma_n+\sigma_p$. Downward (magenta) triangles: inclusive quasi-free cross section from Weiss et al. [29]. Insert: ratio of neutron cross sections. Right hand side: Distribution of deviations between $d\sigma_n/d\Omega$ and $d\sigma'_n/d\Omega = d\sigma_{incl}/d\Omega - d\sigma_p/d\Omega$. Solid (red) curve: fitted Gaussian distribution (width σ=1.25, mean μ=0.034), dashed (blue) curve: standard Gauss: (σ=1, μ=0).

the detection efficiency for the mesons enter into the inclusive cross section. The completely different proton and neutron detection efficiencies enter into σ_p and σ_n so that Eq. 3 in general holds only when all detection efficiencies are correct. This means that the neutron cross section can be measured in two different ways, either as σ_n by the detection of recoil neutrons or as difference $\sigma'_n = \sigma_{incl} - \sigma_p$. The result of such an analysis for η-production from the Crystal Barrel/TAPS experiment [49] is shown in Fig. 5. The left hand side of the figure demonstrates the agreement of the total cross sections. The right hand side of the figure shows the distribution of the deviations $\delta\sigma_i$ normalized by the statistical uncertainties $\Delta\sigma_i$

$$\frac{\delta\sigma_i}{\Delta\sigma_i} \equiv \frac{d\sigma'_n/d\Omega - d\sigma_n/d\Omega}{\sqrt{\Delta^2(d\sigma'_n/d\Omega) + \Delta^2(d\sigma_n/d\Omega)}} \tag{4}$$

for all data points (420 entries) of the angular distributions from production threshold to 2.5 GeV, compared to a Gaussian distribution, which is fairly close to a standard Gaussian distribution. Very similar results have been obtained for η' production [48] and $\pi^0\pi^0$ production (in preparation). In case of reactions with very small cross sections, it may be useful to use σ'_n instead of σ_n, because of better statistical quality. In the extreme, one may even attempt to measure σ''_n, where only events *without* recoil protons are accepted, provided proton acceptance and detection efficiency can be pushed above the $\approx$90% level and the corresponding background contribution to σ''_n can be simulated. Since neutron detection efficiencies are typically below the 30% limit, this would increase count rates by approximately a factor of three. The disadvantage of this indirect methods is, that the kinematic reconstruction discussed below for the elimination of Fermi motion effects is then not possible.

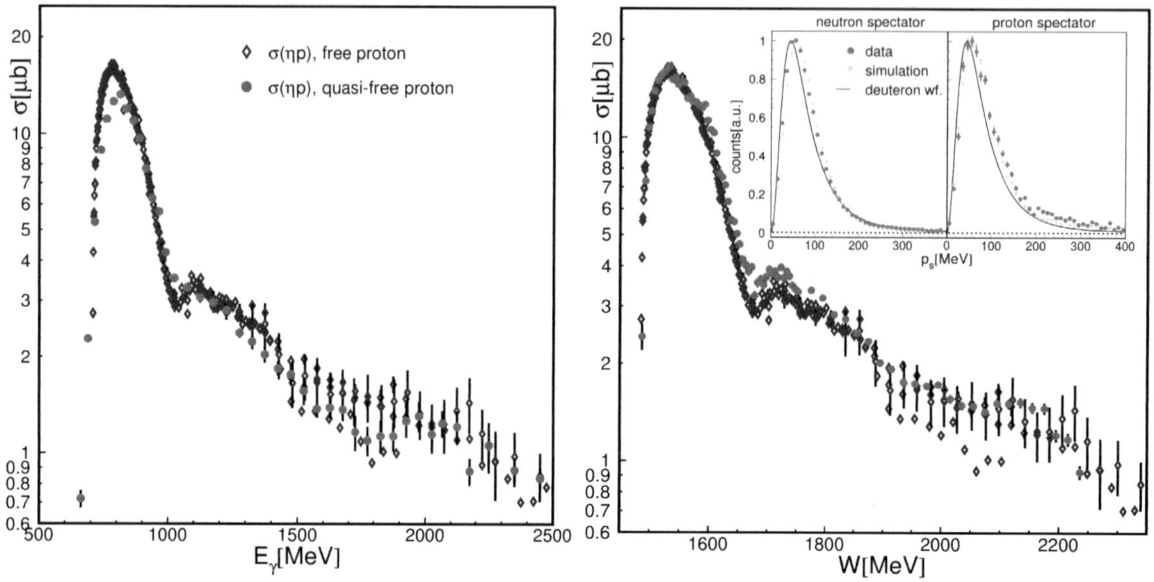

Figure 6. Comparison of free and quasi-free photoproduction of η-mesons off the proton (see text). Left hand side: (Blue) diamonds world data base for $\gamma p \to p\eta$ versus incident photon energy. (Red) dots: quasi-free $\gamma' p' \to' p'\eta$ reaction. Right hand side: Free proton data (blue diamonds) versus W. (Red) dots: quasi-free data from kinematic reconstruction of W. Insert: reconstructed momenta of spectator nucleons compared to momentum distributions from deuteron wave function (neutron spectator corresponds to detection of recoil proton and vice versa). Curve: deuteron wave function, simulation includes detector resolution.

The influence of Fermi motion is shown for the reaction $\gamma p \to p\eta$ in Fig. 6, where at the left hand side the total cross sections for free protons and protons bound in the deuteron are compared as function of incident photon energy. The results are similar, but the slope at threshold is less step, the maximum of the S_{11} peak is underestimated, and the little dip-like structure around 1 GeV is smeared away in the quasi-free data; all typical effects of Fermi motion. These effects can be removed when instead of the incident photon energy (or W calculated from it neglecting Fermi motion), the invariant mass W calculated from the measured four-vectors of the recoil nucleon and the meson is used. In this case, in principle angle and kinetic energy of the recoil nucleon are needed, which is straight forward for the proton, but less so for the neutron. At forward angles it can be measured by time-of-flight methods, but not for neutrons in the Barrel or Ball. However, since the incident photon energy is known and the incident deuteron is at rest, the three-body final state (meson, participant, and spectator nucleon) is already kinematially completely determined by the mass of all three particles, the three-momentum vector of the meson, and the angles of the participant nucleon. The four missing variables (three-momentum of the spectator, kinetic energy of the participant nucleon) can be calculated from energy and momentum conservation. An example is shown at the right hand side of Fig. 6. The insert in the figure shows the momenta of the spectator nucleons derived from the kinematic reconstruction. As long as FSI effects are negligible and the reaction kinematics was quasi-free, they should correspond to the momentum distribution of the nucleons bound in the deuteron, which can be derived from the deuteron wave function [61]. As shown in the figure, the agreement is excellent when the instrumental resolution is taken into account. The main plot then demonstrates the very good agreement between free and quasi-free data, which also implies that effects beyond Fermi smearing (FSI) play no significant role for the η-channel.

In the following we will discuss results from a few selected channels in view of their physical interpretation.

3.1.1. photoproduction of η-mesons Photoproduction off η-mesons off the free proton has been studied during the last two decades in much detail by all laboratories running tagged photon experiments [25, 62, 63, 64, 65, 66, 67, 68, 69, 70, 71, 72, 73, 74, 75, 76, 77]. The main results were the strong dominance of the $S_{11}(1535)$ in the threshold region, with only small contributions from non-resonant background terms, a destructive interference between the $S_{11}(1535)$ and the $S_{11}(1650)$ at somewhat higher incident photon energies, and a small contribution from the $D_{13}(1535)$ resonance. The latter showed up in a detailed analysis of the angular distributions [25, 29] via an interference with the leading S_{11} excitation; but the effect was more pronounced for the beam asymmetry measured with linearly polarized photons [62, 72] and an analysis in the framework of the 'Eta-MAID' model [78] allowed the extraction of the tiny $N\eta$ branching ratio (0.23±0.04 %) [2] of the D_{13} resonance. The situation is less clear above this range, where different analyses like Eta-MAID [78] and Bonn-Gatchina (BoGa) [79] propose different resonance contributions (see e.g. [72]).

Measurements of the $\gamma n \rightarrow \eta n$ reaction were originally mainly motivated as an almost background free study of the isospin composition of the electromagnetic excitation of the $S_{11}(1535)$ resonance. Furthermore, for the energy range above the S_{11}, there were predictions [78] for a strong contribution from the $D_{15}(1675)$ resonance, which is suppressed for the proton due to the Moorehouse selection rules [81] for electromagnetic excitations. Early results [26, 27, 28, 29], covering only the S_{11} range (mainly up to 900 MeV), basically confirmed expectations and yielded a precise value for the ratio of proton and neutron electromagnetic couplings of the $S_{11}(1535)$ [80]. But there was a large surprise when the experiments reached incident photon energies above 1 GeV: all cross section measurements extending into this region [82, 83, 84] found a significant and relatively narrow structure in the excitation function. The result from the ELSA experiment [84, 49] is summarized in Figs. 7,8. The structure is located around an invariant mass of $W \approx 1.65$ GeV with a width below 50 MeV. At the time when it was first seen in experiment, the possible discovery of exotic pentaquark states was hotly debated and models predicted that the P_{11}-like state of the anti-decuplet should couple electromagnetically much more strongly to the neutron than to the proton, have a large branching ratio into $N\eta$, and mass/width comparable to the observations of the structure in η-photoproduction. However, in the meantime evidence for the

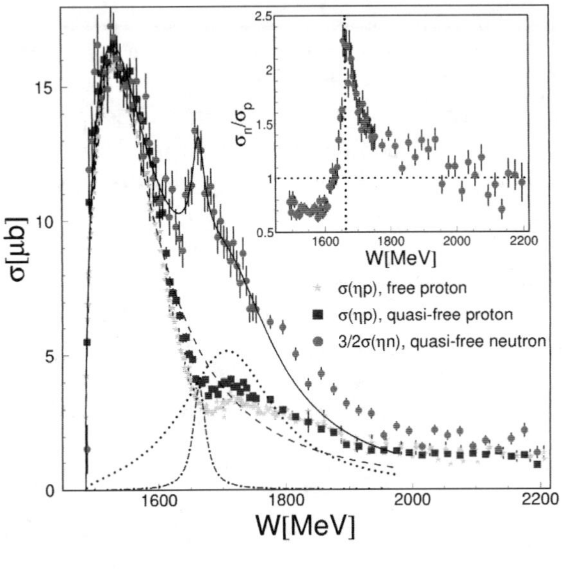

Figure 7. Quasi-free photoproduction of η-mesons off the proton and neutron [49]. Main plot: Total cross sections as function of the final state invariant mass W reconstructed from measured four-vectors (Red) dots: quasi-free neutron, (blue) squares: quasi-free proton, (green) stars: free proton data. The quasi-free neutron data are normalized the to proton data in the S_{11} peak (factor $\approx 2/3$) All curves for neutron data; dashed: fitted S_{11} line shape, dotted: broad Breit-Wigner resonance, dash-dotted: narrow Breit-Wigner, solid: sum of all. Insert: ratio of quasi-free neutron - proton data.

Θ^+ pentaquark has faded and many attempts for alternative explanations of the results from $\gamma n \to \eta n$ have been made [85, 86, 87, 88, 89], ranging from intricate interference effects over threshold cusps to nucleon resonances with different properties. The data reconstructed from the final state kinematics and thus not effected by nuclear Fermi motion have only ruled out an explanation involving only a normally broad (i.e. typically >150 MeV) nucleon resonance. Apart from that the nature of this structure is still unknown.

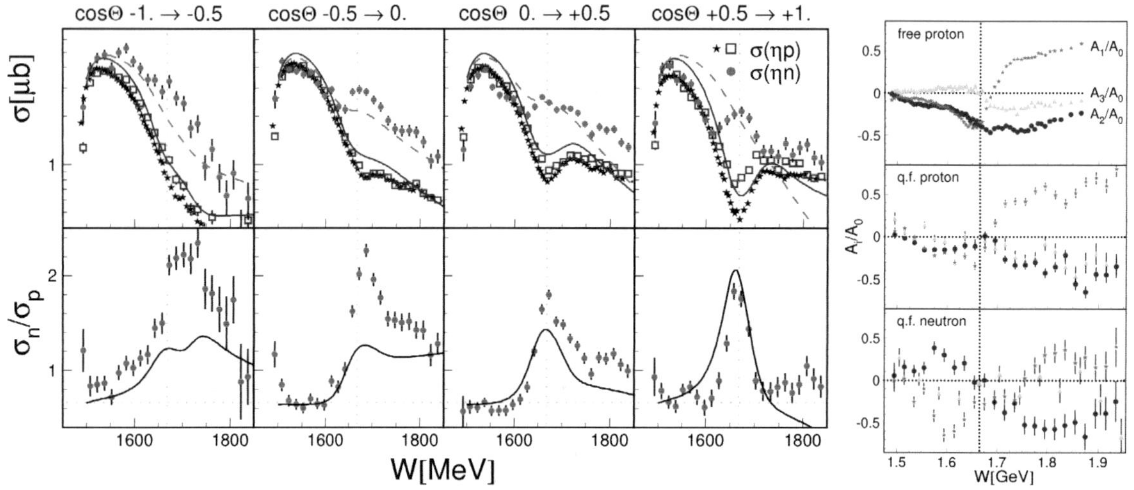

Figure 8. Quasi-free photoproduction of η-mesons off the proton and neutron [49]. Left hand side: First row: excitation functions for different bins of η cm polar angle. (Blue) open squares: quasi-free proton data, (black) stars: free proton data from [77], (red) dots: quasi-free neutron data scaled up by 3/2. (Blue) solid lines: η-MAID [78] for the proton target, (red) dashed lines: η-MAID for the neutron target. Second row: ratio of neutron and proton cross section for data and η-MAID. Vertical dotted lines: position of narrow peak in neutron data, horizontal dotted lines: σ_n/σ_p=2/3. Right hand side: coefficients of Legendre series fitted to angular distributions normalized to A_0. From top to bottom: free proton [77], quasi-free proton, quasi-free neutron [49].

Some additional information comes from the angular dependence shown in Fig. 8 [49]. The excitation functions for different ranges of η cm angle show a clear angular dependence of the effect. It is most prominent for forward angles. Moreover, the peak-like structure in the neutron data corresponds to a dip-like structure in the proton data with similar angular dependence. This might be an indication that at least at some level an interference term with opposite sign for proton and neutron contributes. It is noteworthy, that the MAID model [78], which reproduces at least for forward angles qualitatively the neutron/proton cross section ratio, does so only because it reproduces the 'dip' in the proton data (because it has been fitted to it) but not the peak in the neutron data. This causes also some doubts, whether the parameterization of the proton data reflects the correct physics. The coefficients of the Legendre polynomials fitted to the data and normalized to the leading coefficient A_0, which is proportional to the total cross section, show that for both, proton and neutron there is a dramatic change in the angular dependence around the position of the discussed structures. So far, there are no models available which reproduce this behavior.

The significance of the structure has been further increased with new data from the MAMI facility for η-photoproduction off quasi-free nucleons from the deuteron but also from ^{3}He. The data are still under analysis and the much superior statistical quality of the deuteron data will allow more precise investigations of the angular dependencies. Preliminary excitation functions are shown in Fig. 9. The data taken with the helium target are a further stringent test that no systematic problems influence the data. They behave exactly as expected, although extracted from a much more challenging environment (larger Fermi motion, proton/neutron ratio 2:1). The somewhat worse resolution for helium in the data with reconstructed final state kinematics is due to the fact that the reconstruction is only exact for the three-particle final state from the deuteron target, but has to rely on the approximation of zero relative momentum between the two spectator nucleons in the helium case.

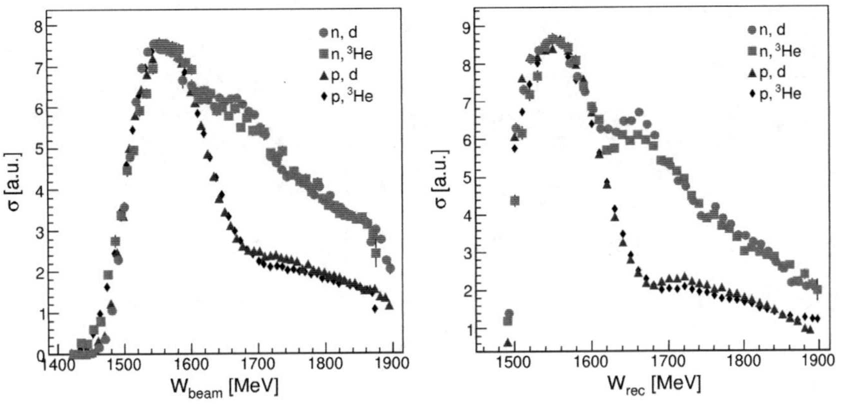

Figure 9. Preliminary excitation functions for quasi-free protons and neutrons from ^{2}H (only small fraction of total statistics) and ^{3}He. Left hand side: W from incident photon energy, right hand side: W from final state invariant mass.

More detailed partial wave analysis of the data will require the measurement of polarization observables. So far, only the beam asymmetry Σ has been measured for $\gamma n \rightarrow \eta n$ by the GRAAL collaboration [90]. More recently, first data have been taken at ELSA to measure the helicity dependence of the cross section with a longitudinally polarized d-buthanol target and a circularly polarized photon beam. At MAMI first data have been taken with a transversely polarized d-buthanol target and a circularly polarized beam to measure the target asymmetry T and the double polarization observable F. These data are still in a very early state of analysis and measurements will continue to collect more counting statistics.

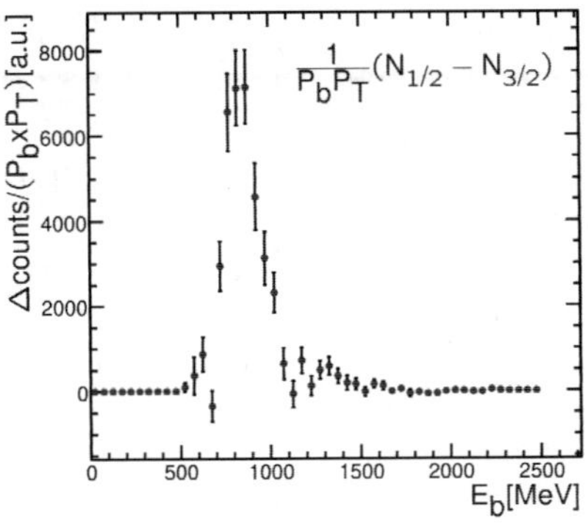

Figure 10. Very preliminary (almost online) result for the count rate difference (normalized to polarization degrees) for $\sigma_{1/2}$ (anti-parallel photon-proton spin) and $\sigma_{3/2}$ (parallel spins) for η photoproduction off quasi-free protons from the deuteron.

Figure 10 shows an example for the measurement of the helicity dependence of the cross section (online data) for quasi-free protons as demonstration of the feasibility of these experiments. Apart from a little effect of Fermi smearing the result is in good agreement with free proton data [91]. It shows the expected dominance of the $\sigma_{1/2}$ component in the S_{11} range. Analysis of the neutron data is expected to reveal whether the structure is related to $J = 1/2$ or to $J > 1/2$ partial waves.

3.1.2. photoproduction of η'-mesons Another channel that has recently attracted much interest in view of nucleon resonances is the $N\eta'$ final state. Like the η, also the isoscalar η' couples only to isospin-1/2 $N^\star$ resonances. Due to its large mass the free-nucleon threshold is at $E_\gamma \approx$1447 MeV, corresponding to $W \approx$1.9 GeV. This means that it is well suited to search for contributions from $N^\star$ resonances around W=2 GeV, i.e. in the nucleon excitation range with a large density of 'missing' $N^\star$ resonances. Since this region is still close to the production threshold one might expect that only few partial waves contribute, which would simplify the interpretation of the data. For the free proton, apart from old bubble chamber data, first results became available from the SAPHIR experiment at ELSA [92]. Subsequently, much more precise data have been reported from the CLAS experiment at Jlab [93, 74] and the Crystal Barrel/TAPS setup at ELSA [75]. Interpretation of these data is still controversial [94, 95]; almost all model fits agree that close to threshold there is an important contribution from an S_{11} resonance and at higher energies t-channel background processes are significant but other resonance contributions are not well constrained by the measured angular distributions alone. Future measurements of polarization observables are mandatory.

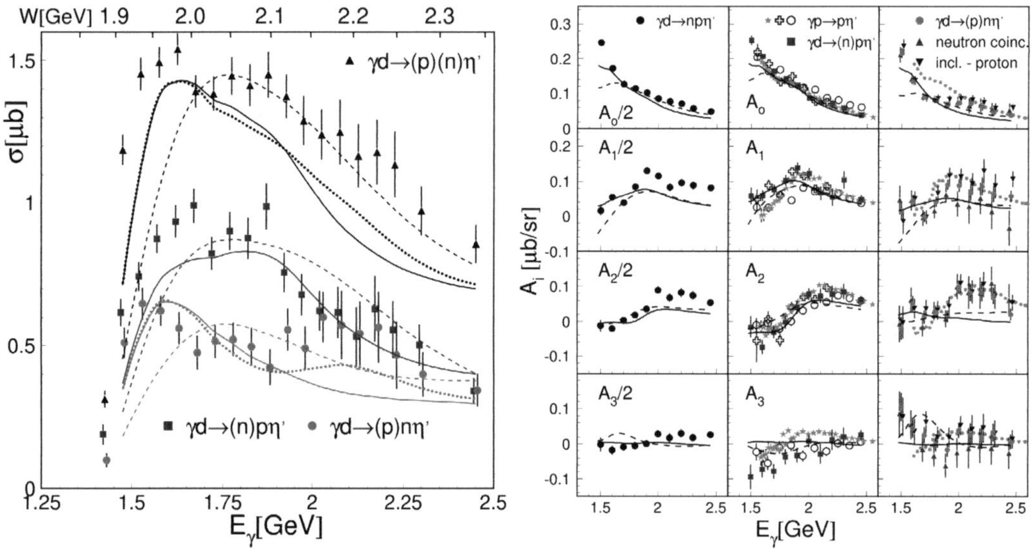

Figure 11. Left hand side: Total cross section for η' production off the deuteron. [48]. Curves: fits with reaction models, solid, dotted: different solutions from [95], dashed: η'-MAID [94]. Right hand side: Coefficients of the Legendre Polynomials for the fitted angular distributions [48]. Left hand column: inclusive reaction scaled down by factor of 2. Center column (proton targets): quasi-free data (blue squares); free proton data: open crosses [93], open circles [75], (magenta) stars [74]. Right hand column (quasi-free neutron data): (blue) upward triangles from neutron coincidence, (black) downward triangles from difference of inclusive and proton data, (red) circles from averaged data. In all plots solid lines: NH model, dashed lines: η'-MAID; for neutron: (magenta) dotted lines CLAS proton data.

The first measurement of the $\gamma n \rightarrow \eta' n$ reaction was recently published [48] from the CBELSA/TAPS experiments, using a liquid deuterium target. Total cross sections and Legendre coefficients of the angular distributions are summarized in Fig. 11. Also in this case good agreement between free and quasi-free proton data was found (i.e. no significant FSI) and the behavior of proton and neutron cross sections points to different resonance contributions. However, also for the neutron target, model solutions are not unique and constraints from polarization observables are needed.

3.1.3. photoproduction of π^0-mesons Photoproduction of pions in the Δ-resonance region and throughout the second and third resonance regions of the nucleon is one of the best studied meson production channels; however also in this case the data base for the all-neutral final state $n\pi^0$ is very sparse [45]. So far, isospin decompositions rely on measurements of the $\gamma n \to p\pi^-$ reaction, but since the non-resonant backgrounds can be very different for charged and neutral mesons, this is unsatisfactory. However, it is known that at least in the range of the Δ-resonance FSI effects for π^0 production off the deuteron are significant [98, 99]. This is not unexpected, already the relatively large contribution from coherent reaction processes in this energy range [98] highlights the importance of nuclear effects. The situation at higher incident photon energies is much less well understood, so far neither exclusive data nor model results including FSI effects were available. The only hint came from a measurement of inclusive π^0 production off the deuteron in the second resonance region [98]. These data were significantly overestimated by simple plane wave approximations using as input the results of the MAID [96] or SAID [97] analyses for the elementary cross sections.

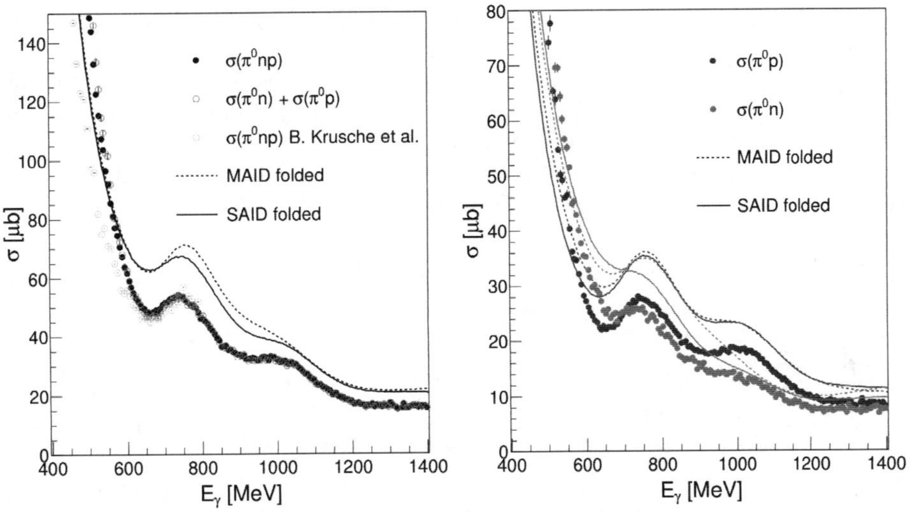

Figure 12. Very preliminary results for quasi-free photoproduction of π^0 mesons off the deuteron in the second and third resonance region. Left hand side: inclusive $\gamma d \to np\pi^0$ data compared to the sum of exclusive cross sections and results from MAID (dashed line) [96] and SAID (solid line) [97]. Both model results are for the incoherent sum of proton/neutron cross sections folded with Fermi motion. Right hand side: Same for exclusive quasi-free $\gamma p \to p\pi^0$ and $\gamma n \to n\pi^0$ reactions.

New exclusive data from the Crystal Ball/TAPS experiment, which are still under analysis, seem to confirm this observation. Preliminary results are summarized in Fig. 12. The left hand side shows the inclusive cross section $\gamma d \to np\pi^0$ constructed in two ways, either by an analysis without any conditions for recoil nucleons or as sum of the quasi-free neutron and proton cross sections measured in coincidence with recoil nucleons (coherent contributions are very small in this energy range). The two results agree and agree in the overlap region also reasonably well with the result from [98]. However, they do not agree with the incoherent sum of the MAID and SAID results for the elementary cross sections folded with Fermi motion. The right hand side of the figure shows the exclusive, quasi-free proton and neutron cross sections. In both cases, they do not agree with the MAID and SAID results. Since those are fitted to the free proton data, this indicates a significant nuclear effect for the quasi-free proton data (and probably a similar effect for the neutron data). Nevertheless the comparison of proton and neutron data confirms the expected differences in resonance excitations, in particular in the third resonance region. These results are still preliminary, further cross checks of the absolute normalization of the data are still under way. The final results will provide precise angular distributions for both quasi-free reactions and thus allow detailed tests of FSI effects and give stringent inputs for the extraction of resonance couplings for the neutron.

3.1.4. further channels under analysis New data sets from ELSA and MAMI are also under analysis for quasi-free production of meson pairs, in particular the double pion channels ($\pi^0\pi^0$, $\pi^0\pi^\pm$) and the $\pi\eta$-channels ($\pi^0\eta$, $\pi^\pm\eta$). First, preliminary results seem to indicate that FSI effects are small for double pion production but substantial for the $\pi\eta$-channels. Since coherent contributions are small for the double pion channels but relatively large for the $\pi\eta$ final states, this seems to be in line with the findings for single η, η', and π^0 production.

3.2. η-mesic nuclei
The search for η-mesic states with meson photoproduction can be done in two different ways. In quasi-free photoproduction of η-mesons kinematic conditions can be selected such that the momentum of the incident photon is carried away by the participant nucleon and the η-meson is produced (almost) at rest in the $A-1$ residual nucleus, or coherent η-production close to threshold off the A initial state nucleus can be used as entrance channels. In the latter case nuclei with quantum numbers that allow coherent η-production (i.e. $I, J \neq 0$) must be used [80]. For light nuclei, the only promising candidates are ^{3}He, ^{3}H, and ^{7}Li, but the radioactive tritium is not considered for practical reasons. In a previous experiment Pfeiffer and collaborators [34] had found some evidence for the formation of η-mesic ^{3}He in two different experimental signals.

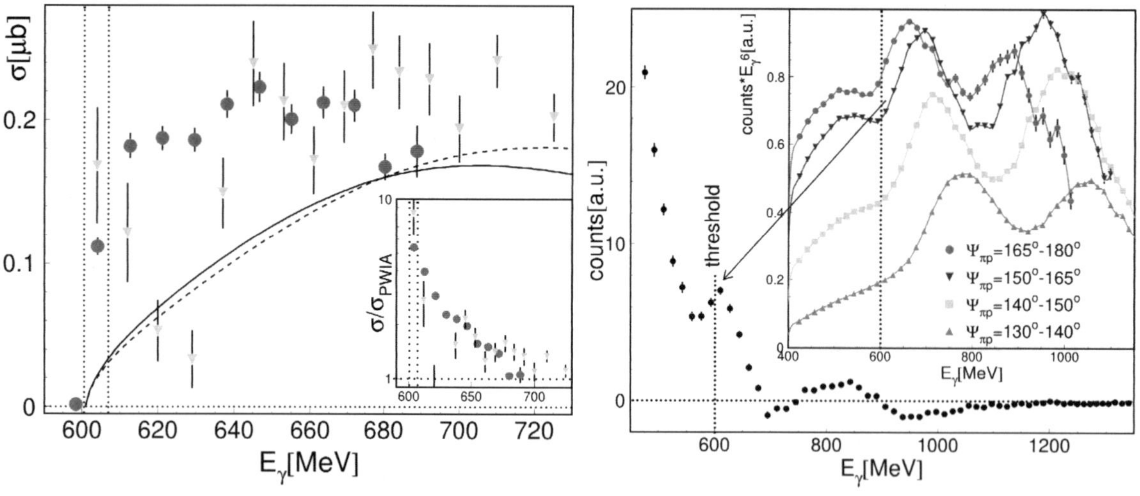

Figure 13. Left hand side: Total cross section for $\gamma^3\text{He}\rightarrow\eta^3\text{He}$ (averaged over 2γ and $3\pi^0$ decays) (red dots) compared to previous data [34] (green triangles). Solid (dashed) curves: PWIA with realistic (isotropic) angular distribution for $\gamma n \rightarrow n\eta$. Insert: ratio of measured and PWIA cross sections. Right hand side: Main plot: difference of excitations functions of $\pi^0 - p$ back-to-back pairs with opening angles between 165° - 180° and 150° - 165°. Insert excitations functions for different ranges of the opening angle $\Psi_{\pi p}$ after removal of the overall energy dependence $\propto E_\gamma^{-6}$. Vertical dotted lines: coherent η-production threshold.

The total cross section of coherent η-prodcution showed an extremely rapid rise at production threshold, far beyond phase-space expection, and the angular distributions close to threshold were much more isotropic than expected from the influence of the nuclear form-factor. Furthermore, a tiny peak-like structure at η-production threshold in the excitation function of $\pi^0 - p$ pairs emitted back-to-back in the photon-nucleus cm system was related to the reaction chain were a quasi-bound η-meson is re-captured by a nucleon into the S_{11} excitation with subsequent decay to $N\pi$. However, both signals suffered from very low conting statistics.

In a follow-up experiment [100] we have remeasured these reactions with Crystal Ball/TAPS at MAMI with much higher statistical precision. In addition, also the systematic uncertainties could be reduced due to two improvements. The almost 4π coverage of the detector allowed a significant reduction of background from quasi-free η-production due to the detection of recoil nucleons. The large solid-angle coverage allowed in addition the measurement of not only the $\eta \rightarrow 2\gamma$ decay but also the $\eta \rightarrow 3\pi^0 \rightarrow 6\gamma$ decay branch could be used. The result for the total cross section is summarized in Fig. 13 and compared to the previous result and phase-space expectations. The overall agreement with the previous data is good, but there is a discrepancy around incident photon energies of 620 MeV. The old data showed a dip-like structure in this region which could not be reproduced. It is very probably due to systematic effects in the separation of the coherent reaction from breakup background in the previous experiment. The data confirm the extremely steep rise of the cross section at the coherent production threshold which is very unlike phase-space behavior. In addition also the behavior of the threshold angular distributions (not shown) disagrees strongly with plane-wave approximations.

The difficulty for the analysis of the $\pi^0 - p$-back-to-back pairs is background from quasi-free single-pion production. The approach for the previous experiment [34] was to subtract from the excitation function of back-to-back pairs (opening angles >170°) a normalized result for opening angles from 150° - 170°, assumed to represent the background. A similar analysis of the present data (main plot at right hand side of Fig. 13) reproduces the previously observed peak with much higher statistical significance. However, closer inspection of the excitation functions for different ranges of opening angle (insert of the figure) shows that this peak-like structure is an artifact, arising from the structures related to quasi-free pion production in the second and third resonance region, which move upwards in photon energy for decreasing opening angles (a purely kinematic effect). Due to this, the difference of back-to-back pairs and pairs with somewhat smaller opening angles, just by chance has a peak-like structure at the η-threshold. Evidence for this second decay channel of an η-mesic resonance could thus not be confirmed.

Data of the same kind for ^{7}Li nuclei have also been taken and are currently still under analysis.

3.3. Photoproduction of pion pairs from nuclei

The main motivation for this experiment [101] was to measure double pion production ($\pi^0\pi^0$ and $\pi^0\pi^\pm$ final states) from very light (deuteron and ^{7}Li) to heavy nuclei (Pb) with such high statistical quality that the invariant mass distributions off the pion pairs can be studied also close to production thresholds where, as discussed in the introduction, FSI effects are small. This can be directly verified again by an investigation of the scaling behavior of the total cross sections as function of the incident photon energy, which is shown in Fig. 14. Very close to threshold the cross sections scale in fact with the nuclear mass number A, indicating almost vanishing absorption effects and thus small FSI. The scaling coefficient α then drops with increasing photon energy and approaches 'surface-scaling' ($\alpha \approx 2/3$) around incident photon energies of 500 MeV. Invariant mass distributions of the pion pairs for both reactions and for different ranges of incident photon energy are summarized in Fig. 15. The shapes of the distributions are very similar for low incident photon energies and develop an enhancement of strength at small invariant masses for heavy nuclei only in the region where the FSI effects become large. Consequently, FSI seems to be a major source of these effects.

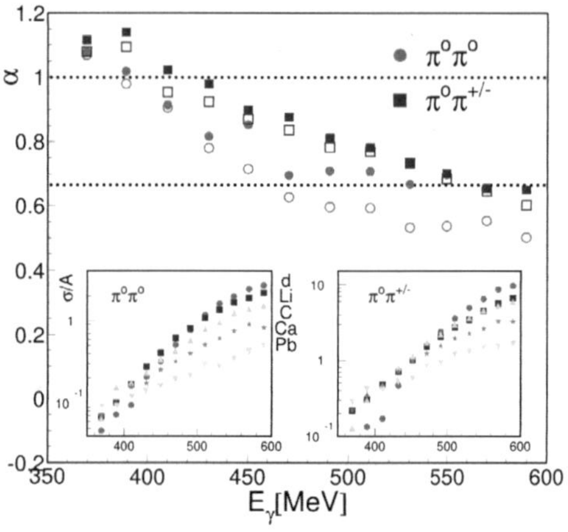

Figure 14. Scaling coefficient α (see Eq. 1) for $\pi^0\pi^0$ and $\pi^0\pi^\pm$ production off nuclei as function of the incident photon energy. Filled symbols include the data from the deuteron, open symbols are from fits to Li, C, Ca, Pb data only. The inserts show for both reaction types the total cross sections in the threshold region normalized to the nuclear mass number A. The change from A-scaling close to threshold to $A^{2/3}$ scaling at higher photon energies is clearly visible.

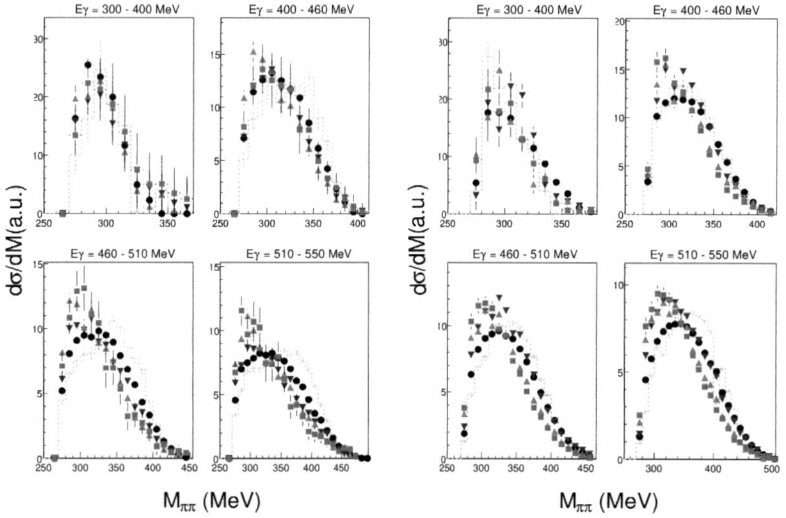

Figure 15. Preliminary invariant mass distributions of $\pi^0\pi^0$ pairs (left hand side) and $\pi^0\pi^+$ pairs (right hand side) off nuclei for different ranges of incident photon energy (all distributions normalized to total cross section). (Green) histograms: deuterium, (black) dots: ^{7}Li, (blue) downward triangles: ^{12}C, (magenta) upward triangles: ^{40}Ca, (red) squares: ^{nat}Pb.

4. Conclusions

An overview was given on recent results from the photoproduction of mesons from nuclei. Quasi-free photoproduction from nucleons bound loosely in the deuteron has been established as a reliable tool for the study of the electromagnetic excitation spectrum of the neutron. The technical/instrumental challenges for the measurement of such reactions are well under control. Problems arising from nuclear effects like FSI can be very different depending on the studied reaction and must be carefully considered separately for each case. In particular for the examples of the η and η' mesons such problems seem to be negligible. Attempts to measure also (double) polarization observables for reactions off quasi-free neutrons with d-buthanol targets are well under way and will in future largely contribute to the interpretation of the data in terms of nucleon resonance contributions. The most interesting current problem is certainly the not yet understood nature of the narrow structure in the excitation function for η-production off the neutron.

In view of η-mesic nuclei, the new data for ^{3}He very convincingly confirm the unusual threshold behavior of coherent η-production off this nucleus, which completely disagrees with expectations from phase-space behavior and, together with the results from hadron induced reactions, points to the existence of a resonance overlapping with the production threshold.

The data for double pion production reactions off heavy nuclei confirm the earlier observed in-medium modifications of the invariant mass of the pion pairs, but suggest that the dominant effect is related to FSI rather than to in-medium modifications of the σ-meson.

Acknowledgements
The results discussed in this paper have been obtained by the CBELSA/TAPS and Crystal Barrel/TAPS collaborations. They are part of the theses works of M. Dieterle, I. Jaegle, Y. Magrhbi, F. Pheron, D. Werthmüller, and L. Witthauer. This work was supported by Schweizerischer Nationalfonds and Deutsche Forschungsgemeinschaft (SFB 443, SFB/TR-16.)

References

[1] Arndt R A *et al.* 2006 *Phys. Rev.* C **74** 045205
[2] Nakamura K *et al.* 2010 *Journal of Physics* G **37** 075021
[3] Edwards R G *et al.* 2011 *arXiv:[hep-ph]* **1104.5152**
[4] Dürr S *et al.* 2008 *Science* **322** 1224
[5] Bianchi N *et al.* 1996 *Phys. Rev.* C **54** 1688
[6] Krusche B *et al.* 2002 *Phys. Lett.* B **526** 287
[7] Krusche B 2005 *Eur. Phys. J.* A **26** 7
[8] Tarbert C M *et al.* 2008 *Phys. Rev. Lett.* **100** 132301
[9] Krusche B *et al.* 2004 *Eur. Phys. J.* A **22** 277
[10] Röbig-Landau M *et al.* 1996 *Phys. Lett.* B **373** 45
[11] Mertens T *et al.* 2008 *Eur. Phys. J.* A **38** 195
[12] Nanova M *et al.* 2012 *submitted to Phys. Lett.* B
[13] Trnka D *et al.* 2005 *Phys. Rev. Lett.* **94** 192303
[14] Kotulla M *et al.* 2008 *Phys. Rev. Lett.* **100** 192302
[15] Nanova M *et al.* 2010 *Phys. Rev.* C **82** 035209
[16] Nanova M *et al.* 2011 *Eur. Phys. J.* A **47** 16
[17] Ishikawa T *et al.* 2005 *Phys. Lett.* B **608** 215
[18] Krusche B *et al.* 1997 *Phys. Lett.* B **397** 171
[19] Liu L C and Haider Q 1986 *Phys. Rev.* C **34** 1845
[20] Chrien R E *et al.* 1988 *Phys. Lett.* B **60** 2595
[21] Johnson J D *et al.* 1993 *Phys. Rev.* C **47** 2571
[22] Sokol G A *et al.* 1999 *Fizika* B **8** 85
[23] Sokol G A and Pavlyuchenko L N 2008 *Phys. of. Atomic Nuclei* **71** 509
[24] Arndt R A *et al.* 2005 *Phys. Rev.* C **72** 045202
[25] Krusche B *et al.* 1995 *Phys. Rev. Lett.* **74** 3736
[26] Krusche B *et al.* 1995 *Phys. Lett.* B **358** 40
[27] Hoffmann-Rothe P *et al.* 1997 *Phys. Rev. Lett.* **78** 4697
[28] Weiss J *et al.* 2001 *Eur. Phys. J.* A **11** 371
[29] Weiss J *et al.* 2003 *Eur. Phys. J.* A **16** 275
[30] Hejny V *et al.* 1999 *Eur. Phys. J.* A **6** 83
[31] Hejny V *et al.* 2002 *Eur. Phys. J.* A **13** 493
[32] Mersmann T *et al.* 2007 *Phys. Rev. Lett.* **98** 242301
[33] Rausmann T *et al.* 2009 *Phys. Rev.* C **80** 017001
[34] Pfeiffer M *et al.* 2004 *Phys. Rev. Lett.* **92** 252001
[35] Bernard V, Meissner U G, Zahed I 1987 *Phys. Rev. Lett.* **59** 966
[36] Hatsuda T, Kunihiro T and Shimizu H 1999 *Phys. Rev. Lett.* **82** 2840
[37] Aouissat Z *et al.* 2000 *Phys. Rev.* C **61** 012201
[38] Chiang H C, Oset E and Vicente Vacas M J 1998 *Nucl. Phys.* A **644** 77
[39] Roca L, Oset E and Vicente Vacas M J 2002 *Phys. Lett.* B **541** 77
[40] Bonutti F *et al.* 1996 *Phys. Rev. Lett.* **77** 603
[41] Grion N *et al.* 2005 *Nucl. Phys.* A **763** 80
[42] A. Starostin A *et al.* 2000 *Phys. Rev. Lett.* **85** 5539
[43] Messchendorp J G *et al.* 2002 *Phys. Rev. Lett.* **89** 222302
[44] Bloch F *et al.* 2007 *Eur. Phys. J.* A **32** 219
[45] Krusche B 2011 *Eur. Phys. J. Special Topics* **198** 199
[46] Husman D, Schwille W J 1988 *Phys. BL.* **44** 40

[47] Hillert W 2006 *Eur. Phys. J.* A **28** 139
[48] Jaegle I *et al.* 2011 *Eur. Phys. J.* A **47** 11
[49] Jaegle I *et al.* 2011 *Eur. Phys. J.* A **47** 89
[50] Aker E *et al.* 1992 *Nucl. Instr. and Meth.* A **321** 69
[51] Novotny R 1991 *IEEE Trans. on Nucl. Science* **38** 379
[52] Gabler A R *et al.* 1994 *Nucl. Instr. and Meth.* A **346** 168
[53] Suft G *et al.* 2005 *Nucl. Inst. Meth.* A **538** 416
[54] Herminghaus H *et al.* 1983 *IEEE Trans. on Nucl. Science.* **30** 3274
[55] Kaiser K H *et al.* 2008 *Nucl. Inst. Meth.* A **593** 159
[56] Anthony I *et al.* 1991 *Nucl. Inst. Meth.* A **301** 230
[57] Starostin A *et al.* 2001 *Phys. Rev.* C **64** 055205
[58] Watts D, in *Calorimetry in Particle Physics, Proceedings of the 11th Internatinal Conference, Perugia, Italy 2004*, edited by C. Cecchi, P. Cenci, P. Lubrano, and M. Pepe (World Scientific, Singapore, 2005, p. 560
[59] Schumann S *et al.* 2010 *Eur. Phys. J.* A **43** 269
[60] Zeitnitz C *et al.* 2001 *The GEANT-CALOR interface user's guide* (http://www.staff.uni-mainz.de/zeitnitz/Gcalor/gcalor.html)
[61] Lacombe M *et al.* 1981 *Phys. Lett.* B **101** 139
[62] Ajaka J *et al.* 1998 *Phys. Rev. Lett.* **81** 1797
[63] Bock A *et al.* 1998 *Phys. Rev. Lett.* **81** 534
[64] Armstrong C S *et al.* 1999 *Phys. Rev.* D **60** 052004
[65] Thompson R *et al.* 2001 *Phys. Rev. Lett.* **86** 1702
[66] Renard F *et al.* 2002 *Phys. Lett.* B **528** 215
[67] Dugger M *et al.* 2002 *Phys. Rev. Lett.* **89** 222002
[68] Crede V *et al.* 2005 *Phys. Rev. Lett.* **94** 012004
[69] Nakabayashi T *et al.* 2006 *Phys. Rev.* C **74** 035202
[70] Batalini O *et al.* 2007 *Eur. Phys. J.* A **33** 169
[71] Bartholomy O *et al.* 2007 *Eur. Phys. J.* A **33** 133
[72] Elsner D *et al.* 2007 *Eur. Phys. J.* A **33** 147
[73] Denizli H *et al.* 2007 *Phys. Rev.* C **76** 015204
[74] Williams M *et al.* 2009 *Phys. Rev.* C **80** 045213
[75] Crede V *et al.* 2009 *Phys. Rev.* C **80** 055202
[76] Sumihama M *et al.* 2009 *Phys. Rev.* C **80** 052201(R)
[77] McNicoll E F *et al.* 2010 *Phys. Rev.* C **82** 035208
[78] Chiang W T *et al.* 2002 *Nucl. Phys.* A **700** 429
[79] Anisovich V A *et al.* 2005 *Eur. Phys. J.* A **25** 427
[80] Krusche B and Schadmand S 2003 *Prog. Part. Nucl. Phys.* **51** 399
[81] Moorehouse R G 1996 *Phys. Rev. Lett.* **16** 772
[82] Kuznetsov V *et al.* 2007 *Phys. Lett.* B **647** 23
[83] Miyahara F *et al.* 2007 *Prog. Theor. Phys. Suppl.* **168** 90
[84] Jaegle I *et al.* 2007 *Phys. Rev. Lett.* **100** 252002
[85] Shklyar V, Lenske H, Mosel U 2007 *Phys. Lett.* B **650** 172
[86] Fix A, Tiator L, and Polyakov M V 2007 *Eur. Phys. J.* A **32** 311
[87] Shyam R and Scholten O 2007 *Phys. Rev.* C **78** 065201
[88] Anisovich V A *et al.* 2009 *Eur. Phys. J.* A **41** 13
[89] Döring M and Nakayama K 2008 *Phys. Lett.* B **683** 145
[90] Fantini A *et al.* 2008 *Phys. Rev.* C **78** 015203
[91] Elsner D 2010 *Int. J. Mod. Phys.* E **19** 869
[92] Plötzke R *et al.* 1998 *Phys. Lett.* B **444** 555
[93] Dugger M *et al.* 2006 *Phys. Rev. Lett.* **96** 169905
[94] Chiang W T *et al.* 2003 *Phys. Rev.* C **68** 045202
[95] Nakayama K, Haberzettl H 2006 *Phys. Rev.* C **73** 045211
[96] Drechsel D *et al.* 1999 *Nucl. Phys.* A **645** 145
[97] Arndt R A *et al.* 2002 *Phys. Rev.* C **66** 055213
[98] Krusche B *et al.* 1999 *Eur. Phys. J.* A **6** 309
[99] Darwish E M, Arenhövel H, Schwamb M 2003 *Eur. Phys. J.* A **16** 111
[100] Pheron F *et al.* 2012 *submitted to Phys. Lett.* B
[101] Magrhbi Y *et al.* 2012 *to be submitted to Phys. Lett.* B

Overview of the recent results from CLAS

Marco Mirazita
I.N.F.N - Laboratori Nazionali di Frascati
E-mail: marco.mirazita@lnf.infn.it

Abstract.
An overview of the recent results obtained at the Thomas Jefferson Laboratory on the study of the nucleon internal structure is presented, with main focus on the CLAS measurements of the Transverse Momentum Dependent partonic functions.

1. The Jefferson Laboratory facility and CLAS

The Thomas Jefferson National Accelerator Facility (TJNAF or JLab) is located at Newport News, in Virginia (USA). It houses the Continuous Electron Beam Accelerator Facility (CEBAF), a high-current, high-duty electron beam machine with three experimental halls. A schematic of the accelerators is shown in Fig. 1.

The CEBAF is basically composed by two linear accelerator, each delivering to the electrons up to 0.4 GeV per pass, and two recirculating arcs. The electron beam can be circulated up to five times in the accelerator, so that the maximum achievable energy is about 6 GeV. The primary beam can then be split and sent to the three experimental halls (Hall A, B and C), with different energies and intensities. The polarization of the electron beam can be as high as 85% and, in order to minimize systematic effects, it is pseudo-randomly flipped between positive and negative values every about 30 ms.

The Hall A and C are both instrumented with two small acceptance, high precision spectrometers and routinely runs with beam currents sufficient to achieve a luminosity of the order of several times $10^{38} cm^{-2}s^{-1}$.

The Hall B is instrumented with the CEBAF Large Acceptance Spectrometer (CLAS), shown schematically in Fig. 2. The CLAS design is based on a toroidal field, generated by six superconducting coils arranged around the beam line, so that the field is pointing primarily in the azimuthal direction. The six coils divide the detector in six independent spectrometers, each of them containing:

- three regions of drift chambers, the second of which is immersed in the magnetic field, for particle's track and momentum measurement;
- an array of scintillator counters for time of flight measurement and particle identification;
- a gas Cerenkov counter, for electron/pion separation;
- a forward electromagnetic calorimeter (EC), for neutral particles detection and used also to improve electron identification;
- a large angle calorimeter (in two sectors only), to extend neutral particle detection.

The geometry of the detector close to the beam line allows the installation of polarized as well as unpolarized targets. The Hall B instrumentation is completed by a tagger system, installed upstream of the CLAS detector, allowing the production of a real photon beam with intensity of up to several times 10^7 photons per second.

For charged particles, CLAS covers all the polar angles θ between 8° and 142° in the laboratory frame and between 60 and 80% of the azimuthal angles ϕ. The momentum resolution is between 0.5% (forward) and 1.5% (backward), with angular resolution of about 1 mrad in θ and about 4 mrad in ϕ.

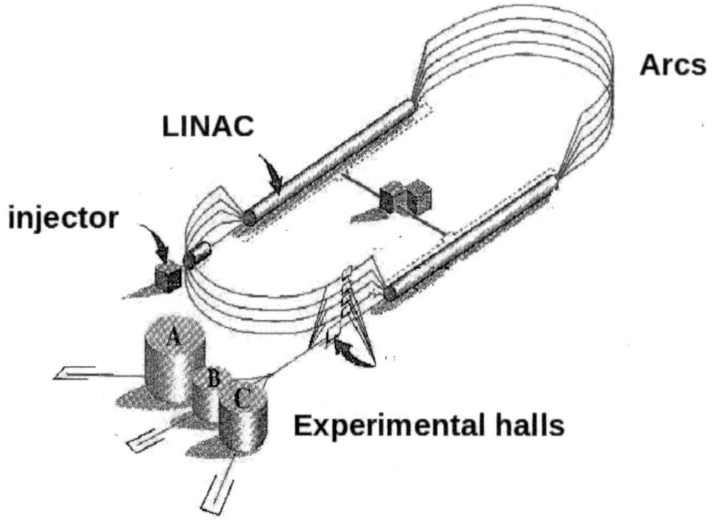

Figure 1. Layout of the CEBAF accelerator.

2. The CLAS physics program

The high intensity electron and photon beams, together with the availability of three experimental halls with different and complementary detectors, offer the unique opportunity to study the QCD from the non-perturbative regime to the microscopic partonic structure of the hadrons and nuclei, making of JLab a laboratory for the hadronic physics. In fact, the JLab physics program, started in 1996, covers many aspects of the hadronic physics, namely:

- measurement of the charge and magnetization distributions in the nucleons and in the nuclei;
- two-body correlations in nuclei;
- meson and baryon excitation spectrum;
- production of exotic hadrons;
- strangeness in nuclei;
- structure functions in the deeply inelastic regime:
- parity violation in electron-proton scattering;

Some of those physiscs items were present in the JLab program since the beginning of its operation, but others have emerged as exciting new developments only in the last years. In particular, the combination of high polarization electron beam with highly polarized targets have

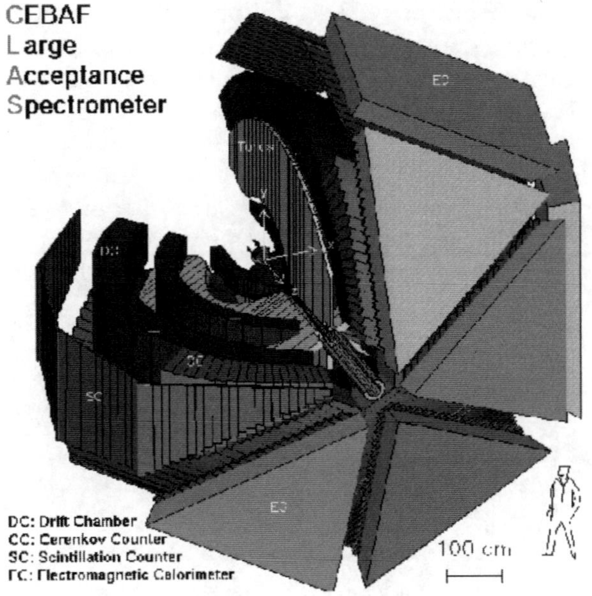

Figure 2. Schematic view of the CLAS detector.

open the possibility to study the internal structure of the nucleon not only in the inclusive regime, but also in the semi-inclusive and exclusive final states, giving access to new elementary functions like the Transverse Momentum Dependent (TMD) partonic distributions and the Generalized Parton Distribution (GPD) functions. In the following, a review of the recent results obtained at JLab on the TMD measurements will be presented.

3. Structure functions in Deeply Inelastic Scattering

The existence of point-like quarks inside the nucleon was inferred for the first time from the study of the Deeply Inelastic Scattering (DIS) performed at SLAC in the late 1960's [1]. The DIS reaction, in which high energy electrons scatter off a proton target and the emerging electron is detected, is described by two kinematic variables: the virtuality of the exchanged photon $Q^2 = -q^2$ (where $q = p - p'$ is the 4-momentum of the virtual photon and p and p' are the initial and final electron 4-momenta), and the Bjorken variable $x_B = Q^2/2Pq$ (where P is the target nucleon 4-momentum). The parity-conserving electromagnetic part of the DIS cross section is described by four structure functions, usually denoted by F_1, F_2, g_1 and g_2. The first two contribute when both the electron and the target nucleon are unpolarized, while the last two arise when the beam and the target are polarized. Although they depend on the two variables Q^2 and x_B, in the Bjorken limit, i.e. for $\nu, Q^2 \to \infty$ (where ν is the virtual photon energy) but with x_B fixed, the structure functions are expected to scale approximately, that is, to depend only on x_B.

In the parton model the nucleon is composed by elementary constituents moving collinearly and in the DIS reaction the virtual photon is assumed to scatter incoherently off these constituents, so that the structure functions can be decomposed in a sum over the contributions coming from its partons. For the unpolarized structure function one gets

$$F_2 = \sum_q e_q^2 x\,(q + \bar{q}) = \frac{4}{9}x\,(u + \bar{u}) + \frac{1}{9}x\,(d + \bar{d}) + \frac{1}{9}x\,(s + \bar{s}) + \cdots \quad (1)$$

and the so-called Callan-Gross relation $F_2(x) = 2xF_1(x)$. Here q and $\bar{q}$ are the quark and antiquark Parton Distribution Functions (PDFs), expressed as a function of the fraction $x = x_B$ of the nucleon longitudinal momentum carried by the parton.

Measurements of the unpolarized structure function have been performed over the last decades at SLAC, CERN, DESY and Fermilab over a kinematic range that spans several order of magnitudes in Q^2 and x_B. A summary of these experimental data is shown in Fig. 3. We see that the Bjorken scaling is well satisfied for the larger values of x_B, that is in the region where valence quark contribution dominates the structure function. As x_B goes down, sea quark contributions start to be relevant and logarithmic scaling violation are observed. It was a great success of the QCD that these scaling violations could be correctly calculated perturbatively.

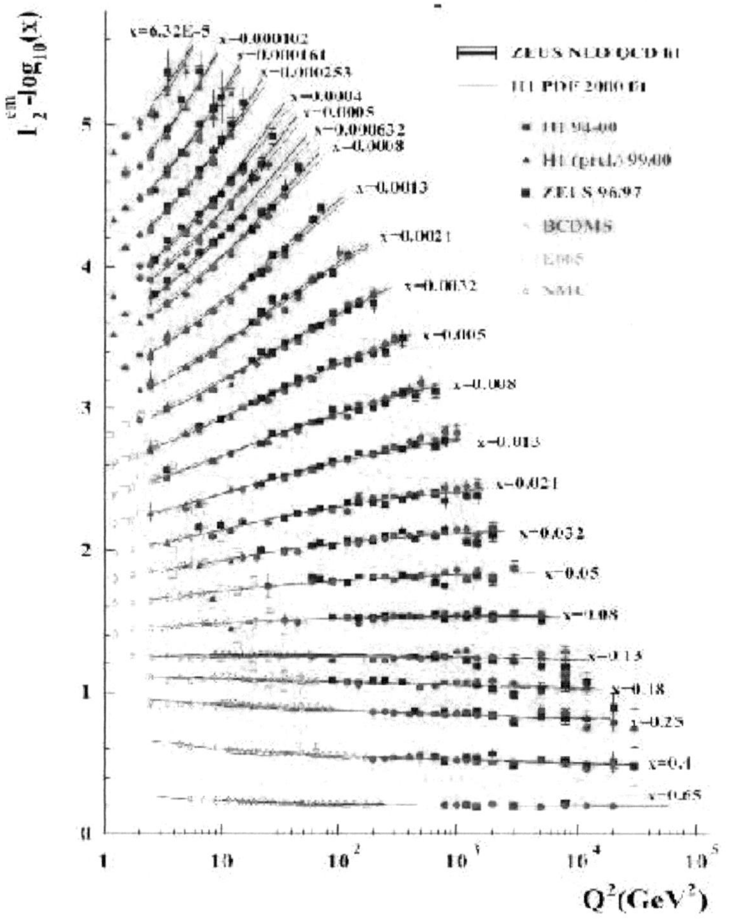

Figure 3. World data on F_2 for the proton as a function of Q^2 for fixed values of x_B.

The parton model expansion of the polarized structure function g_1 is given by

$$g_1 = \sum_q e_q^2 (\Delta q + \Delta \bar{q}) = \frac{1}{2}\left[\frac{4}{9}x(\Delta u + \Delta \bar{u}) + \frac{1}{9}x(\Delta d + \Delta \bar{d}) + \frac{1}{9}x(\Delta s + \Delta \bar{s}) + \cdots\right] \quad (2)$$

where Δq and $\Delta \bar{q}$ are the polarized quark and antiquark PDFs. They represent the difference between the probabilities to find a parton with helicity aligned minus that with helicity antialigned to the nucleon momentum.

The second spin structure function g_2, related to the transverse spin structure, has no simple probabilistic interpretation, nevertheless it carries important information on the quark-gluon correlations. Its contribution to the cross section is generally small and only in the last years its measurement became possible.

Double spin experiments on the proton in the DIS domain have been performed at CERN, SLAC, DESY and in the last years also with COMPASS [2] and CLAS at JLab [3] and the results are shown in Fig. 4. Although not at the same level of the unpolarized data, those data provide a good knowledge of g_1. We also see that in the high-x_B region information comes almost exclusively from CLAS data.

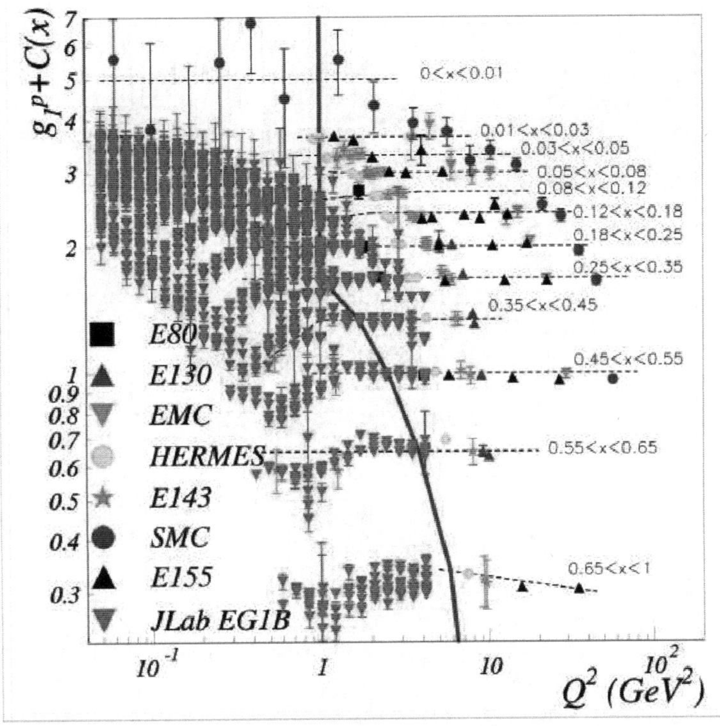

Figure 4. World data on g_1 for the proton as a function of Q^2 for fixed values of x_B. The solid line is the conventional DIS frontier ($Q^2 = 1$ GeV^2 and $W = 2$ GeV.). Note that Compass data [2] are not shown.

4. Orbital angular momentum in DIS

The four structure functions F_1, F_2, g_1 and g_2 are the observables that can be directly measured in DIS experiments. However, physics information is enclosed in the PDFs and the extraction of those information from the data requires complicated fit procedures that involve some *a priori* assumption. Extractions of the polarized PDFs from the experimental data of polarized structure function on the proton and on the neutron have been performed by SLAC [4], Hermes [5], Compass [2] and by CLAS [3] and Hall A [6]. Global fits of the data[5, 2], including valence and sea quark contributions, have revealed surprising results, as for example the sizeable breaking of the isospin symmetry in the light quark sector and the difference between the strange quark and antiquark distributions.

Extractions of the helicity distributions for u and d quarks are shown in Fig. 5. While $\Delta u/u$ seems to approach the unity at high x as predicted by a variety of models (see for example [9]),

$\Delta d/d$ remains negative over the whole explored region and shows no tendency to turn toward one.

In Fig. 5, the data are compared with a leading order pQCD [9] (dashed curve) and a pQCD fit including quark orbital angular momentum corrections [10] (solid curve). While both calculations are in agreement with the data for $\Delta u/u$, the leading order pQCD model assuming quark helicity conservation is not reproducing the $\Delta d/d$ data. Better agreement is found with the fit including quark orbital angular momentum, suggesting that it may play an important role in this kinematic regime. Still this fit predicts $\Delta d/d \to 1$ for x approaching unity. In contrast, different models [8], [7] predict $\Delta d/d \to -1/3$ at high x, in agreement with the trend that the experimental data seem to show. It is clear that new data at high x are needed to disentangle between the various models.

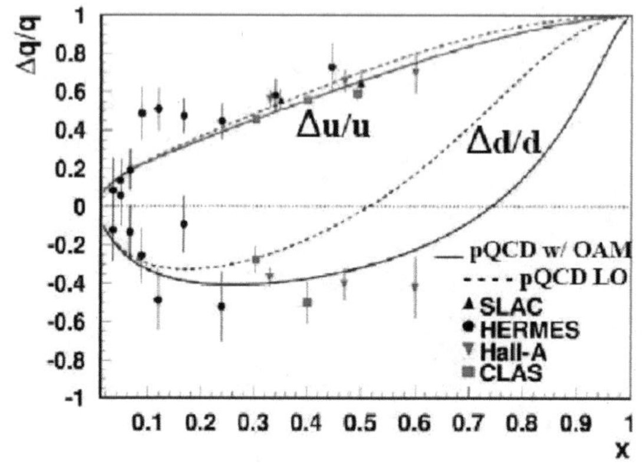

Figure 5. Extraction of $\Delta u/u$ and $\Delta d/d$ from SLAC [4], Hermes [5] and from CLAS [3] and Hall A [6] at JLab, compared with theoretical predictions.

The doubly polarized experiments in the DIS domain are the cornerstone of our understanding of the nucleon spin. Naively, in the parton model one can write down the sum rule for the nucleon spin $S_N = 1/2 = 1/2\Delta\Sigma$, where $\Delta\Sigma$ is the quark spin contribution and is obtained by integrating the structure function $g_1(x)$. From the experimental data a value $\Delta\Sigma \approx 0.3$ is estimated, thus more contributions must considered in the spin sum rule:

$$S_N = \frac{1}{2}\Delta\Sigma + L_q + \Delta G + L_G \tag{3}$$

with L_q being the quark orbital angular momentum and ΔG and L_G the gluon spin and orbital momentum. Measurement of the polarized gluon distribution $\Delta G/G$ at $x \approx 0.1$ are small, compatible with zero [11]. Large contributions might come from the low x region, which is not yet explored exerimentally, but in any case it is clear that orbital angular momenta play an important role in solving the spin puzzle.

At the parton model level, introduction of the orbital angular momentum requires that quarks have non zero transverse momentum, thus the PDF definition should be extended beyond the collinear approximation used to parametrize the DIS structure fuctions.

5. Transverse Momentum Dependent PDFs

In the approximation of partons moving collinearly with the nucleon, three PDFs can be defined: the two measured in DIS experiments, f_1 (using now the SIDIS notation of the DIS F_1) and g_1, plus a third one, h_1, describing transversely polarized quarks in a transversely polarized nucleon.

N \ q	U	L	T
U	$\mathbf{f_1}$		$h_1^\perp$
L		$\mathbf{g_1}$	$h_{1L}^\perp$
T	$f_{1T}^\perp$	g_{1T}	$\mathbf{h_1}$ $h_{1T}^\perp$

Table 1. Leading-twist transverse TMD functions: U, L, and T stand for transitions of unpolarized, longitudinally polarized, and transversely polarized nucleons (rows) and quarks (columns).

The latter is not accessible in inclusive measurements and very little is known so far about this function.

When the partons are allowed to have also a transverse momentum, the PDFs aquire a more rich and complicated structure [12, 13]. At leading twist, there are eight independent Transverse Momentum Dependent (TMD) parton distribution functions, as shown in Tab. 1, where each row (column) refers to unpolarized (U), longitudinally (L) or transversely (T) polarized nucleon (quark). Each TMD depends not only on the fraction of longitudinal nucleon momentum x carried by the quark, but also on the quark transverse momentum $\mathbf{k}_\perp$. Along the diagonal of Tab. 1, we find $f_1(x, \mathbf{k}_\perp)$, $g_1(x, \mathbf{k}_\perp)$ and $h_1(x, \mathbf{k}_\perp)$ that, after integrating over the treansverse momentum, reduce to the corresponding collinear PDFs. The remaining five TMDs are new objects, describing all the possible correlations between the momentum and the spin of the quark with those of the nucleon and originate from interference between wave functions with different angular momentum. They vanish after integration over $\mathbf{k}_\perp$, thus their measurement requires the tagging of the transverse momentum of the struck quark.

6. Experimental access to TMDs

Because of the need to tag the quark transverse momentum, experimental access to TMDs requires the presence of more than one hadron in the initial or final state. In electron-proton scattering, this can be achieved by studying Semi-Inclusive Deeply Inelastic Scattering (SIDIS) reactions $ep \to e'hX$, in which the struck quark fragments into a final hadron h with transverse momentum $\mathbf{P_T}$. In a well defined kinematic regime, a factorization theorem for SIDIS processes has been proved [15], showing that the cross section of the reaction in Fig. 6 can be decomposed as

$$\sigma^{SIDIS} = \sum_q DF^q \otimes \sigma^{eq\to e'q'} \otimes FF^{q'} \tag{4}$$

where DF^q is a TMD distribution function and $\sigma^{eq\to e'q'}$ is the elementary cross section for electron-quark scattering. A new object appears in this expression, the Fragmentation Function FF^q, describing amplitudes that a quark q fragments into the final observed hadron. The FFs depend on the energy fraction z of the final hadron and on the transverse momentum $\mathbf{P}_\perp$ of the fragmenting quark. Restricting to spinless mesons, two of such FF are relevant in SIDIS processes: the unpolarized $D_1(z, \mathbf{P}_\perp)$ and the Collins $H_1^\perp(z, \mathbf{P}_\perp)$, the latter parametrizing the fragmentation of a transversely polarized quark into an unpolarized hadron.

Being a convolution of a two functions, the SIDIS process allows the extraction of information on the TMD distribution functions only if the FFs are known from other measurements. This is in fact possible because of the general universality of the TMDs, which ensures that the same functions entering in the SIDIS cross section can be measured also in a large variety of experiments, namely:

- Drell-Yan (DY) processes $HH \to l^+l^-X$, involving two distribution functions for the initial hadrons;

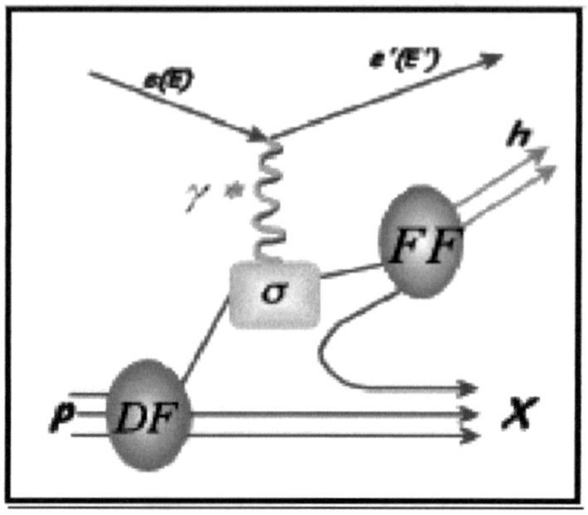

Figure 6. SIDIS reaction $ep \to e'hX$.

- hadron production in collider experiment $l^+l^- \to h_1h_2X$, involvin two fragmentation functions for the final hadrons;
- hadron production in proton proton scattering $pp \to hX$, involving two distribution functions for the initial hadrons and one fragmentation function for the final hadron.

It must be noted that for some of these processes, the factorization has not been proved yet but it is simply assumed to analyze the data.

It is clear that a full comprehension of the internal nucleon structures in terms of TMDs requires the combined efforts of many laboratories and results from many different experiments. Also, disentangling between PDFs and FFs in the data calls for new analysis techniques [14] fully differential in the relevant kinematic variables.

A crucial test of the universality of the TMDs involves the so-called Sivers and Boer-Mulders functions. The Sivers function $f_{1T}^{\perp}$ describes unpolarized quarks inside a transversely polarized nucleon and it arises from correlation between the quark transverse momentum and the nucleon spin. Conversely, the Boer-Mulders function $h_1^{\perp}$ describes unpolarized quarks inside a transversely polarized nucleon and it is produced by correlation between the quark transverse spin and the nucleon momentum. Both functions are non-zero because of initial or final state interaction and, because of that, they must change sign between SIDIS and DY processes, namely:

$$f_{1T}^{\perp}(SIDIS) = -f_{1T}^{\perp}(DY) \tag{5}$$

and

$$h_1^{\perp}(SIDIS) = -h_1^{\perp}(DY) \tag{6}$$

7. Measuring TMDs in SIDIS at JLab and CLAS

The SIDIS cross section contains 18 structure functions, each appearing as modulation in the azimuthal angle ϕ [16] of the final hadron. Decomposition of the structure functions in terms of TMDs distribution and fragmentation functions have been derived at leading order (twist-2, involving the eight PDFs af Tab. 1 and the unpolarized and Collins FFs) and next-to-leading order (twist-3, that is suppressed by one power of M_p/Q) [13, 12], although for the latter the factorization has not been completely proven. Contributions of the various structure functions can be isolated by computing azimuthal asymmetries or, equivalently, ϕ moments of the cross section.

Pion SIDIS measurements have been performed at JLab using the CLAS detector and in Hall A and C. Having different equipments, the three experiments provided different and complementary data sets. In particular, CLAS measured asymmetries over a large kinematic range, while Hall A and C measured asymmetries and total cross sections in a limited kinematic range but with high statistical precision.

7.1. Results from CLAS
In the collinear limit, the helicity distribution g_1 is well known (see Fig. 4). The dependence of g_1 on the transverse momentum has been studied by CLAS [17] by measuring the double spin asymmetry A_{LL} in the scattering of longitudinally polarized electron beam on a longitudinally polarized proton target and with a pion in the final state. Results for charged and neutral pions are shown in Fig. 7 as a function of the pion transverse momentum P_T. Although the data are consistent with flat P_T dependence, the asymmetry moderately decreases with P_T for charged pions. A possible interpretation of these results may involve different widths of the transverse momentum distribution of g_1 and f_1, arising from different orbital motion of quarks polarized in the direction of the proton spin or in the opposite one. Assuming a gaussian shape of the transverse momentum dependence, the convolution integrals over the involved transverse momenta can be analytically performed and the structure functions reduce to a product of the collinear DFs and FFs. In Fig. 7, the full, dashed and dotted curves are the results of calculations [18] with three different values for the ratio of the transverse momentum widths for g_1 and f_1 (with fixed f_1 width of 0.25 GeV^2).

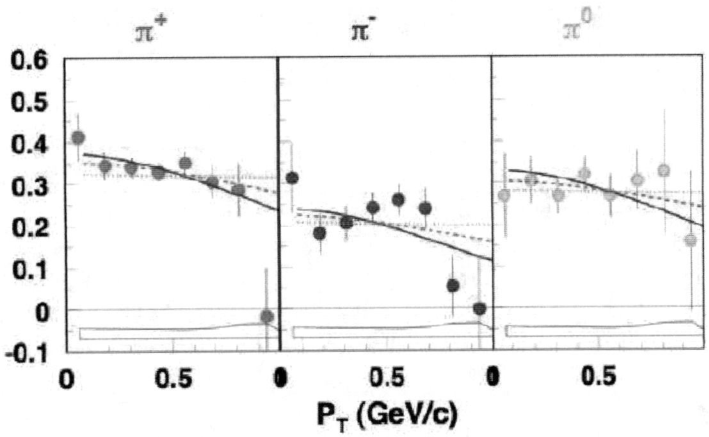

Figure 7. Double spin asymmetry as a function of the transverse momentum P_T measured by CLAS [17] for the three pions. The full, dashed and dotted curves are calculations for different values for the ratio of the transverse momentum widths for g_1 and f_1 [18].

Single target spin asymmetries A_{UL} have also been measured by CLAS [17] in SIDIS production of pions. The relevant structure function presents a leading twist contribution with a $\sin 2\phi$ modulation and a higher twist term with a $\sin\phi$ modulation. The leading twist term involves the Collins fragmentation function $H_1^\perp$ and the $h_{1L}^\perp$ DF, describing the transverse polarization of quarks in a longitudinally polarized nucleon. The measured P_T dependence of the amplitude of the two azimuthal modulations are shown in Fig. 8. The data have a significant $\sin 2\phi$ modulation for the two charged pions with opposite sign, and a relatively small one for π^0, in agreement with the expectation that the Collins function for π^0 may be suppressed, as already showed by Hermes data [19]. While π^+ data are roughly consistent with calculation of $h_{1L}^\perp$ [20, 21] and the Collins [22] function extracted from Hermes [19] and Belle [23] data, the interpretation of the π^- is less straightforward. It may require accounting for additional contribution as for example interference effects from exclusive $\rho\pi^0$ and $\pi^-\Delta^{++}$ channels. The CLAS data [17] also show a non-zero $\sin\phi$ modulation for all pions, which indicates that higher twist term can be relevant at the JLab energies.

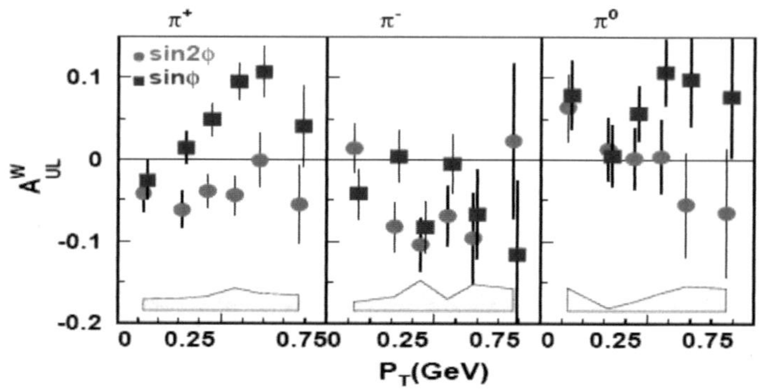

Figure 8. The amplitude of the $\sin 2\phi$ (circles) and $\sin\phi$ (squares) modulations of the target single spin asymmetry as a function of the transverse momentum P_T measured by CLAS [17] for the three pions.

CLAS measured also single beam asymmetry A_{LU} in pion SIDIS production [24, 25]. The relevant structure function is modulated by a $\sin\phi$ dependence and presents four terms, involving higher twist distribution or fragmentation functions. The P_T dependence of the asymmetry for π^0 is shown in Fig. 9 in four x_B bins. It increases at low P_T and reaches a plateau at about $P_T \approx 0.3$ GeV. There is also an indication of a possible decrease at large P_T, as expected from perturbative QCD.

The x_B dependence of A_{LU} for π^+ and π^0 is compared in Fig. 10. It is of the same order of magnitude for both pions over the whole measured range. This may suggest that the Collins term in the structure function may be negligible, because Collins function for π^+ is non-zero while it is small for π^0. Thus contributions related to spin-orbit correlations in the PDFs can be studied without a significant background from Collins mechanism.

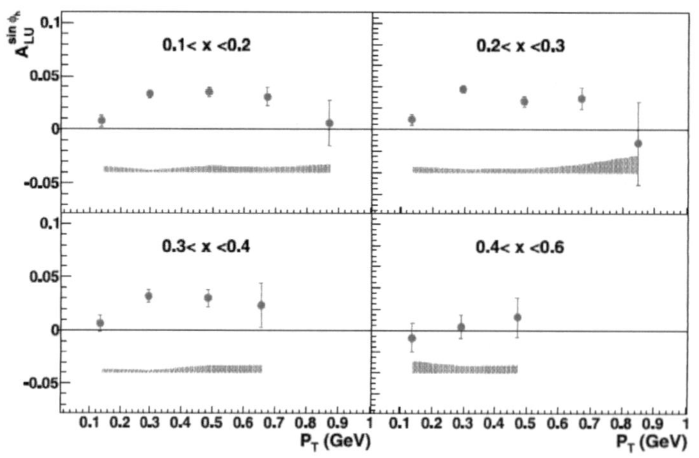

Figure 9. The single beam spin asymmetry measured by CLAS [24, 25] for π^0 as a function of P_T in four x_B bins.

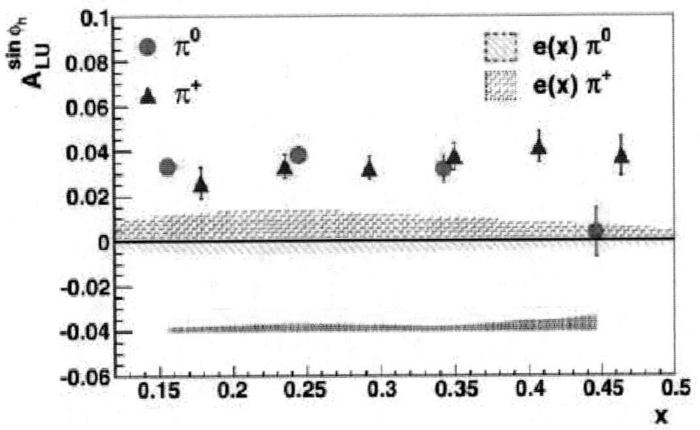

Figure 10. The single beam spin asymmetry measured by CLAS [24, 25] for π^+ (circles) and π^0 (triangles) as a function of x_B.

7.2. Results from Hall A and C

The Hall C accurately measured the unpolarized cross section for SIDIS production of charged pions on hydrogen and deuterium target [26]. The P_T^2 dependence of the cross section is shown in Fig. 11 with the full circle. These four data sets have been fitted by assuming different widths of the transverse momentum dependencies (assumed to be gaussians) of u and d quarks in both PDFs and FFs and neglecting the sea quarks (typical global fits give less than 10% sea contributions at $x_B \approx 0.3$). The fit results are shown in Fig. 11 by the full lines. They provide larger widths for u quark than for d in both distributions and fragmentation functions. They also seems to indicate a preference for larger widths in the fragmentation than in the distributions functions.

In Hall A, the first measurement of transverse target spin asymmetry A_{UT} on the neutron has been performed [27]. The CEBAF unpolarized electron beam is scattered off a ^3He transversely polarized target and charged pions are detected in the final state. The asymmetry A_{UT} has contributions from Collins and Sivers effects. The Collins effect involves convolution of the transversity DF h_1 and the Collins FF $H_1^\perp$ and presents a $\sin(\phi + \phi_S)$ modulation (here ϕ_S is the azimuthal angle of the target spin). The Sivers effect involves the convolution of the Sivers DF $f_{1T}^\perp$ and the unpolarized FF D_1, with a $\sin(\phi - \phi_S)$ modulation. The two contributions can be separated by calculating the Collins $A_C = 2 < \sin(\phi + \phi_S) >$ and Sivers $A_S = 2 < \sin(\phi - \phi_S) >$ moments of the cross section. Moreover, the neutron asymmetry is obtained from the measured data on the ^3He target by deconvolving the proton contribution [28] and taking into account final state interaction corrections [29]. The results are shown in Fig. 12 as a function of x_B and compared with several models [30, 31, 32, 33]. The Collins moments are small, compatible with zero (in agreement with theoretical expectations) except for the highest x_B data point for π^+, showing a non-zero value at the 2σ level. The data also favor a negative Sivers moment for π^+, while the π^- one is close to zero. This behaviour supports a negative Sivers function for d quark, as already suggested by fenomenological fits [30, 34] to Hermes [36] and Compass data.

7.3. Experimental results: summary and open isues

TMDs studies have been performed in recent years or are underway not only at JLab but also in many other Laboratories, in SIDIS processes (EMC, Hermes, Compass) as well as in polarized proton-proton collisions (PHENIX, STAR, BRAHMS) and electron-proton annihilation (Belle,

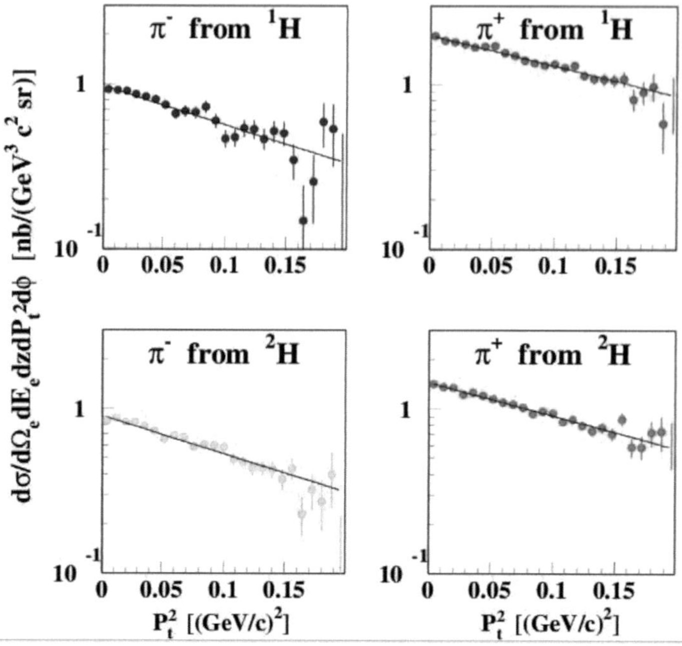

Figure 11. The P_T^2 dependence of the unpolarized cross section for charged pion SIDIS production on proton (upper plots) and deuteron (lower plots) target measured at Hall C. The curves are fenomenological fits of the data, as explained in the text.

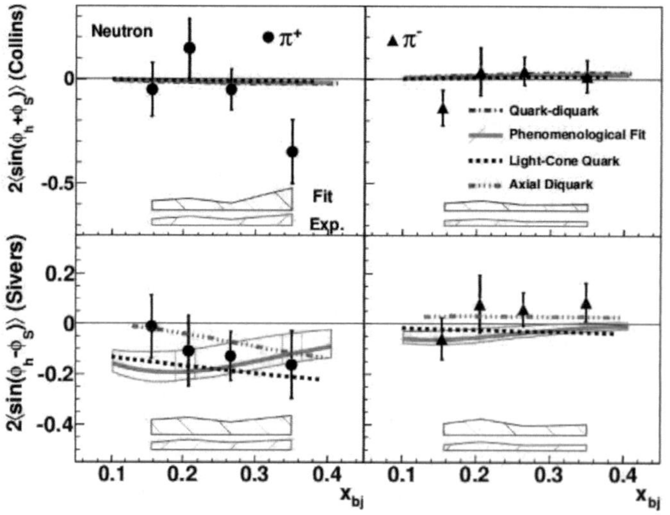

Figure 12. The x_B dependence of the Collins (upper plots) and Sivers (lower plots) moments for charged pions SIDIS production measured by Hall A.

BaBar). All these data have demonstrated that measurements of polarized hadronic cross sections in hard processes are a powerful tool to probe the nucleon structure.

Many of the DFs in Tab. 1 have been measured to be non-zero. In particular, the recent Belle data [35] on two pion pairs production together with the Collins moment measurements in double pion SIDIS done at Hermes [36] have allowed the first extraction of the transversity [37] in the framework of collinear factorization, the less known of the three PDFs surviving after transverse momentum integration.

Higher order (twist-3) contributions have been also measured at JLab, Hermes and Compass. They are not only important to correctly extract the twist-2 terms from the data, but they also carry important information on the largely unknown quark-gluon correlations.

Besides the need for more accurate experimental data to better understand the dynamics of quarks inside the nucleon, a number of problems remains still unclear. Extraction of the strange unpolarized quark distribution have been performed using DIS data under the assumption that it is proportional to the average of the non-strange quarks [38]. However, relaxing the proportionality requirement and including in the fit semi-inclusive data [39] showed that the shape of the s quark is very different from the average of non-strange sea quarks.

The helicity distribution Δs has also been extracted from DIS data, yielding to negative polarization. However, from more recent SIDIS data the strange sea polarization resulted, within statistic and systematic uncertainties, consisten with zero. It is important to independently verify this result with higher accuracy.

Hermes measurements of the Collins asymmetry for pions and kaons [36] have indicated that the differences between pions and kaons may be very significant. The asymmetry for K^+ is about double in magnitude than the π^+, while from u-quark dominace very similar amplitudes are expected. This indicates possible significant role of sea quarks. Similar behaviour can be seen also in the Collins measurements from Compass [40, 41] and from the Hermes data on the $\cos 2\phi$ moment of the unpolarized cross section [42], which involves the Boer-Mulders function $h_1^\perp$ and the Collins FF.

It is clear that the strange sector of TMD requires a strong effort and dedicated measurements in order to fully address these open issues.

8. The JLab upgrade at 12 GeV

The 6 GeV physics program of JLab will end in May 2012, when the accelerator complex will be shut down to undergo to an upgrade that will bring the maximum available energy of the electron beam to 12 GeV. The luminosity will also be increased by an order of magnitude. All the three experimental Hall equipments will be upgraded (Hall A and C) or completely renewed (Hall B), in order to better match the increased energy and a new experimental Hall D will be built. The start up of the new facility is expected in may 2013 with the commissioning of the new accelerator, followed by the commissioning of the experimental halls.

The physics program of the upgraded JLab will represent the natural continuation of the 6 GeV program, with the benefit of the increased beam energy and luminosity. In particular, for the nucleon structure study, the higher energy implies a better insight into the Deeply Inelastic Scattering region, allowing to reach higher values of Q^2 (where the factorization is in more solid grounds) and x_B (i.e., extending the valence quark explored region).

In Hall B, the new CLAS12 detector, shown in Fig. 13, will be built. It is divided in a forward part to detect leading particles and a central part for recoil particles detection. The central detector will cover the angles between 40° and 135° and will be housed in a compact superconducting solenoidal magnet, wich will serve for particle track measurement and as shielding of the detectors from electromagnetic backgrounds as well.

The general structure of the forward part of CLAS12 is similar to that of the CLAS, being based on a toroidal field and azimuthally segmented in six independent spectrometers. It will detect charged and neutral particles with polar angles between 5° and 40°. Charged particle identification will rely on a time of flight system and on two Cerenkov counters with different thresholds. These system are well suited for electron/pion separation and for proton identification over the whole accessible momentum range, but, taking into account the relative

Journal of Physics: Conference Series **349** (2012) 012004 doi:10.1088/1742-6596/349/1/012004

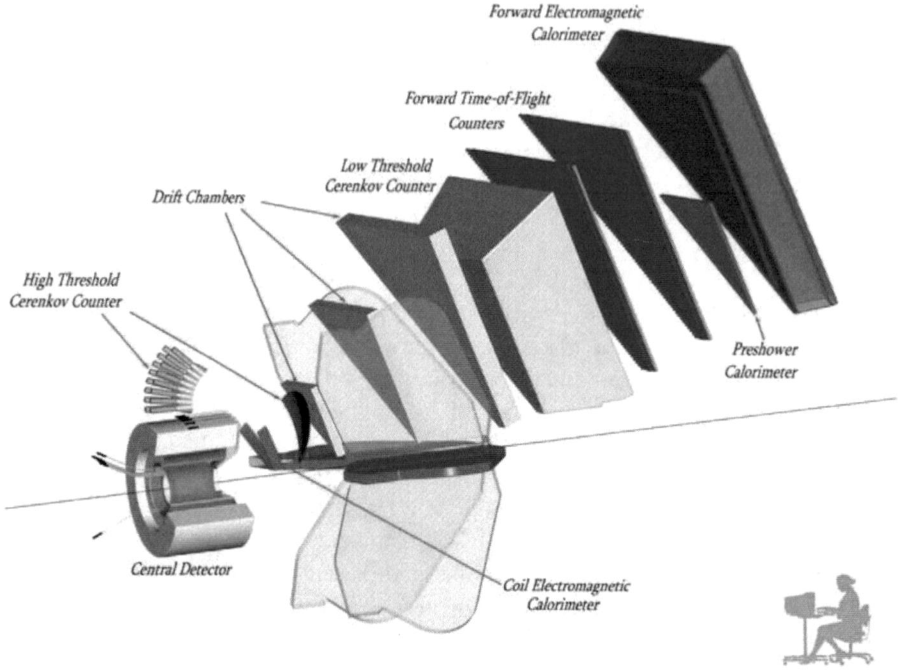

Figure 13. Schematic view of the CLAS12 detector.

population of pions and kaons, it is not sufficient for good kaon identification above 3 GeV/c.

As discussed in the previous sections, the more recent experimental results have pointed out as many of the open issues in the TMD studies are connected to strange channels and in fact a relevant part of the CLAS12 physics program will involve kaon detection. For this reason, a project to replace part of the low threshold Cerenkov counter with a RICH detector is under development. The current project (still under discussion) foresees a first phase when only one RICH sector will be instrumented and in which unpolarized and longitudinal polarization measurements will be performed. In the next stage, a second (or possibly more) RICH sector will be installed, allowing to perform measurements that require either more statistics or a symmetric coverage (as for Collins and Sivers asymmetries).

9. Conclusions

The JLab with its CEBAF accelerator and the three experimentals halls is a Laboratory dedicated to the study of the hadronic physics with electromagnetic probes in a large variety of aspects. The physics program extends from typical nuclear effects like long range correlations in nuclei to the study of the short distance vacuum structure in parity violating experiments.

In recent years, a big part of the physics program was devoted to the study of the Transverse Momentum Dependent distribution functions, new functions introduced to describe the internal structure of the nucleon. Studies of TMDs at JLab and in other laboratories have shown sizeable effects due to transverse motion of the quarks inside the nucleon, but also have open questions. These questions need to be addressed in a new generation of experiments, poviding higher precisions experimental data, and with new analysis techniques, necessary to unfold fundamental properties from the measured obsevables.

The 6 GeV era of JLab, started in 1996, will end in spring 2012, and after about one year of shut down, the 12 GeV operations of JLab will begin. The physiscs program of the 12 GeV JLab will be the natural continuation of the studies performed at the laboratory in the last 15 years. In particular, the increased beam energy will allow to extend the TMD studies at higher Q^2 and x_B and the luminosity increased by about an order of magnitude will provide experimental data with an unprecedented statistical precision.

References

[1] Panofsky W H K 1968 *Proceedings of the 14th International Conference on High Energy Physics, Vienna, Austria*; Bloom E D *et al.* 1969 *Phys. Rev. Lett.* **23** 930; Breidenbach M *et al.* 1969 *Phys. Rev. Lett.* **23** 935);
[2] Alekseev M G *et al.* 2010 *Phys. Lett.* B **693** 227;
[3] Fatemi R *et al.* 2002 *Phys. Rev. Lett.* **91** 222002; Yun J *et al.* 2003 *Phys. Rev.* C **67** 055204; Dharmawardane K *et al.* 2006 *Phys. Lett.* B **641** 11; Bosted P *et al.* 2007 *Phys. Rev.* C **75** 035203; Prok Y *et al.* 2009 *Phys. Lett.* B **672** 12.
[4] Baum G *et al.* 1980 *Phys. Rev. Lett.* **43** 2000; Abe K *et al.* 1997 *Phys. Rev. Lett.* **78** 815; 1998 *Phys. Rev.* D **58** 112003.
[5] Airapetian A *et al.* 2003 *Phys. Rev. Lett.* **90** 092002.
[6] Zheng X *et al.* 2004 *Phys. Rev. Lett.* **92** 012004; Zheng X *et al.* 2004 *Phys. Rev.* C **70** 065207.
[7] Close F E 1974 *Nucl. Phys.* B **80** 269.
[8] Isgur N 1999 *Phys. Rev.* D **59** 034013; Close F A and Thomas A 1988 *Phys. Lett.* B **212** 227.
[9] Farrar G R, Jackson D R 1975 *Phys. Rev. Lett.* **35** 1416; Brodsky S, Burkardt M and Schmidt I 1995 *Nucl. Phys.* B **441** 197;
[10] Avakian H, Brodsky S, Deur A, and Yeng F *Phys. Rev. Lett.* 2007 **99** 082001.
[11] Adeva B *et al.* 1998 *Phys. Rev.* D **58** 112001; Airapetian A *et al.* 2000 *Phys. Rev. Lett.* **84** 2584; Liebig P 2007 *AIP Conf. Proc.* **915** 331; Ageev E S *et al.* 2006 *Phys. Lett.* B **633** 25; Pretz J 2010 *Proceeding of the Photon 09 Conference , Hamburg.*
[12] Mulders P J and Tangerman R D 1996 *Nucl. Phys.* B **461** 197.
[13] Bacchetta A *et al.* 2007 *JHEP* **0702** 093.
[14] Boer D, Gamberg L, Musch B U and Prokudin A *arXiv:1107.5294.*
[15] Collins J C and Metz A 2004 *Phys. Rev. Lett.* **93** 252001; Ji X D, Ma J P and Yuan F 2005 *Phys. Rev.* D **71** 034005; Bacchetta A, Boer D, Diehl M and Mulders P J 2008 *JHEP* 023.
[16] Bacchetta A, D'Alesio U, Diehl M and Miller C A 2004 *Phys. Rev.* D **70** 117504.
[17] Avakian H *et al.* 2010 *Phys. Rev. Lett.* **105** 262002.
[18] Anselmino M, Efremov A, Kotzinian A and Parsamyan B 2006 *Phys. Rev.* D **74** 074015.
[19] Airapetian A *et al.* 2005 *Phys. Rev. Lett.* **94** 012002.
[20] Avakian H *et al.* 2008 *Phys. Rev.* D **77** 014023.
[21] Efremov A V, Goeke K and Schweitzer P 2003 *Phys. Rev.* D **67** 114014.
[22] Efremov A V, Goeke K and Schweitzer P 2006 *Phys. Rev.* D **73** 094025.
[23] Abe K *et al.* 2006 *Phys. Rev. Lett.* **96** 232002.
[24] Aghasyan M *et al.* 2011 *Phys. Lett.* B **704** 397.
[25] Avakian H and Elouadrhiri L 2003 *High Energy Spin Physics Proc., Editors Efremov A V and Teryaev O* 239.
[26] Asaturian R *et al.* *arXiv:1103.1649.*
[27] Qian X *et al.* 2011 *Phys. Rev. Lett.* **107** 072003.
[28] Scopetta S 2007 *Phys. Rev.* D **75** 054005.
[29] Airapetian A *et al.* 2007 *Nucl. Phys.* B **780** 1; Accardi A *et al.* 2003 *Nucl. Phys.* **A720** 131.
[30] Anselmino M *et al.* 2007 *Phys. Rev.* D **75** 054032.
[31] She J and Ma B Q 2011 *Phys. Rev.* D **83** 037502.
[32] Ma B Q 2002 *Phys. Rev.* D **65** 034010.
[33] Pasquini B *et al.* 2008 *Phys. Rev.* D **78** 034025; Boffi S *et al.* 2009 *Phys. Rev.* D **79** 094012.
[34] Anselmino M *et al.* *arXiv:0812.4366.*
[35] Vossen A *et al.* 2011 *Phys. Rev. Lett.* **107** 072004.
[36] Airapetian A *et al.* 2005 *Phys. Rev. Lett.* **94** 012002.
[37] Bacchetta A, Courtoy A and Radici M 2011 *Phys. Rev. Lett.* **107** 012001.
[38] Ball R D *et al.* 2010 *Nucl. Phys.* B **838** 136.
[39] Airapetian A *et al.* 2008 *Phys. Lett.* B **666** 446.
[40] Alekseev M *et al.* 2009 *Phys. Lett.* B **673** 127.
[41] Alekseev M *et al.* 2010 *Phys. Lett.* B **692** 240.
[42] Giordano F and Lamb R 2010 *arXiv:1011.5422.*

Journal of Physics: Conference Series **349** (2012) 012005 doi:10.1088/1742-6596/349/1/012005

Hadronic Light-by-Light contribution to the muon anomalous magnetic moment

Dario Moricciani on behalf of KLOE-2 and BGO-OD collaborations
INFN - Sez. Roma "Tor Vergata", Via della Ricerca Scientifica 1, I-00133 Rome Italy
E-mail: Dario.Moricciani@roma2.infn.it

Abstract. The muon anomalous magnetic moment is one of the most precisely measured quantities in particle physics. The E821 recent experiment at Brookhaven it measured with a remarkable precision of 0.54 ppm. Since the first results were published, a persisting "discrepancy" between theory and experiment of about 3 standard deviations is observed. The dominating uncertainty of the prediction, caused by strong interaction effects, should be reduced substantially in the next future as requested by the "New g-2" experiment planned at FNAL, where a precision of 0.14 ppm should be reached . A review of the the role of the hadronic light–by–light scattering to $a_\mu = (g-2)_\mu/2$ contribution is discussed focussing on the experimental point of view.

1. Introduction

The systematic comparison of Standard Model (SM) predictions with precise experimental data served in the last decades as an invaluable tool to test this theory at the quantum level. It has also provided stringent constraints on new physics scenarios. The remarkable agreement between the measurements of the electroweak observables and their SM predictions is a striking experimental confirmation of the theory, even if there are a few observables like the muon anomalous magnetic moment a_μ where the agreement seems to be not so satisfactory.

The comparison between the experimental [1] value of $a_\mu^{\mathrm{EXP}} = 116592080(0.54)(0.33)[0.63] \times 10^{-11}$ and the theoretical [2] $a_\mu^{\mathrm{SM}} = 1165917900(0.65) \times 10^{-11}$ produce a difference between them of $\Delta a_\mu = a_\mu^{\mathrm{EXP}} - a_\mu^{SM} = (290.0 \pm 90.3) \times 10^{-11}$ while a different theoretical analysis [3] produces a similar results : $\Delta a_\mu = (255 \pm 80) \times 10^{-11}$.

In order to clarify the nature of the observed discrepancy between theory and experiment, and eventually reinforce its statistical significance, new direct measurements of the muon $g-2$ with a fourfold improvement in accuracy have been proposed at Fermilab [4] and J-PARC [5].

2. The Hadronic Ligh-by-Light contribution

The main contributions to the error on the Standard Model prediction of a_μ^{SM} are shown in Table 1 for three recent estimates [2, 6, 3] [1], where the two dominant contributions to the uncertainty, namely a_μ^{HLO} and a_μ^{HLbL} terms are shown separately:

The most problematic set of hadronic corrections to a_μ are those related to hadronic light–by–light scattering (HLbL), which show up at order $O(\alpha^3)$ via the diagrams of Fig. 1.

[1] Ref. [6] includes the recent BaBar 2π data; [2] uses a more conservative error analysis.

Error	[2]	[6]	[3]	prospect
$\Delta a_\mu^{\mathrm{SM}}$	65	49	48	35
$\Delta a_\mu^{\mathrm{HLO}}$	53	41	40	26
$\Delta a_\mu^{\mathrm{HLbL}}$	39	26	26	25
$\Delta(a_\mu^{\mathrm{SM}} - a_\mu^{\mathrm{EXP}})$	88	80	79	40

Table 1. Estimated uncertainties Δa_μ in units of 10^{-11} according to Refs. [2, 6, 3] and (last column) prospects in case of improved precision in the e^+e^- hadronic cross-section measurement (the prospect on $\Delta a_\mu^{\mathrm{LbL}}$ is an *educated guess*). Last row: Uncertainty on Δa_μ assuming the present experimental error of 63 from BNL-E821 [1] (first two columns) and of 14 (last column) as planned by the future $g-2$ experiments [4, 5].

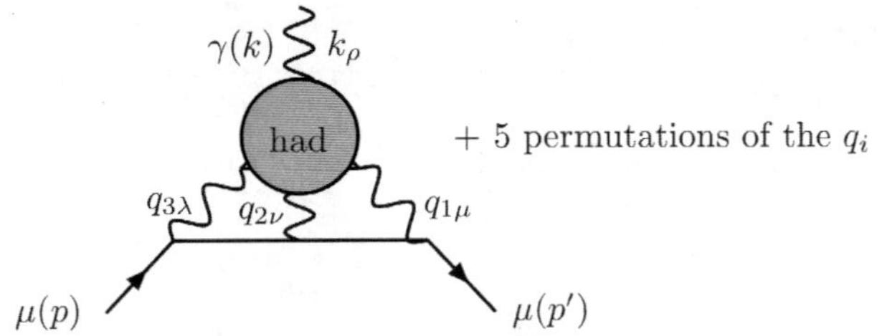

Figure 1. Assignment of momenta for the calculation of the hadronic contribution of the light–by–light scattering to the muon electromagnetic vertex.

As a contribution to the a_μ, three of the four photons in Fig. 1 are virtual and to be integrated over all four–momentum space, such that a direct experimental input for the non–perturbative dressed four–photon correlator is not available. In this case one has to resort to the low energy effective descriptions of QCD-like *chiral perturbation theory* (CHPT) extended to include vector–mesons. Note that early evaluations assumed that the main contribution to hadronic light-by-light scattering comes from momentum regions around the muon mass.

In order to clarify the importance of the HLbL contribution, let us take into account some numerical estimation [2] for the "total" HLbL contribution, only the part coming from the light pseudo-scalar (π^0, η, η'), and only the part coming from π^0:

$$\begin{aligned} a_\mu^{\mathrm{HLbL;tot}} &= (116 \pm 39) \times 10^{-11} \\ a_\mu^{\mathrm{HLbL;PS}} &= (99 \pm 16) \times 10^{-11} \\ a_\mu^{\mathrm{HLbL;}\pi^0} &= (72 \pm 12) \times 10^{-11} \end{aligned}$$

By a comparison of the above errors with the proposed error of the new $(g-2)$ experiment (± 14 in the unit of 10^{-11}), we notice that a method for reduce the error on HLbL is needed both from theory, where lattice QCD seems promising [7], and experimental side. Here we discuss the dominating hadronic contributions which are due to the neutral pseudoscalar–exchange diagrams shown in Fig. 2.

We first concentrate on the exchange of the neutral pion. The key object which enters the Feynman diagrams is the off-shell $\pi^0\gamma\gamma$ form factor $\mathcal{F}((q_1+q_2)^2, q_1^2, q_2^2)$, which is defined up to small mixing effects with the states η and η'. In this framework the $a_\mu^{\mathrm{HLbL;}\pi^0}$ can be evaluated by

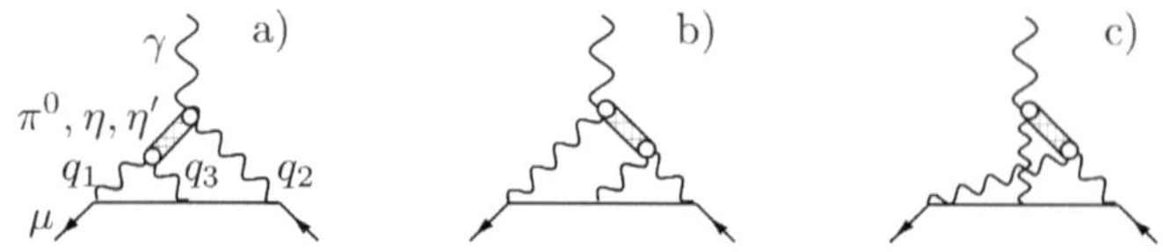

Figure 2. Leading hadronic light–by–light scattering diagrams. In accord with Eq. (1), here all photon momenta are chosen incoming to the pion transition form factors. Internal photon lines are dressed by $\rho - \gamma$ mixing.

the following integral:

$$\begin{aligned} a_\mu^{\mathrm{HLbL};\pi^0} &= -e^6 \int \frac{d^4q_1}{(2\pi)^4} \frac{d^4q_2}{(2\pi)^4} \frac{1}{q_1^2 q_2^2 (q_1+q_2)^2[(p+q_1)^2 - m_\mu^2][(p-q_2)^2 - m_\mu^2]} \\ &\times \left[\frac{\mathcal{F}_{\pi^{0*}\gamma^*\gamma^*}(q_2^2, q_1^2, q_3^2)\ \mathcal{F}_{\pi^{0*}\gamma^*\gamma}(q_2^2, q_2^2, 0)}{q_2^2 - m_\pi^2} T_1(q_1, q_2; p) \right. \\ &+ \left. \frac{\mathcal{F}_{\pi^{0*}\gamma^*\gamma^*}(q_3^2, q_1^2, q_2^2)\ \mathcal{F}_{\pi^{0*}\gamma^*\gamma}(q_3^2, q_3^2, 0)}{q_3^2 - m_\pi^2} T_2(q_1, q_2; p) \right], \end{aligned} \tag{1}$$

where the $T_1(q_1, q_2; p)$ and $T_2(q_1, q_2; p)$ terms are defined in [2]. We have to notice that the only unknown parts in (1) are the Transition Form Factors (TFF) $\mathcal{F}_{\pi^{0*}\gamma^*\gamma^*/\gamma}((q_1+q_2)^2, q_1^2, q_2^2)$.

3. Experimental options

From the experimental side we have to face that the unique observable that we can measure in order to evaluate the equation (1) is the TFF where the PS is on-shell : $\mathcal{F}_{\mathrm{PS}\gamma^*\gamma^*/\gamma}(M_{\mathrm{PS}}^2, q_1^2, q_2^2)$. In Fig. 3 it is shown our present experimental knowledge of the TFF for the reactions $\gamma\gamma^* \to \pi^o$, $\gamma\gamma^* \to \eta$, and $\gamma\gamma^* \to \eta'$ from [8]. In this case the space-like region $Q^2 = -q_1^2$ and $q_2 = 0$ are explored.

Two are the key points which we have to mention :

- the π^o TFF doesn't follow the Brodsky-Lepage [11] asymptotic limit i.e. $\mathcal{F}_{\pi^0\gamma^*\gamma} \sim 1/Q^2$ for $Q^2 \to \infty$, while the TFF for η and η' follow it;
- for all the pseudo-scalars there are no experimental data for $Q^2 \leq 0.5$ GeV^2.

The point at $q^2 = 0$ plays an important role because in it the TFF is proportional to the width of the decay of PS $\to \gamma\gamma$:

$$\mathcal{F}^2_{\pi^0\gamma^*\gamma^*}(q_1^2 = 0, q_2^2 = 0) = \frac{4}{\pi\alpha^2 m_\pi^3} \Gamma_{\pi^0 \to \gamma\gamma}. \tag{2}$$

A precise knowledge of width decay represents an absolute normalization of the TFF and then permits to reduce error on the evaluation of the a_μ^{HLbL}.

Due to the recent advances, the decay width $\Gamma_{\pi^0\to\gamma\gamma}$ is now predicted with a 1.4% accuracy: $\Gamma^{\mathrm{theor}}_{\pi^0\to\gamma\gamma} = 8.09 \pm 0.11$ eV [14, 15]. The major experimental information on this decay comes from the photo-production of pions on a nuclear target via the Primakoff effect [16]. The most precise value of the pion lifetime cited by PDG [17] is that of a direct decay measurement [18] and, via the precisely known $\pi^0 \to \gamma\gamma$ branching ratio, it can be related to the two-photon width.

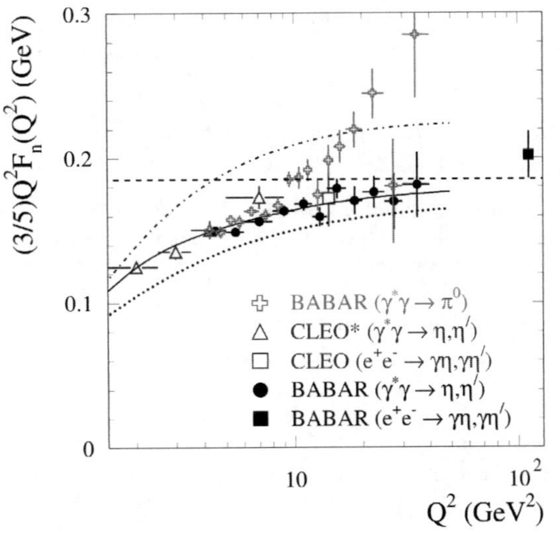

Figure 3. The $\gamma\gamma^* \to$ PS TFF multiplied by $3Q^2/5$ in comparison with the $\gamma\gamma^* \to \pi^0$ TFF [9]. The dashed line indicates the asymptotic limit for the π^0 form factor. The dotted, dash-dotted, and solid curves show predictions of Ref. [10] for the asymptotic DA [11], the Chernyak-Zhitnitsky π^0 DA [12], and the π^0 DA from Ref. [13], respectively.

Until recently, the experimental world average of $\Gamma^{\rm PDG}_{\pi^0\to\gamma\gamma} = 7.74 \pm 0.48$ eV [17] was only known to 6.2% precision. Due to the poor agreement between the existing data, the PDG error of the width average is inflated (scale factor 2.6) and it gives an additional motivation for new precise measurements. The PrimEx Collaboration, using a Primakoff effect experiment at JLab, has achieved 2.8% precision, reporting, see Table 2, the value $\Gamma_{\pi^0\to\gamma\gamma} = 7.82 \pm 0.14 \pm 0.17$ eV [19], but this result is not yet included in the PDG average. There are plans to further reduce the uncertainty to the per cent level.

Though theory and experiment are in a fair agreement, a better experimental precision is needed to really test the theory predictions. The Primakoff effect-based experiments suffer from model dependence due to the contamination by the coherent and incoherent conversions in the strong field of a nucleus [20].

Target	$\Gamma(\pi^0 \to \gamma\gamma)$ [eV]	$\Delta\Gamma/\Gamma$
^{12}C	7.79±0.18	2.31 %
^{208}Pb	7.85±0.23	2.93 %

Table 2. The fit values extracted from the measured cross sections on ^{12}C and ^{208}Pb [19]. The values for the decay widths are the weighted averages from two analyses. The uncertainties shown here are statistical only (see text for notations).

4. KLOE-2 option

In the second data taking campaign of the KLOE experiment (KLOE-2[21]) new small angle tagging detectors are installed, which offer the possibility to study a program of $\gamma\gamma$ physics through the process $e^+e^- \to e^+\gamma^*e^-\gamma^* \to e^+e^-X$. Here we refer to the High Energy Tagger (HET) [22] A detailed study of the reaction $e^+e^- \to e^+e^-\pi^o$ based on the EKHARA [23] Monte

Journal of Physics: Conference Series **349** (2012) 012005 doi:10.1088/1742-6596/349/1/012005

Carlo generator has been done [24] in order to evaluate the possibility to measure both the $\Gamma_{\pi^o\to\gamma\gamma}$ at 1 % level and the $\mathcal{F}_{\pi^0\gamma^*\gamma}(M^2_{\pi^o}, q^2, 0)$ with $q^2 \in (0, 0.1)$ GeV^2.

Two different acceptance regions are considered :

$H \otimes H$) both leptons in the final state are detected in the HET taggers and the two γs from the decay of π^o are detected in KLOE;

$H \otimes K$) one lepton in the final state is detected in the corresponding HET while the second one in KLOE, the two γs from the decay of π^o are detected in KLOE;

In the first region since the q^2 coverage of the HET detector as show in Fig. 4 is surely below 10^{-3} GeV^2 we can make the safe approximation that $\sigma \propto \Gamma_{\gamma^*\gamma^*\to\pi^o}\Gamma_{\pi^o\to\gamma\gamma} \simeq \Gamma^2_{\pi^o\to\gamma\gamma}$ then measuring the cross section we can extract from it the π^o width. The simulation shows that a statistical error of 1 % can be obtained with an integrated luminosity of 5 fb^{-1}.

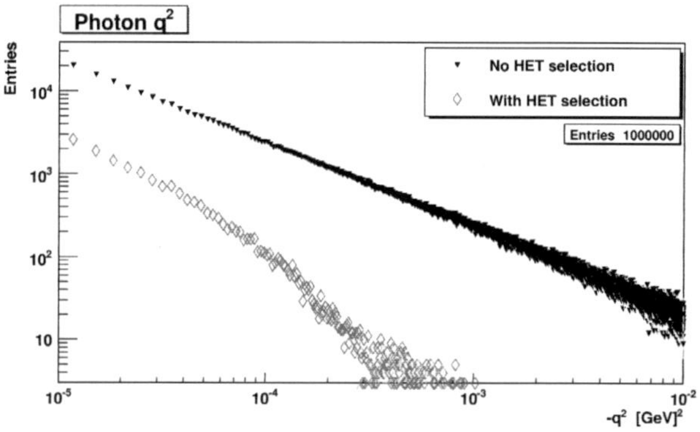

Figure 4. Distribution of the photon virtuality in $\gamma^*\gamma^* \to \pi^0$. The lepton double tagging (HET-HET) selects the events (red diamonds) with small virtuality of the photons.

In the second region considered, thanks to the acceptance of the KLOE detector for one of the final leptons, a q^2 coverage in the unexplored region below 0.1 GeV^2 is allowed, as show in Fig. 5. The value of the pion-exchange part of the $a_\mu^{\mathrm{HLbL};\pi^0}$ contribution to a_μ is currently obtained using hadronic models and any experimental information on the TFF is important in order to constrain the models. However, having a good description for the TFF is only necessary, not sufficient, in order to uniquely determine $a_\mu^{\mathrm{HLbL};\pi^0}$. After combining both regions an evaluation of the impact on $a_\mu^{\mathrm{HLbL};\pi^0}$ in the framework of JN [2, 28] and MV [29], VMD-inspired, models shows that a reduction of a factor 4 for **on-shell** part is obtainable. A further data taking period don't improve so much the knowledge of $a_\mu^{\mathrm{HLbL};\pi^0}$ because in this case the error on the mass of vector meson in the JN and MV models starts to play a role.

5. BGO-OD option

The TFF could also be studied in the time-like region through the PS Dalitz decay $PS \to \gamma\gamma^* \to \gamma e^+e^-$. Electromagnetic form factors are regarded as an important tool to study the intrinsic structure of hadrons. Of particular interest are the decays of the narrow pseudoscalar states π,η and η' into two real or virtual photons and of the narrow vector states ω and ϕ into a pseudoscalar and a (real or virtual) photon. The pseudoscalar decays are even of some relevance for searches for physics beyond the standard model of elementary particles. Their size influences for the anomalous magnetic moment of the muon [30].

On the other hand, the mentioned decays of pseudoscalar mesons and of vector mesons are even interrelated, since the neutral vector mesons have the same quantum numbers as the

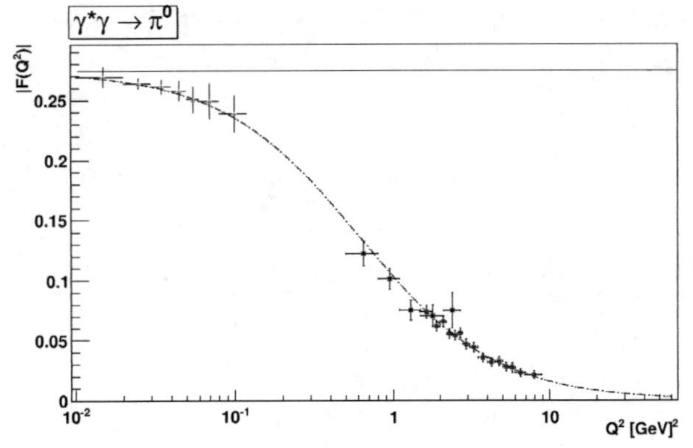

Figure 5. Simulation of KLOE-2 measurement of $\mathcal{F}(Q^2)$ (red triangles) with statistical errors for 5 fb^{-1}. Dashed line is the $\mathcal{F}(Q^2)$ form factor according to LMD+V model [25], solid line is $F(0)$ given by Wess-Zumino-Witten term $\mathcal{F}(0) = 1/4\pi^2 F_\pi$; CELLO [26] (black crosses) and CLEO [27] (blue squares) data at high Q^2 are also shown for illustration.

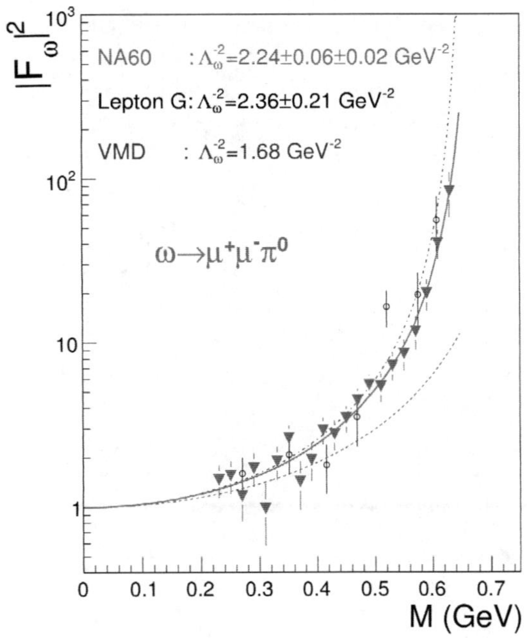

Figure 6. Experimental data on the ω-meson electromagnetic TFF (red triangles), compared to the previous measurement by the Lepton-G experiment (open circles) and to the expectation from VMD (blue dashed line). The solid red and black dashed- dotted lines are results of fitting the experimental data with the pole dependence $|F|^2 = (1 - M^2/\Lambda^2)^{-2}$. The normalization is such that $|F_\omega(M = 0)| = 1$.

photon. In fact, for many reactions, where dileptons couple (via a virtual photon) to hadrons, the vector-meson dominance (VMD) assumption [31] turned out to be phenomenologically very successful. One example where standard VMD dramatically fails, however, is the TFF of the ω-meson [32] as show in Fig. 6. In the same paper the TFF of the reaction $\eta \to \gamma\mu^+\mu^-$ is also studied, in this case the VMD is not satisfactory but dons't fail dramatically like in the ω case. The naive VMD model suggests for the TFF that:

$$\mathcal{F}(Q^2) \simeq \frac{m_V^2}{m_V^2 - Q^2} \tag{3}$$

where the vector mass m_V is in the range m_ρ, m_ω, m_ϕ but obviously this model needs to be more sophisticated in order to reproduce η and ω data.

Recently new models appeared in the literature, as the one in the Ref. [33] where the decays of narrow light vector mesons into pseudoscalar mesons and dileptons are calculated to leading order in a recently proposed scheme which treats pseudoscalar and vector mesons on equal footing. A more recent model [34] uses the chiral perturbation theory with resonances for the description of the conversion transition of the vector meson V into the pseudoscalar meson $\mathcal{P}$ and the lepton pair l^+l^- ($V \equiv \rho^0$, ω, ϕ and $\mathcal{P} \equiv \pi^0$, η, η').

All this experimental information seems to suggest that the VMD work well for π start to fail for the η and fail dramatically for the ω for that reason seems to be very interesting investigate the η' case because its mass is very close to m_V.

In the BGO-OD experiment the η' can be photo-produced by the reaction $\gamma p \to \eta' p$ and we can study the Dalitz decay $\eta' \to \gamma e^+ e^-$, the recoil proton will be detected in the forward MRPC detector $\Delta_{tof} \simeq 30 \div 50$ ps which covers the polar angle between 8° and 25°; the decay photon in the BGO calorimeter and the two decay lepton in the forward region by the tracking system in the OD magnet. The performance of this apparatus is even better than GRAAL apparatus where the forward magnet was absent. Using the GRAAL a similar analysis has been already done showing the excellent resolution on the invariant mass of the system: one γ and a pair of charged particles (e^+e^-) see Fig. 7[35].

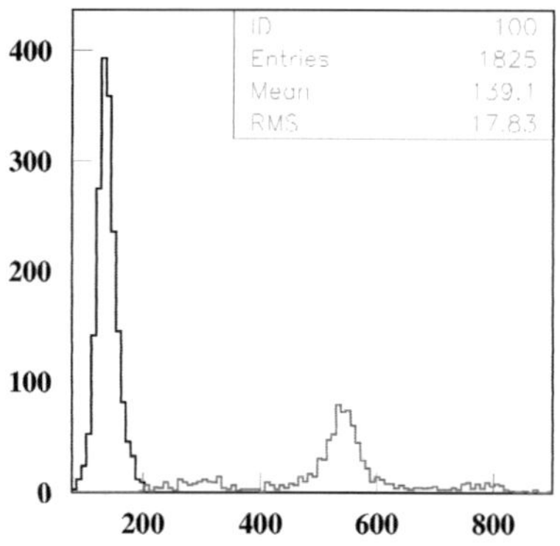

Figure 7. Invariant mass of the PS meson

References

[1] G. W. Bennett et al. [Muon g-2 Collaboration], Phys. Rev. D **73**, (2006), 072003.
[2] F. Jegerlehner and A. Nyffeler, Phys. Rept. **477**, (2009), 1.
[3] T. Teubner, *et al.* arXiv:1001.5401.
[4] R.M. Carey et al. (2009), FERMILAB-PROPOSAL-0989.
[5] J. Imazato, Nucl. Phys. Proc. Suppl. **129**, (2004), 81.
[6] M. Davier, *et al.* Eur. Phys. J. C **66** (2010), 1.
[7] T. Blum *at al.*, PoS LAT2005 (2005) 353.
[8] P. del Amo Sanchez, *et al.* Phys. Rev. D **84**, (2011), 052001.
[9] B. Aubert *et al.* , Phys. Rev. D **80**, (2009), 052002.
[10] A. P. Bakulev, S. V. Mikhailov, and N. G. Stefanis, Phys. Rev. D **67**, (2003), 074012, Phys. Lett. B **578**, (2004).
[11] G. P. Lepage and S. J. Brodsky, Phys. Lett. B **87**, (1979), 359.
[12] V. L. Chernyak and A. R. Zhitnitsky, Nucl. Phys. B **201**, (1982), 492, [Erratum-ibid. B **214**, (1983), 547].
[13] A. P. Bakulev, S. V. Mikhailov, and N. G. Stefanis, Phys. Lett. B **508**, (2001), 279, [Erratum-ibid. B **590**, (2004), 309].
[14] K. Kampf, B. Moussallam, Phys. Rev. D **79**, (2009), 076005.
[15] J. Bijnens, K. Kampf, Nucl. Phys. Proc. Suppl. **207-208**, (2010), 220.
[16] H. Primakoff, Phys. Rev. **81**, (1951), 899.
[17] K. Nakamura, J. Phys. G **37**, (2010), 075021.
[18] H. W. Atherton, *et al.*, Phys. Lett. B **158** (1985), 81.
[19] I. Larin, et al., Phys. Rev. Lett. **106**, (2011), 162303.
[20] M. M. Kaskulov, U. Mosel (2011). arXiv:1103.2097.
[21] G. Amelino-Camelia *at al.* Eur. Phys. Jour. C **68**, (2010), 619.
[22] F. Archilli *at al.* Nucl. Instrum. Meth. A **617**, (2010), 266.
[23] H. Czyz and S. Ivashyn, Comp. Phys. Com. **182**, (2011), 1338.
[24] D. Babusci *et al.* (2011). arXiv:1109.2461.
[25] M. Knecht, A. Nyffeler, Eur. Phys. J. C **21**, (2001), 659.
[26] H. J. Behrend, et al., Z. Phys. C **49**, (1991), 401.
[27] J. Gronberg, et al., Phys. Rev.D **57**, (1998), 33.
[28] A. Nyffeler, Phys. Rev. D **79**, (2009), 073012.
[29] K. Melnikov, A. Vainshtein, Phys. Rev. D **70**, (2004),113006.
[30] J. Bijnens and J. Prades, Mod. Phys. Lett. A **22**, (2007), 767.
[31] J. J. Sakurai, *Currents and Mesons* (University of Chicago Press, 1969).
[32] R. Arnaldi et al. (NA60), Phys.Lett. B **677**, (2009), 260.
[33] C. Terschlsen, S. Leupold, Phys.Lett. B **691**, (2010), 191.
[34] S. Ivashyn (2011). arXiv:1111.1291.
[35] G. Mandaglio and M. Romaniuk (2011) private comunication.

The $\gamma p \to K^0\Sigma^+$ photoproduction reaction[1]

Hartmut Schmieden

Physikalisches Institut, University of Bonn, Nussallee 12
D-53115 Bonn, Germany

E-mail: schmieden@physik.uni-bonn.de

Abstract. At the electron accelerator facility ELSA of the University of Bonn the photoproduction reaction $\gamma p \to K^0\Sigma^+$ was investigated using the CBELSA/TAPS experimental setup. A pronounced structure in the cross section is found at the K^* threshold. It is speculated whether this is due to the formation of a K^*-hyperon quasibound state below the K^* threshold. The impact of recent polarisation measurements is discussed.

1. Introduction

Excitations of baryons carry information about their inner structure. The effective degrees of freedom are not necessarily just quarks and gluons as may be "simply" expected from the short range nucleon structure revealed, e.g., in deep inelastic lepton scattering. Due to the closeness of the chiral symmetry breaking scale to the nucleon mass/size scale, also the rsp. Goldstone bosons enter as effective "elementary" objects [1, 2]. It is hence no surprise that in some aspects, e.g. the parity ordering of the lowest nucleon excitations [1], models which include the interaction of the light mesons with quarks are more successful than genuine three quark models which attempt to parameterize pure colour interactions. If already meson-quark interactions play a significant role, then meson-baryon interactions appear as a natural extension. Some models indeed assign a leading role in hadronic excitations to meson-baryon interactions [3, 4, 5, 6, 7], and some of the states which persistingly resisted a conventionial three-quark explanation in the meanwhile are speculated to predominantly exhibit meson-baryon dynamics, e.g. the $\Lambda(1405)$ [8]. In addition to the interaction of pseudoscalar mesons with baryons also vector mesons are considered for dynamical resonance generation [9, 10, 11]. Degenerate states of $J^P = 1/2^-, 3/2^-$ are expected, in particular in the mass region around 2 GeV [12].

K^0_s photoproduction off the proton provides a very interesting channel for several reasons [8]. For the study presented here it is of particular interest that, compared to the extensively measured charged kaon channels, t-exchange is suppressed in the production mechanism. Hence, a cleaner probe to s-channel contributions is provided. On the other hand, t-exchange is not strictly forbidden despite the zero charge of the K^0, because the photon may magnetically couple to a K^{0*} vector meson as is shown in Figure 1(e). If vectormeson-baryon dynamics was significant in the production process, then diagrams of the type Figure 1(f) should play a role and differently affect K^0 photoproduction above and below the K^{0*} threshold, unobstructed by the strong charge-dominated t-exchange in K^+ production. While previous $K^0\Sigma^+$ data of Crystal Barrel [13] and SAPHIR [14] agree rather well in general, differences show up just in the

[1] Supported by the Deutsche Forschungsgemeinschaft (DFG) within SFB/TR-16.

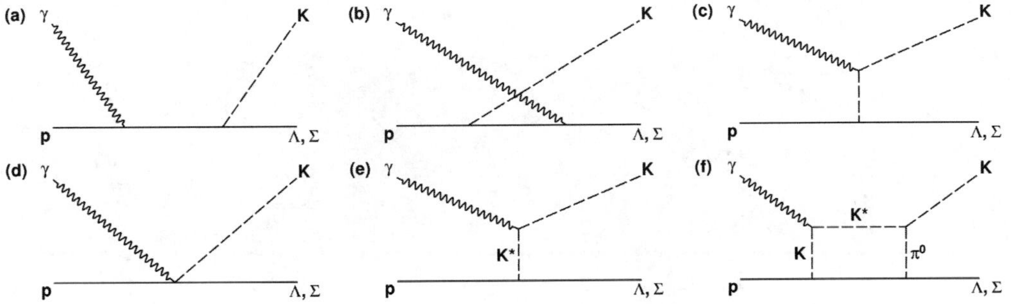

Figure 1. Diagrams contributing to charged and uncharged kaon photoproduction. The Born diagrams are shown in (a) - (d). Non-strange resonances may contribute as intermediate states in the s-channel (a) and u-channel (b). The t-channel meson exchange (c) and the seagull term (d) are proportional to the meson charge, hence contribute only to the charged kaon channel. Vector meson exchange (e) is allowed also for K^0 photoproduction. Diagram (f) visualizes subthreshold K^* production with subsequent coupling into the kaon channel through π^0 rescattering, as explained in the text.

energy region of the K^* threshold. This prohibits a clear conclusion on the role of K^*-hyperon dynamics. The new data discussed here improve this unsatisfactory situation.

2. Experiment

The experiment was performed at the Electron Stretcher Accelerator ELSA [15] of the University of Bonn using the combined Crystal Barrel [16] and TAPS [17, 18] detector system. At an electron beam energy of $E_0 = 3.2$ GeV tagged photon beams were generated by coherent bremsstrahlung from a 500 μm thick diamond radiator [19]. The bremsstrahlung electrons were momentum analysed in a magnetic "tagging" spectrometer. A photon energy range of $E_\gamma = 0.18$–$0.92 E_0$ was covered. The tagging system provided an energy resolution between 10 and 25 MeV. It was run at electron rates up to 10^7 Hz. The absolute photon flux was determined from the tagged electron spectrum in combination with a fast total absorbing $PbWO_4$ detector to measure the energy dependent tagging efficiency.

The photon beam impinged on liquid hydrogen contained in a 5.3 cm long target cell with 80 μm Kapton windows. The reaction products were observed in the Crystal Barrel and TAPS spectrometers, complemented by a three layer scintillating fibre detector [20]. In total, the detector system covered a polar angular range of 5.8 – 165 degree. Further details of setup and readout are given in Ref. [21].

The $K^0\Sigma^+$ reaction was investigated in the neutral decay modes $K^0 \rightarrow \pi^0\pi^0$ (B.R. 31.4 %) and $\Sigma^+ \rightarrow p\pi^0$ (B.R. 51.6 %), yielding 6 photons and the proton. The identification of the $K^0\Sigma^+$ reaction channel was based on a kinematic fit to the $\gamma p \rightarrow p 3\pi^0$ reaction. Energy and angle parameters for the fit were provided by the final state photon hits, the photon energy by the tagging spectrometer. Due to the hermetic and even acceptance, cross sections and polarisation observables were obtained over the full angular range. Total cross sections could be extracted without any extrapolations.

3. Cross sections

After the kinematic fit two-dimensional invariant mass distributions are obtained as shown in Figure 2. The $K^0\Sigma^+$ events are selected from the appropriate region in Figure 2. Background is subtracted through simulation of various channels which may lead to the $6\,\gamma$ event topology

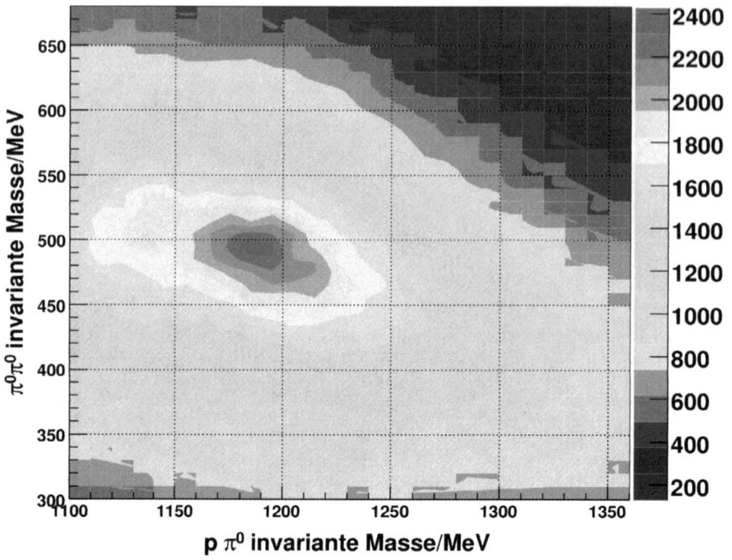

Figure 2. $\pi^0\pi^0$ against $p\pi^0$ invariant mass distribution after event preselection and kinematic cut, showing a concentration of events in the $K^0\Sigma^+$ final state.

and tuning to the observed one-dimensional invariant mass distributions. The most severe background is found to come from uncorrelated $3\,\pi^0$ photoproduction. The deacy $\eta \rightarrow 3\,\pi^0$ is suppressed by an appropriate anti-cut in the $6\,\gamma$ invariant mass distribution. Upon flux normalisation and acceptance correction then differential cross sections are obtained (for details see [21]).

Figure 3 shows the results in comparison to previous measurements of Crystal Barrel [13] and SAPHIR [14]. The data sets agree rather well in general. Discrepancies are however observed in forward directions and in the energy range of the K^* threshold ($E_\gamma = 1750 - 1850\,\text{MeV}$ bin). These discrepancies are resolved by the new data. Below the K^* threshold mostly the SAPHIR data is favored. At the K^* threshold and higher energies the previous Crystal Barrel data is supported.

Directly above the $K^0\Sigma^+$ threshold a differential cross section of $\simeq 0.02\,\mu\text{barn/sr}$ is obtained with flat angular dependence, typical for s-wave production. The cross section rises with increasing photon energy and also develops an increasing forward peaking, suggesting increasing t-channel contributions. This forward peaking is most pronounced in the $E_\gamma = 1700 \pm 50\,\text{MeV}$ bin. The next energy bin exhibits an entirely flat distribution again and the cross section drops back to $\simeq 0.02\,\mu\text{barn/sr}$. This is at $E_\gamma = 1800\,\text{MeV}$, i.e. between the thresholds of $K^*\Lambda$ and $K^*\Sigma$ photoproduction. The effect is strong enough to be clearly visible also in the total cross section, cf. Figure 4, but it is most pronounced in the most forward angular bin where the cross section drops by a factor of four. This is shown in Figure 5. Above the K^* threshold the differential cross section remains almost flat and practically constant up to the highest measured energies.

4. Polarisation Observables

In order to achieve full experimental account of meson photoproduction reactions the measurement of polarisation observables is mandatory. The most basic one is the photon beam

Journal of Physics: Conference Series **349** (2012) 012006 doi:10.1088/1742-6596/349/1/012006

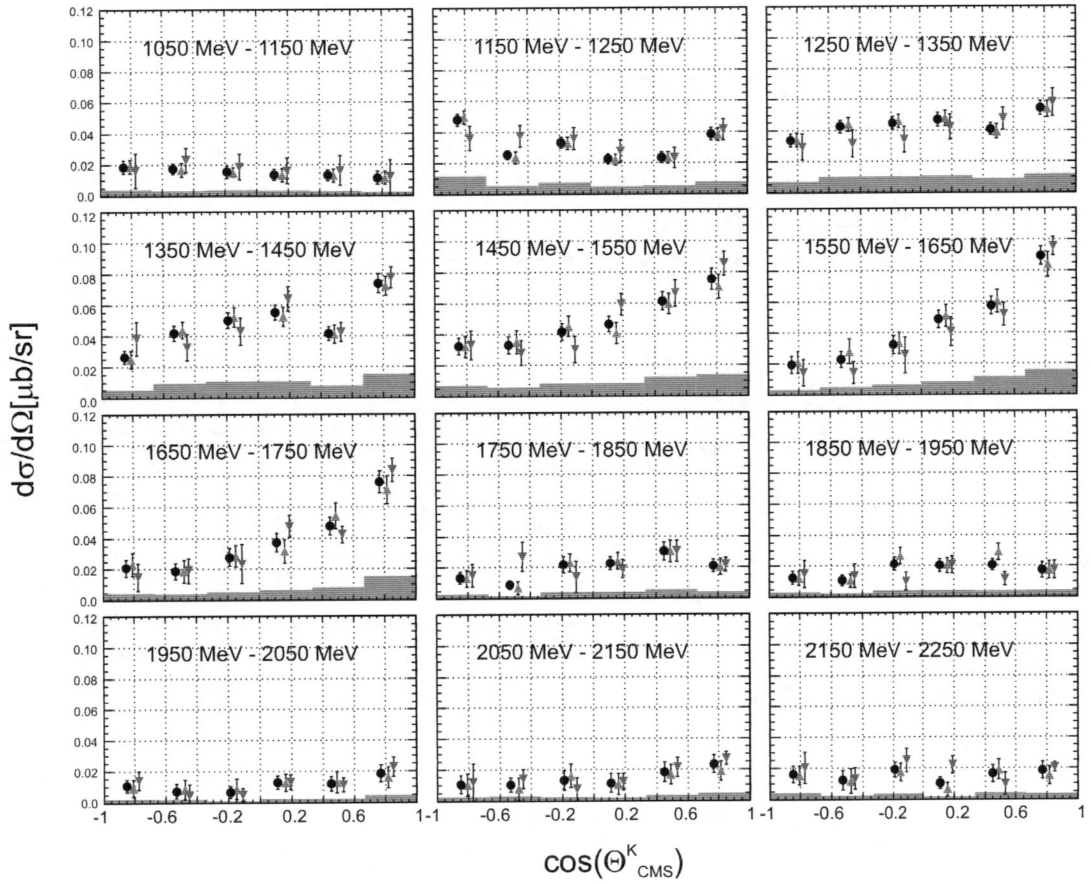

Figure 3. Measured differential cross sections for $K^0\Sigma^+$ photoproduction as a function of the kaon center-of-mass angle in ± 50 MeV wide bins of photon energy from 1100 to 2200 MeV. The present results (full squares) are compared to previous measurements of Crystal Barrel (open squares) [13] and SAPHIR (triangles) [14]. The error bars are purely statistical. An estimate of the systematic uncertainty is given by the bars on the x-axis.

asymmetry. It measures the asymmetry of the azimuthal orientation of the reaction plane relative to the plane of photon linear polarisation. In associated photoproduction of strange mesons with hyperons the polarisation of the recoiling baryon is also easily accessible due to the self analysing weak deacy. Both observables were analysed in addition to the cross sections.

4.1. Recoil polarisation

In meson photoproduction the recoiling spin 1/2 baryon generally carries polarisation. The origin is twofold: (i) polarisation transfer from the linearly or circularly polarised photon field, and (ii) spin-orbit coupling in final state interactions [25]. The present experiment used linearly polarised photon beams. In this case, angular momentum transfer results in $\sin 2\phi$ and $\cos 2\phi$ modulations of the recoil polarisation *within* the reaction plane [26], which is spanned by the ejected kaons and hyperons. The angle ϕ denotes the azimuth between the plane of linear polarisation and the reaction plane. Integrating over all possible orientations which, effectively, corresponds to unpolarised photons, solely leaves the polarisation component (ii) non-zero. That

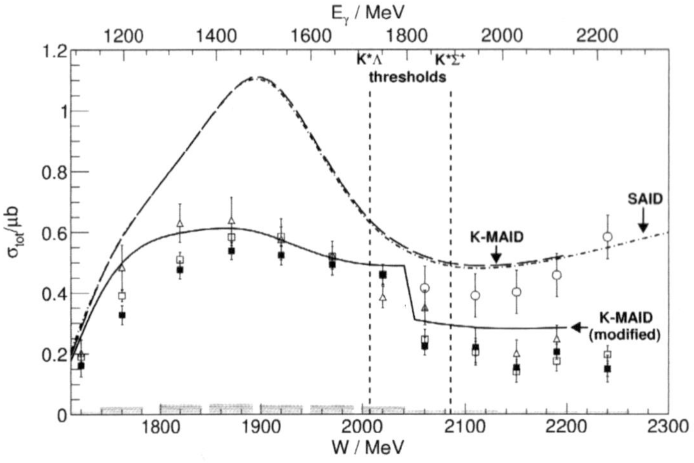

Figure 4. Total cross section for $K^0\Sigma^+$ photoproduction as a function of the centre-of-mass energy from the present experiment (full squares) in comparison to the previous Crystal Barrel (open squares) [13] and SAPHIR (triangles) [14] data. The vertical lines indicate the $K^*\Lambda$ and $K^*\Sigma^+$ thresholds at $W = 2007.4$ and 2085.5 MeV, respectively. The SAID parameterisation [29] is represented by the dashed-dotted curve. A K-MAID calculation [23] with standard parameters yields the dashed curve. The full curve is obtained with the modifications described in the text. Above the K^* threshold the grey circles represent the sum of the $K^0\Sigma^+$ cross section of the present experiment and the $K^{0*}\Sigma^+$ cross section of the work of Nanova et al. [24]. The vertical bars on the abscissa again indicate the experimental systematic errors, the errors plotted with the data symbols are purely statistical.

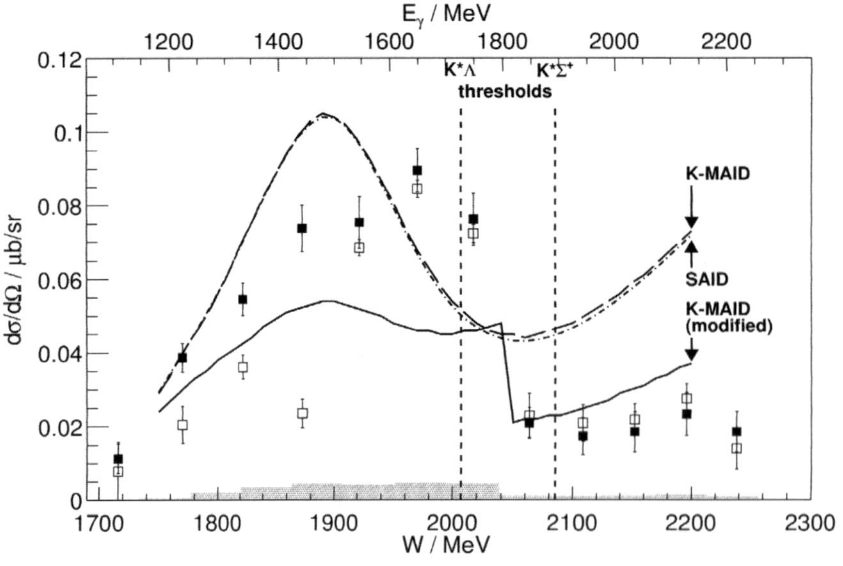

Figure 5. Cross section for $K^0\Sigma^+$ photoproduction as a function of the centre-of-mass energy from the present (full squares) and a previous (open squares) [13] Crystal Barrel experiment in the most forward angular bin of Figure 3.

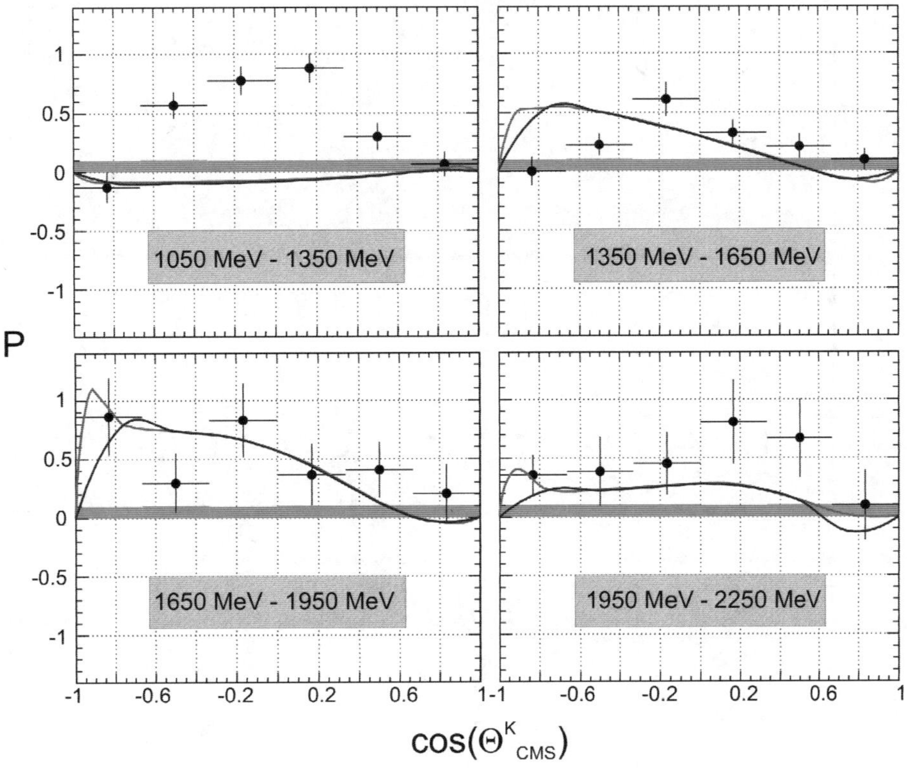

Figure 6. Recoil polarisation of the Σ^+ in the four bins of photon energy indicated in the diagrams. The results of the present measurement are compared to the parameterisations of SAID [29] (blue curve) and K-MAID [23] (red curve). Attached to the data points are the statistical errors and, as horizontal lines, the bin widths in $\cos\theta$. The systematic errors are indicated by the grey bars on the abscissa.

one is usually called *the* recoil polarisation, P. Due to parity conservation it is oriented normal to the reaction plane.

In the studied reaction, the weak decay of the final state Σ^+ enables the reconstruction of the recoil polarisation P of the hyperon. The decay angular distribution has the form

$$W(\theta_p) = \frac{1}{2}(1 + \alpha_0 P \cos\theta_p) \tag{1}$$

with α_0 denoting the so-called decay parameter, and θ_p the angle between the decay proton direction and the normal of the reaction plane. In consequence, the count rate asymmetry

$$\frac{N_\uparrow - N_\downarrow}{N_\uparrow + N_\downarrow} = \frac{1}{2}\alpha_0 P \tag{2}$$

is obtained relative to the reaction plane, where $N_\uparrow$ and $N_\downarrow$ represent the event numbers above and below the reaction plane, respectively. A particular benefit of the $\Sigma^+ \to \pi^0 p$ decay observed in the present experiment is the large decay parameter $\alpha_0 = -0.980$ [27]. According to Eq. 2 it results in large asymmetries from which the recoil polarisation is then determined.

In order to facilitate the comparison to previous measurements, the results are presented in two different sets of binnings, Fig. 6 and Fig. 7. In both cases the errors attached to the data

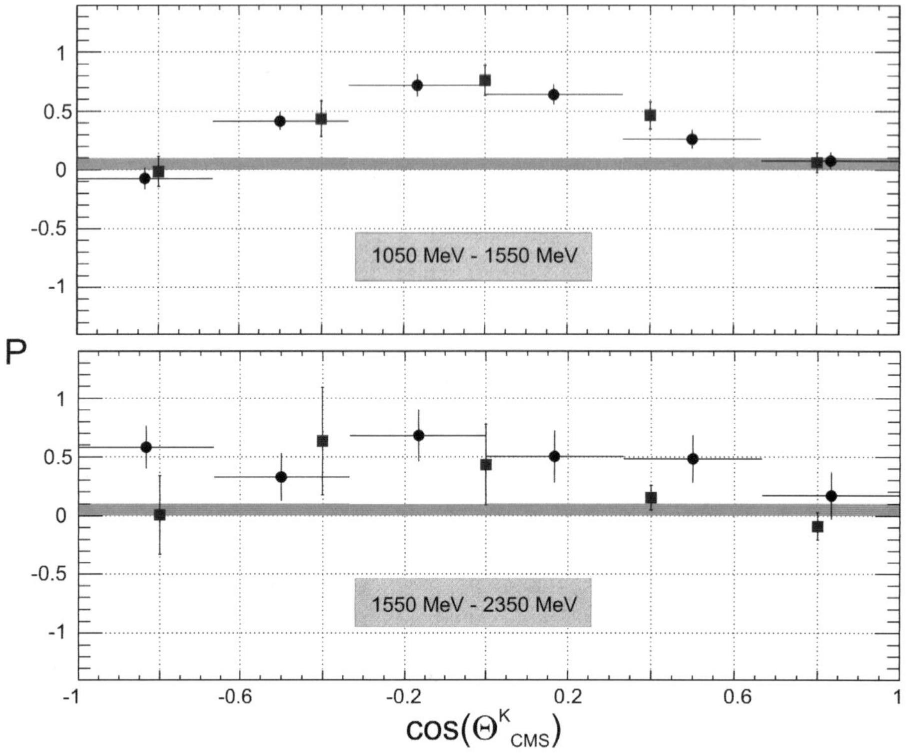

Figure 7. The dots represent the same recoil polarisation data as in Fig. 6 now rebinned into two bins of photon energy to allow better comparison to the previous SAPHIR data (squares). Errors and bin widths are indicated as in Fig. 6.

points are purely statistical. The grey bars on the abscissa give an estimate of the systematic uncertainties. Since according to Eq. 2 the recoil polarisation is determined from a ratio of event rates, some of the systematic effects cancel out which may affect cross section or beam asymmtry measurements. Among those are photon flux, detector inefficiencies and beam polarisation. Remaining systematic errors were studied by variations of the cuts applied in the analysis as described in Ref. [21] for the cross section analysis.

Fig. 6 shows the polar angle dependence of the recoil polarisation in four $\pm 150\,\mathrm{MeV}$ wide bins from $E_\gamma = 1050$ to $2250\,\mathrm{MeV}$ along with a previous measurement of the CBELSA/TAPS collaboration [13] where a similar detector setup but unpolarised beam was used. The general agreement between the data sets seems fair. The errors of the present measurement represent a significant improvement in particular at lower energies. Fig. 7 shows the same data binned in the way of the previous SAPHIR measuremets [14]. Within the errors excellent agreement is obtained.

4.2. Photon beam asymmetry

Accounting for a linearly polarised photon beam, the cross section of photoproduction of pseudoscalar mesons off a nucleon can be written in the form [28]

$$\frac{d\sigma}{d\Omega} = \left(\frac{d\sigma}{d\Omega}\right)_0 (1 - P_{lin}\Sigma\cos 2\phi)\,. \tag{3}$$

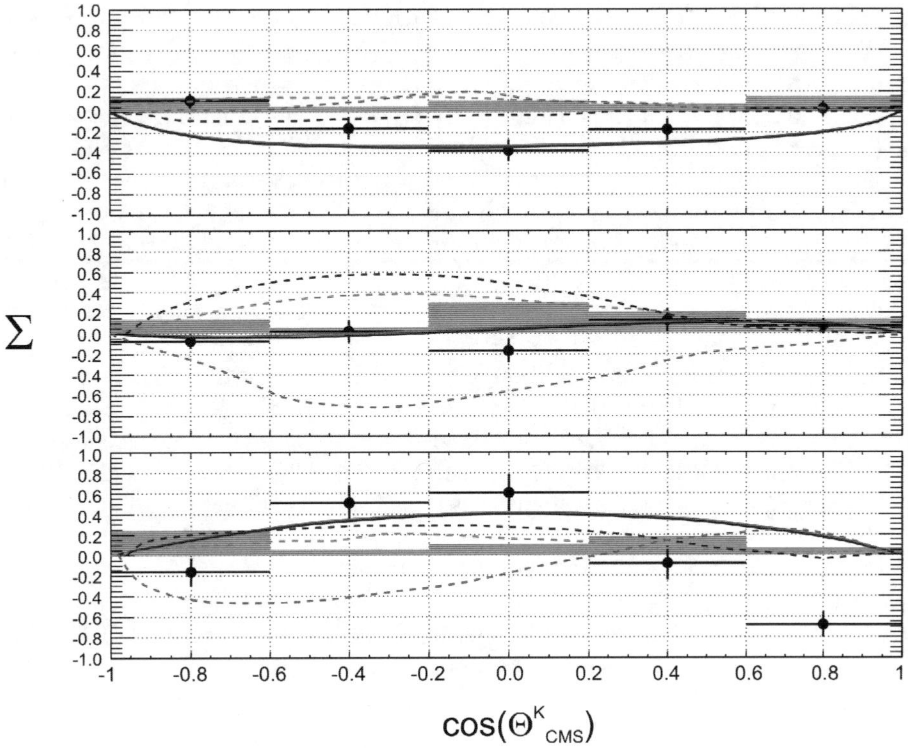

Figure 8. Angular distribution of the photon beam asymmetry Σ in the three bins of photon energy indicated in the diagrams. Attached to the data points are the statistical errors and, as horizontal lines, the bin widths in $\cos\theta$. The systematic errors are indicated by the grey bars on the abscissa. The curves represent: SAID [29] (full blue), K-MAID [23] (full red), and variants of the Gent model [31] (dashed).

Here, $(d\sigma/d\Omega)_0$ is the polarisation independent cross section, P_{lin} the degree of linear polarisation, and Σ denotes the photon beam asymmetry. The product $P_{lin}\Sigma$ determines the magnitude of modulation of the cross section with the azimuthal angle ϕ. Vice versa, it is obtained from the measured modulation by a fit of the function

$$f(\phi) = A\left(1 - \frac{B}{2A}\cos 2\phi)\right) \tag{4}$$

to the K^0 yield. Σ can then be determined once the beam polarisation is determined as described in Ref. [19]. The method is exactly the same as described in Ref. [22] for η photoproduction. The results for the beam asymmtries are shown in Fig. 8.

5. Discussion

Both K-MAID [23] and SAID [29] disagree with the measured cross section data. This can clearly be seen in Figures 4 and 5. Neither parameterisation describes the observed dip in the total cross section. Due to the vicinity of the dip to the K^* threshold, it seems likely related to K^* exchange in the t-channel according to Figure 1(e).

The role of this t-exchange was tested with the K-MAID parameterisation, where it is possible to manually change the K^* exchange strength. In addition to some modifications of couplings below the K^* thresholds to improve the overall description of th data (cf. [21]), mainly the K^{0*} exchange was retained below the $K^*\Sigma^+$ threshold but manually set to zero above. As is demonstrated by the full curve in Figure 4, with these modifications K-MAID yields a significantly improved description of the cross section, including the structure at the K^* thresholds. In forward directions (cf. Figure 5) the modified K-MAID still compares only modestly with the data at smaller energies. The drop of the cross section at the K^* thresholds is however rather well reproduced by switching off the K^{0*} exchange contribution.

From this it is concluded that below K^* threshold there may happen a transition in t-exchange from K^* vector to pseudoscalar K^0. Close to the K^* threshold the vector meson is almost on mass shell in a relative s-wave. It subsequently decays into a pion, which rescatters with the hyperon, and the final state K^0. Such a process is depicted in Fig. 1 (f). It resembles the formation of a vectormeson-baryon quasibound state. Above K^* threshold the vector meson is produced as a free particle. Hence, this process no longer feeds the K^0 final state, leading to the sharp drop of the $K^0\Sigma^+$ cross section. It speaks in favour of this hypothesis that corresponding strength is indeed found in the K^{0*} total cross section [24]. This is illustrated in Figure 4. With the sum of K^0_S and K^{0*} cross sections a smooth transition is obtained at the K^* threshold.

Besides the cross sections, both photon beam asymmetry and recoil polarisation are indispensible observables to unambiguously extract the reaction amplitudes [30], and hence partial wave amplitudes, in a reliable manner. The curves in Figures 6 and 8 demonstrate the level of agreement with the present polarisation data which can be obtained within the K-MAID and SAID parameterisations, as well as with the Gent model [31] for the photon beam asymmetry.

The recoil polarisation neither is satisfactorily described by SAID nor by K-MAID (cf. Figure 6). Since, through interfences, the recoil polarisation is very sensitive to even small resonances, this points to a yet incomplete resonance basis for $K^0\Sigma^+$ photoproduction in these parameterisations. Either there are other s-channel resonances involved than included in the models, or the decomposition into s- and t-channel is different than assumed.

In contrast to the Gent model, the K-MAID and SAID parameterisations of the photon beam asymmetry show reasonable agreement with the data, except in forward directions in the highest energy bin. Similar to the recoil polarisation, also the beam asymmetry is sensitive to interferences of partial waves, and in s-channel production therefore a large sensitivity is obtained to even small resonance contributions. In t-channel processes, however, the photon beam asymmetry reflects the character of the t exchange. In the transition amplitudes the dominant pieces of the electromagnetic coupling with the t-exchange can be written [32]

$$T \propto (\vec{\epsilon} \times \vec{q}) \cdot \vec{p}_f \quad (5)$$

$$T \propto \vec{\epsilon} \cdot (\vec{q} + \vec{p}_f) \quad (6)$$

for K^* and K exchange, respectively. Here, $\vec{\epsilon}$ denotes the photon polarisation vector, and $\vec{q}$ and $\vec{p}_f$ the momenta of the exchanged (intermediate) and final-state kaons, respectively. The scattered kaon and hyperon in the final state are produced most likely perpendicular to the plane of photon polarization when a K^* is exchanged, and parallel to it in the case of an intermediate K. The photon beam asymmetry is determined by

$$\Sigma = \frac{\sigma_\perp - \sigma_\parallel}{\sigma_\perp + \sigma_\parallel} \quad (7)$$

with $\sigma_\perp$ and $\sigma_\parallel$ denoting the cross sections perpendicular and parallel to the photon polarisation plane, respectively. Therefore, in case of a K^* exchange the asymmetry is expected to be positive and, oppositely, negative for a pseudoscalar K exchange.

In this light the beam asymmetry shows an interesting behaviour in the highest energy bin where the differential cross sections already exhibit forward peaking, suggesting significant t-channel contributions. In the case of K^0_s production the exchange particle is a K^{*0} as depicted in Fig. 1(e). The reaction mechanism suggested above requires a flip-over from K^* to K exchange in the vicinity of the K^* thresholds. The generally positive beam asymmetry is compatible with K^*-exchange. In forward directions of the 1450–1650 MeV bin, however, it turns significantly negative. This indeed seems especially here to indicate the predominant exchange of a pseudoscalar K – as it would be expected, if a K^*-hyperon molecular state was generated via diagram Fig. 1(f). Unfortunately, the energy range of the present beam asymmetry data is still restricted to somewhat below the K^* threshold. It is necessary to extend the data base over and beyond the K^* threshold, in order to further clarify the reaction mechanism.

In chiral unitary approaches quasibound K^*-hyperon states are indeed theoretically expected close the K^* threshold through the interaction of the nonet of vector mesons with the octet of baryons [33].

6. Summary and outlook

Cross sections and single polarisation observables of $K^0\Sigma^+$ photoproduction off the proton were discussed. The differential and total cross sections exhibit a dip structure at the K^* threshold which isn't described in any of of the standard parameterisations of the reaction. It is speculated whether a transition is seen from t-channel K^*-exchange to K^0 exchange. In the latter case a K^* is produced off mass shell, subsequently decaying into a π^0, which rescatters with the hyperon, and the K^0 which is observed in the final state. A K^*-Σ intermediate state may be dynamically generated.

The recoil polarisation P could be determined from threshold to $E_\gamma = 2250\,\text{MeV}$. It agrees well with previous measurements. The model descriptions remain unsatisfactory as well, pointing to yet unresolved resonance contributions.

The photon beam asymmetry was measured for the first time in $K^0\Sigma^+$ photoproduction. Again, SAID and K-MAID are not able to reproduce the data. The beam asymmetry shows the interesting feature that it turns strongly negative in forward directions just below K^* threshold. This is expected if, close to K^*-threshold, a dominating t-exchange mechanism flips from K^* to pseudoscalar K exchange. Hence, the beam asymmetry is in agreement with the reaction mechanism hypothesised to explain the observed anomaly of the K^0 photoproduction cross section at the K^* threshold. However, before definite conclusions can be drawn it appears important to extend the beam asymmetry measurements across the K^* threshold. This will be subject to future investigations. The reduced photon flux and beam polarisations at higher photon energies can be compensated by exploiting the charged decay channels in addition to the neutral ones used here. This will be made possible by the new BGO-OD detector setup [34] at ELSA.

Acknowledgements

The data discussed are obtained within the frame of the CBELSA/TAPS collaboration and I thank all collaborators for their support. Ralf Ewald deserves the biggest share of credit for performing the data analysis within his doctoral thesis. The project is supported by the *Deutsche Forschungsgemeinschaft* DFG within the SFB/TR-16 and by the Federal State of *Northrhine Westphalia*.

Journal of Physics: Conference Series **349** (2012) 012006 doi:10.1088/1742-6596/349/1/012006

References

[1] L.Ya. Glozman and D.O. Riska, Physics Reports **268** (1996) 263
[2] A. Manohar and H. Georgi, Nucl. Phys. **B 234** (1984) 189
[3] R.H. Dalitz and J.G. McGinley, in *Low and Intermediate Energy Kaon-Nucleon Physics*, ed. by E. Ferrari and G. Violini, Reidel, Boston (1981) 381; R. H. Dalitz, T.C. Wong, and G. Rajasekaran, Phys. Rev **153** (1967) 1617
[4] P.B. Siegel, and W. Weise, Phys. Rev **C 38** (1988) 2221
[5] N. Kaiser, T. Waas, and W. Weise, Nucl. Phys, **A 612** (1997) 297
[6] C. Garcia-Recio, M.F.M. Lutz, and J. Nieves, Phys. Lett B 582 (2004) 49
[7] M.F.M. Lutz and E.E. Kolomeitsev, Phys. Lett. **B 585** (2004) 243
[8] T. Jude, these proceedings
[9] P. Gonzalez, E. Oset and J. Vijande, Phys. Rev **C 79** (2009) 025209
[10] S. Sarkar et al., Eur. Phys. J. **A 44** (2010) 431
[11] E. Oset and A. Ramos, Eur. Phys. J. **A 44** (2010) 445
[12] E. Oset et al., arXiv:1103.0807v1 [nucl-th]
[13] R. Castelijns et al., Eur. Phys. J. **A 35** (2008) 39
[14] R. Lawall et al., Eur. Phys. J. **A 24** (2005) 275
[15] W. Hillert, Eur. Phys. J. **A 28**, s01 (2006) 139
[16] E. Aker et al., Nucl. Instrum. Methods **A 321** (1992) 69
[17] R. Novotny et al., IEEE Trans. Nucl. Sci. **38** (1991) 379
[18] A.R. Gabler et al., Nucl. Instrum. Methods **A 346** (1994) 168
[19] D. Elsner et al., Eur. Phys. J. **A 39** (2009) 373
[20] G. Suft et al., Nucl. Instrum. Methods **A 538** (2005) 416
[21] R. Ewald et al., arXiv:1112.0811v1 [nucl-ex]
[22] D. Elsner *et al.*, Eur. Phys. J. **A33** (2007) 147-155
[23] D. Drechsel et al., http://www.kph.uni-mainz.de/MAID/ (Version 29.3.2007)
[24] M. Nanova et al. (Crystal-Barrel/TAPS Collab.), Eur. Phys. J. **A35** (2008) 333
[25] Chinese Physics **C 33** (2009) 1146 (Proceedings of NSTAR09, Beijing)
[26] D. Drechsel and L. Tiator, Journal of Physics **G 18** (1992) 449
[27] K. Nakamura et al. (Particle Data Group), Journal of Physics **G 37** (2010) 075021
[28] G. Knöchlein, D. Drechsel and L. Tiator, *Z. Phys.* **A352** (1995) 327
[29] R.A. Arndt et al., http://gwdac.phys.gwu.edu
[30] W.-T. Chiang and F. Tabakin, *Phys. Rev.* **C55** (1997) 2054
[31] S. Janssen et al., *Phys. Rev.* **C66** (2002) 035202
[32] S. Nam et al., Phys. Rev. **D 75** (2007) 014027
[33] E. Oset and A. Ramos, Eur. Phys. J. **A 44** (2010) 445
[34] H. Schmieden, Int. J. Mod. Phys. **15** (2010) 1043

Real Photon Experiments with a Transverse Polarized Target at MAMI

A.Thomas[1]

E-mail: thomas@kph.uni-mainz.de

Abstract. The A2-collaboration at the Mainz Microtron MAMI is measuring photon absorption cross section using circularly and linearly polarized photons up to the energy of 1.5GeV. The photons are produced in the 'Bremsstrahlungs' process. In the years 2005/2006 the Crystal Ball detector with its unique capability to cope with multi photon final states was set up in Mainz. Since 2010 the experimental apparatus has been completed by a polarized target. The horizontal dilution refrigerator of the Frozen-Spin Target has been constructed and is operated in close cooperation with the Joint Institute for Nuclear Research in Dubna, Russia. The system offers the opportunity to provide longitudinally and transversely polarized protons and deuteron. In this paper experiments with the transversely polarized Frozen-Spin Target that was used for the runs in 2010 and 2011 are discussed.

1. Introduction

Since more than 20 years the international A2 collaboration has been measuring photo absorption cross sections of circularly polarized photons on unpolarized and polarized protons and deuterons to determine the total cross section and partial reaction channels in a large kinematical range, which provides new information about the excitation spectrum of the nucleon. The A2-Glasgow-Mainz tagging facility stands out due to its high photon intensity. In the framework of the GDH-experiment there was used a longitudinally polarised Frozen-Spin Target [1] from the University Bonn in combination with the DAPHNE detector. In the year 2010 a new Frozen-Spin Target, produced in collaboration with the Joint Institute for Nuclear Research, Dubna, Russia, came into operation and was used with transverse polarized protons and deuteron for more than 5000 hours in beam.

2. The A2 Real Photon Facility

The A2 Real Photon Facility is one of the three major experiments using the electron beam from the MAMI accelerator, see figure 1. The experiments A1 and A4 use the direct electron beam, while in A2 the electron beam is converted in a beam of photons in the 'Bremsstrahlungs' process.

[1] *Institut für Kernphysik, Universität Mainz, J.-J. Becherweg 45, D55128 Mainz, Germany*

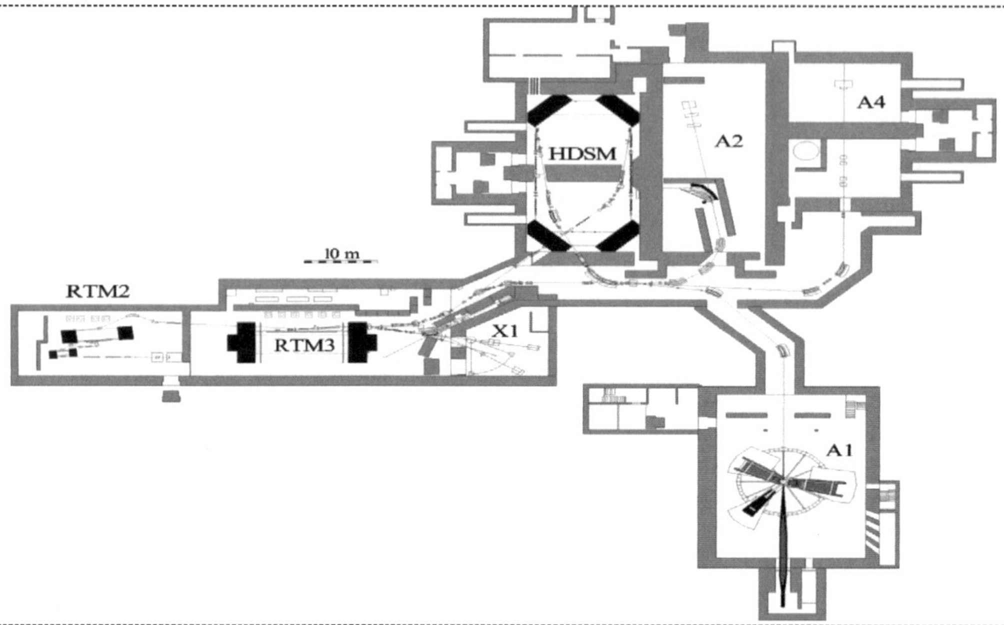

Figure 1. Floor plan of the MAMI accelerator with experimental halls.

The MAMI accelerator with its source of polarized electrons, based on the photo-effect on a strained GaAs crystal, routinely delivers polarized beams with a maximum energy of 1608 MeV. We typically have a degree of polarization of about 85%. Details about the new machine type can be found in reference [1].

The electrons are used to produce a secondary beam of real photons in the 'Bremsstrahlungs' – process. The energy of these photons is detected in the Glasgow-Mainz tagging-system [2].

The resulting photons can be circularly polarized, with the application of a polarized electron beam, or linearly polarized, in the case of a crystalline radiator. The degree of polarization achieved is dependent on the energy of the incident photon beam E_0 and the energy range of interest, but currently peaks at ~75% for linear polarization and ~85% for circular polarization.

The Glasgow Photon Tagger (see figure 2) provides energy tagging of the photons by detecting the post-radiating electrons and can determine the photon energy with a resolution of 2 to 4 MeV depending on the incident beam energy, with a single-counter time resolution of 0.117ns. Each counter can operate reliably to a rate of ~1 MHz, giving a photon flux of $2.5 \cdot 10^5$ photons per MeV. Photons can be tagged in the momentum range from 4.7 to 93.0% of E_0.

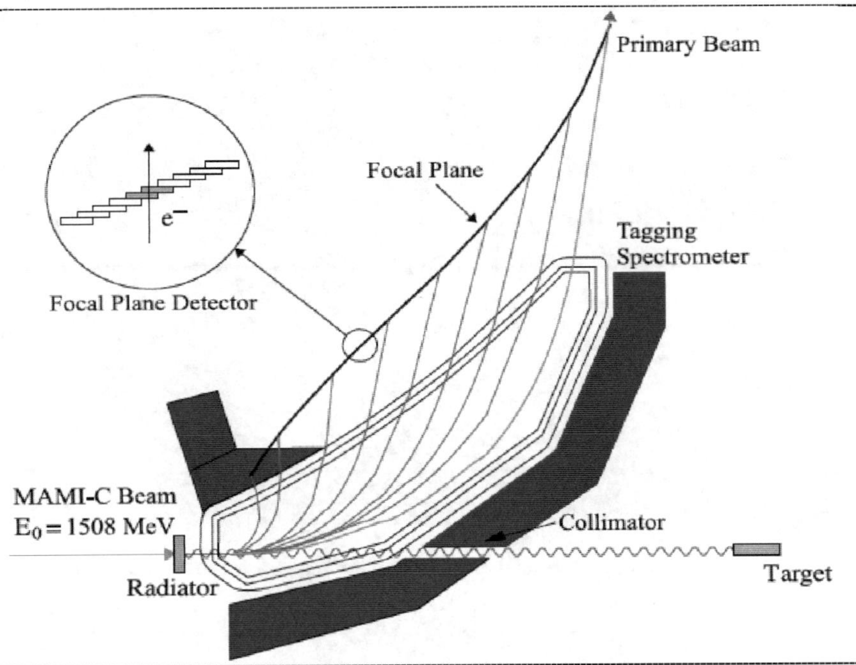

Figure 2. The A2 Glasgow/Mainz tagging system.

2.1. The detector system

The central detector system consists of the Crystal Ball calorimeter combined with a barrel of scintillation counters for particle identification and two coaxial multi-wire proportional counters for charged particle tracking. This central system provides position, energy and timing information for both charged and neutral particles in the region between 21° and 159° in the polar angle and over almost the full azimuthal range. At forward angles, less than 21°, reaction products are detected in the TAPS forward wall. The full, almost hermetic, detector system is shown schematically in figure 3.

The full angular coverage of this detector system sets very rigorous condition for the construction of the polarized target.

The Crystal Ball detector (CB) is a highly segmented 672-element NaI(Tl), self triggering photon spectrometer constructed at SLAC in the 1970's. Each element is a truncated triangular pyramid, 41cm (15.7 radiation lengths) long. The readout electronics for the Crystal Ball were completely renewed in 2003, and it now is fully equipped with SADCs which allow for the full sampling of pulse-shape element by element. In normal operation, the onboard summing capacity of these ADCs is used to enable dynamic pedestal subtraction and the provision of pedestal, signal and tail values for each element event-by-event. Each CB element is also newly equipped with multi-hit CATCH TDCs.

The readout of the CB is effected in such a way as to allow for flexible triggering algorithms. There is an analogue sum of all ADCs, allowing for a total energy trigger, and also an OR of groups of sixteen crystals to allow for a hit-multiplicity second-level trigger - ideal for use when searching for high multiplicity final states.

In order to distinguish between neutral and charged particles species detected by the Crystal Ball, the system is equipped with PID2, a barrel detector of twenty-four 50mm long, 4mm thick scintillators, arranged so that each PID2 scintillator subtends an angle of 15^0 in ϕ. By matching a hit in the PID2 with a corresponding hit in the CB, it is possible to use the locus of the ΔE, E combination to identify the particle species. This is primarily used for the separation of charged pions, electrons and protons.

Journal of Physics: Conference Series **349** (2012) 012007 doi:10.1088/1742-6596/349/1/012007

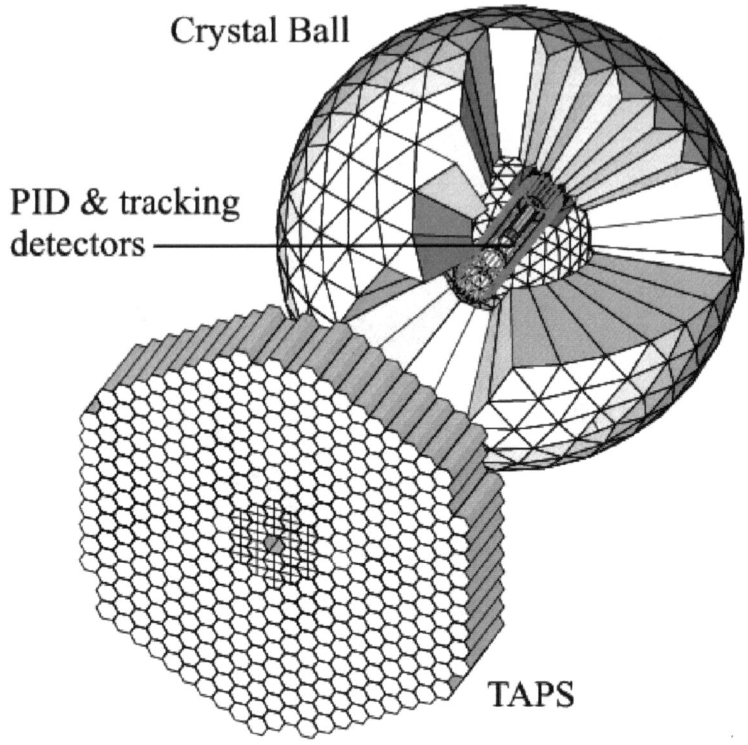

Figure 3. The Crystal Ball calorimeter, with cut-away section showing the inner detectors, and the TAPS forward wall.

3. The Polarized target

The new Frozen-Spin target was designed to retain the high angular acceptance of the detector system. The main boundary condition for the outer diameter of the target cryostat was the most inner particle identification detector PID2 with a diameter of 104 mm. The internal holding coils had to be as thin as possible to allow particles to punch through.

The core of the Frozen-Spin target for the Crystal Ball detector is a specially designed, large, roughly 2m long, horizontal ^{3}He/^{4}He dilution refrigerator (see figure 4) that was built in cooperation with the Joint Institute for Nuclear Research (JINR) Dubna.

The cryostat has a separator working at 3 K and an evaporator working at 1.2 K in the pre-cooling stages. These are pumped by rotary pumps with pumping speed of 60 m^3/h, 100 m^3/h and 250 m^3/h (company Busch). The beam axis is equal to the cryostat axis and the target material has to be loaded along the beam axis using a specially adapted, twofold target-insert. This target-insert needs to seal the cavity against the beam pipe vacuum. It has minimum limitations for the particle detection and fits into the central core of the inner Particle Identification Detector (PID2). This was achieved by using the Frozen-Spin technique in combination with the new concept of placing a thin superconducting holding coil on the thermal radiation protection shields of the refrigerator.

The cryostat could provide a very low operation temperature of 25mK in the target chamber. The butanol is filled into a PTFE-cylinder of 2cm length and diameter. A temperature stability of better than $\pm$0.2mK over a time scale of a week was reached. This corresponds to very long relaxation times of the target nuclei in the order of some thousands of hours. Typically the target had to be re-polarized once a week. Longitudinal and transverse polarizations are possible.

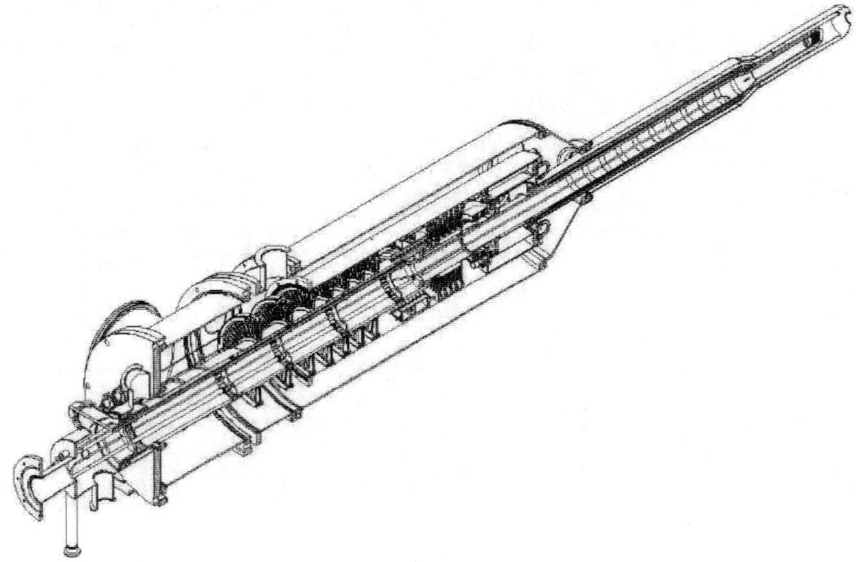

Figure 4. The ^{3}Helium-^{4}Helium dilution refrigerator.

3.1. The transverse holding coil

The transverse holding coil was designed as a 4-layer dipole magnet, allowing for high homogeneity. The coil was wound from a single wire to avoid soldering joints and to provide a high critical current.

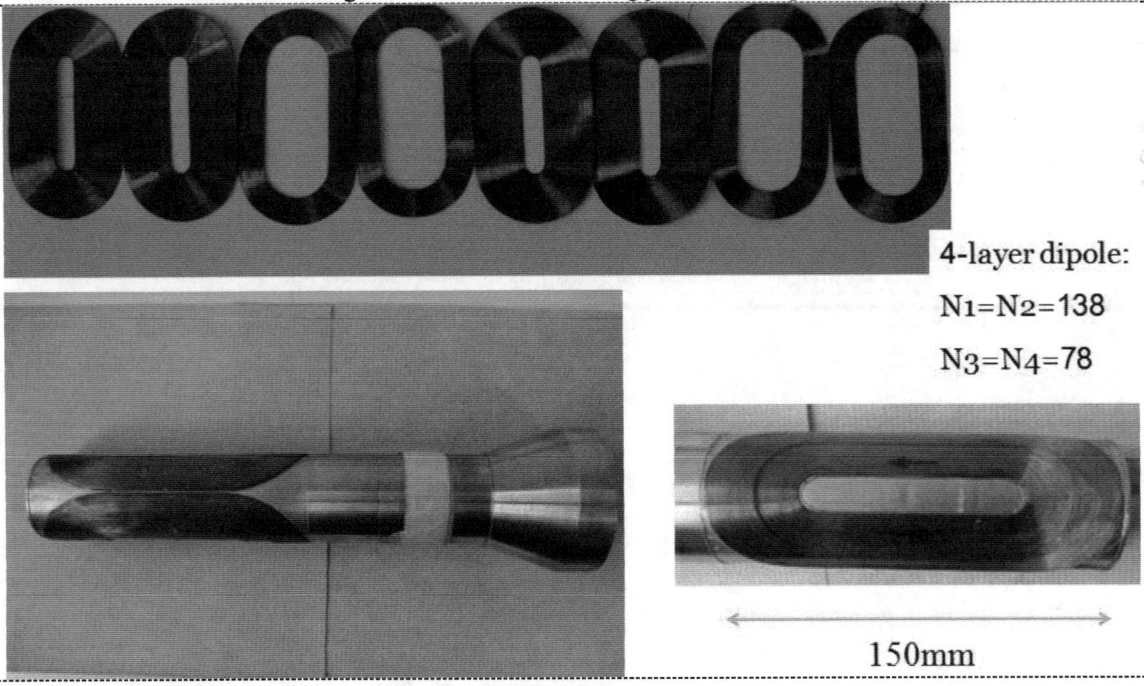

Figure 5. The transverse holding coil.

The transverse coil has been in operation for more than 4000 hours in the years 2010 and 2011 at a current of 35 A, corresponding to a field of 0.45 Tesla. The current leads are optimized for low thermal input: In the first step we enter into the cryostat with thick, normal conducting copper wire to a temperature of 70 K, next step high temperature superconducting band is used. The helium consumption for stable operation of the complete cryostat (target material at 25 mK and holding coil at 0.45 T) was below 2.5 liters per hour, showing a very economic design of all heat exchangers.

4. First Measurement of the T- and F-observables in π- and η-production

In the first measurement campaign with the new Frozen-Spin target we have used a circularly polarized photon beam in combination with transversely polarized protons and deuterons. The analysis is still preliminary [4] and final calibrations have to be done. Figure 6 from a recent publication on an unpolarized hydrogen target shows the quality of the data to be expected from our polarized measurements.

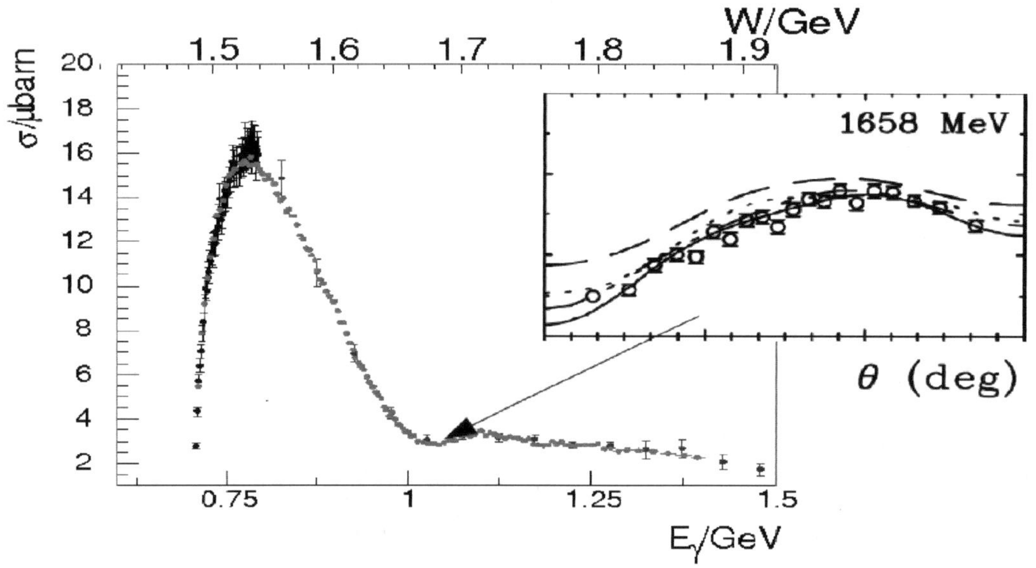

Figure 6. Total cross section for the reaction $\gamma\pi \rightarrow p\pi^0$.

5. References

[1] C. Bradtke et al., NIM A **436**, p.430 (1999).
[2] A. Jankowiak, *EPJA*, **28**Suppl.1, p. 149-160 (2006).
[3] J.C. McGeorge et al., EPJ **A 37**, p.129 (2008).
[4] V. Kashevarov, private communication (2011).
[5] E. F. McNicoll et al., Phys.Rev.C82:035208, 2010.

Acknowledgments
This work was supported by the Deutsche Forschungsgemeinschaft (SFB 443, SFB/TR16), DFG-RFBR (Grant No. 09-02-91330), the European Community-Research Infrastructure Activity under the FP6 “Structuring the European Research Area” program (Hadron Physics, contract number RII3-CT-2004-506078).

Latest on g-2 from experiment

Graziano Venanzoni
Laboratori Nazionali di Frascati dell' INFN, Frascati
E-mail: graziano.venanzoni@lnf.infn.it

Abstract. I will review the latest experimental achievements on the hadronic cross section measurements at low-energy which are of fundamental importance for a precise evaluation of the hadronic contribution to the g-2 of the muon. Then I will discuss the new proposed muon g-2 experiments, with particular emphasis on E989 at Fermilab which plans to improve the experimental uncertainty of a factor 4 respect to the previous E821 experiment at BNL.

1. The Muon anomaly as a precision test of the Standard Model

The muon anomaly $a_\mu = (g-2)/2$ is a low-energy observable, which can be both measured and computed to high precision [1]. Therefore it provides an important test of the Standard Model (SM) and it is sensitive search for new physics [2]. Since the first precision measurement of a_μ from E821 experiment at BNL in 2001 [3], there has been a discrepancy between its experimental value and the SM prediction. This discrepancy, has been slowly growing due to impressive theory and experiment recent achievements. Figure 1 (from Ref. [4]) shows an up-to-date comparison of the SM predictions of different groups and the BNL measurement for a_μ. Evaluation of different groups are in very good agreement, showing a persisting $3\,\sigma$ discrepancy (as, for example, $26.1 \pm 8.0 \times 10^{-10}$[4]), despite many changes in the recent history. It should be noted that both theory and experiment uncertainties have been reduced by more than a factor two in the last ten years[1].

The accuracy of the theoretical prediction ($\delta a_\mu^{\rm SM}$, between 5 and 6 $\times 10^{-10}$) is limited by the strong interaction effects which cannot be computed perturbatively at low energies. Table 1 shows their contribution to the error for three recent estimates [5, 6, 4][2].

The leading-order hadronic vacuum polarization contribution, $a_\mu^{\rm HLO}$, gives the main uncertainty (between 4 and 5 $\times 10^{-10}$). It can be related by dispersion integral to the measured hadronic cross sections, and it is known with a fractional accuracy of 0.7%, i.e. to about 0.4 ppm. The $O(\alpha^3)$ hadronic light-by-light contribution, $a_\mu^{\rm HLbL}$, is the second dominant error to the theoretical evaluation. It cannot at present be determined from data, and relies on same specific models. Although its value is almost one order of magnitude smaller than $a_\mu^{\rm HLO}$, it is much worse known (with a fractional error of the order of 30%) and therefore it still give a significantly contribution to $\delta a_\mu^{\rm SM}$ (between 2.5 and 4 $\times 10^{-10}$).

From the experimental side, the error achieved by the BNL E821 experiment is $\delta a_\mu^{\rm EXP} = 6.3 \times 10^{-10}$ (0.54 ppm) [7]. This impressive result is still limited by the statistical errors, and

[1] In 2001 this discrepancy was $(23.1 \pm 16.9) \times 10^{-10}$.
[2] Ref. [5] uses a more conservative error analysis.

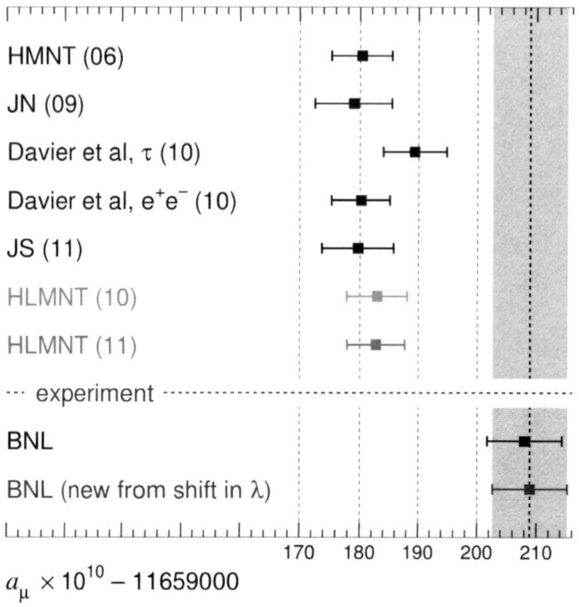

Figure 1. Standard Model predictions of a_μ by several groups compared to the measurement from BNL. (From Ref. [4]).

Error	[5]	[6]	[4]	prospect
$\delta a_\mu^{\rm SM}$	65	49	49	35
$\delta a_\mu^{\rm HLO}$	53	42	43	26
$\delta a_\mu^{\rm HLBL}$	39	26	26	25
$\delta(a_\mu^{\rm SM} - a_\mu^{\rm EXP})$	88	80	80	40

Table 1. Estimated uncertainties δa_μ in units of 10^{-11} according to Refs. [5, 6, 4] and (last column) prospects in case of improved precision in the e^+e^- hadronic cross-section measurement (the prospect on $\delta a_\mu^{\rm LbL}$ is an *educated guess*). Last row: Uncertainty on Δa_μ assuming the present experimental error of 63 from BNL-E821 [7] (first two columns) and of 16 (last column) as planned by the future (g-2) experiments [8, 9].

proposals to measure the muon g-2 with a fourfold improvement in accuracy have been proposed at Fermilab [8] and J-PARC [9].

2. Recent progress on the hadronic contribution to a_μ

Differently from the QED and Electroweak contributions to a_μ, which can be calculated using perturbation theory, and therefore are well under control, the hadronic ones (LO VP and HLbL) cannot be computed reliably using perturbative QCD. The hadronic contribution $a_\mu^{\rm HLO}$ can be computed from hadronic e^+e^- annihilation data via a dispersion relation, and therefore its uncertainty strongly depends on the accuracy of the experimental data. For the hadronic Light-by-Light contribution $a_\mu^{\rm HLbL}$ there is no a direct connection with data and therefore only model-dependent estimates exist. As the hadronic sector dominates the uncertainty on the theoretical prediction $a_\mu^{\rm SM}$, considerable effort has been put by on it by the experimental and theoretical groups, reaching the following main achievements:

- A precise determination of the hadronic cross sections at the e^+e^- colliders (VEPP-2M, DAΦNE, BEPC, PEP-II and KEKB) which allowed a determination of $a_\mu^{\rm HLO}$ with a fractional accuracy below 1%. These effort led to the development of dedicated high precision theoretical tools, like Radiative Corrections (RC) and the non-perturbative hadronic contribution to the running of α (i.e. the vacuum polarisation, VP) in Monte Carlo (MC) programs used for the analysis of the data [10];
- the use of '*Initial State Radiation* (ISR) [11, 12, 13] which opened a new way to precisely obtain the electron-positron annihilation cross sections into hadrons at particle factories operating at fixed beam-energies [14, 15].
- a dedicate effort on the evaluation of the Hadronic Light-by-Light contribution, where two different groups [16, 5] obtained results consistent on the size of the contribution (with slightly different errors), and therefore strengthening our confidence in the reliability of these estimates;
- an impressive progress on lattice, where an accuracy of 2% were reached on on the two-flavor QCD correction to $a_\mu^{\rm HLO}$ [17];
- a better agreement between the e^+e^-– and the τ– based evaluation of $a_\mu^{\rm HLO}$, thanks to improved isospin corrections [6]. These two sets of data are eventually in agreement (with τ data moving towards e^+e^-– data) after including vector meson and $\rho-\gamma$ mixing [18, 19].

3. σ_{had} measurements at low energy

In the last years big efforts on e^+e^- data in the energy range below a few GeV led to a substantial reduction in the hadronic uncertainty on a_μ^{had}. Figure 2 shows an up-to-date compilation of this data. The main improvements have been achieved in the region below 5 GeV: between 2 and 5 GeV (where the data are now closer to the prediction of pQCD), the BESII collaboration reduced the error to ~7% [20] (before it was ~15%); between 1 and 4.5 GeV BaBar measured various final states with more than two hadrons with a systematic accuracy between 5% and 15%, as shown in Tab. 2; below 1 GeV, the CMD-2 and SND collaborations at Novosibirsk, KLOE at Frascati and BaBar at Stanford measured the pion form factor in the energy range around the ρ peak with a systematic error of 0.8%, 1.3%, 0.9%, and 0.5%, respectively. The CMD-2 and SND collaborations at Novosibirsk and BESII in Beijing were performing the hadronic cross section measurements in a traditional way, i.e., by varying the e^+e^- beam energies. KLOE, BaBar, and more recently Belle used the method of radiative return as reviewed in Refs. [10, 14, 15]. Figure 2 shows that, despite the recent progress, the region between 1 and 2 GeV is still poorly known, with a fractional accuracy of ~10%. Since about 50% of the error squared, $\delta^2 a_\mu^{\rm HLO}$ comes from this region (see Fig. 3), it is evident how desirable an improvement on this region is.

3.1. Measurement of $\sigma_{\pi\pi}$ below 1 GeV

The region below 1 GeV is dominated by the two-pion channel which accounts for 70% of the contribution to a_μ^{HLO}, and for 40% to the total error $\delta^2 a_\mu$ (see Fig. 3). Therefore due to its particular importance, it has been studied by different experiments as shown in Fig. 4. CMD-2 and SND have performed an energy scan at the e^+e^- collider VEPP-2M ($\sqrt{s} \in$ [0.4–1.4] GeV) with ~10^6 and ~4.5×10^6 events respectively, and systematic fractional errors from 0.6% to 4% in the cross sections, depending on $\sqrt{s}$. The pion form factor has also been measured by KLOE using ISR, and more recently by BaBar. KLOE collected more than 3.1 million events, corresponding to an integrated luminosity of 240 pb^{-1}, leading to a relative error of 0.9% in

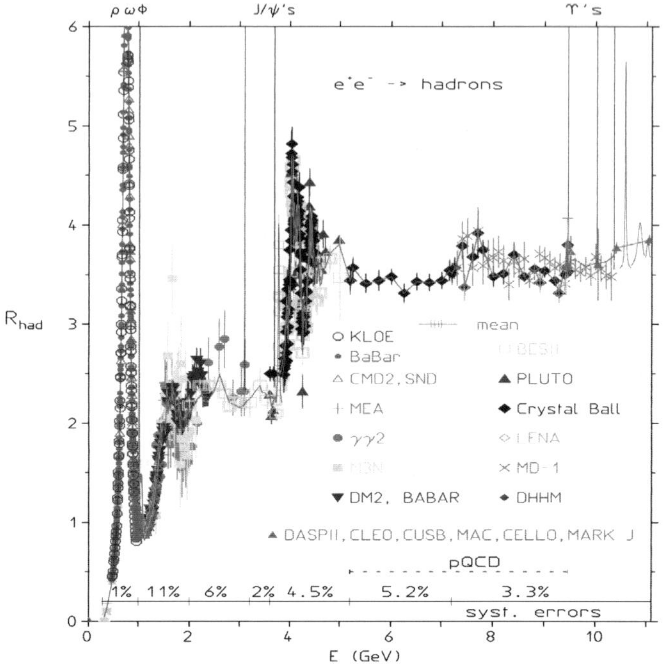

Figure 2. An updated compilation of R measurements. In the bottom line the overall uncertainties of the different regions are reported (*Courtesy of Fred Jegerlehner*).

Table 2. Systematic accuracy on several processes studied by BaBar in the energy range $\sqrt{s} <$ 4.5 GeV using ISR. Integrated luminosity is 232 fb^{-1} for all processes but 3π and $2\pi^+2\pi^-$ where only 89 fb^{-1} were used.

Process	Systematic accuracy
$\pi^+\pi^-\pi^0$	(6-8)%
$2\pi^+2\pi^-$	5%
$2\pi2\pi^0$	(8-14)%
$2(\pi^+\pi^-)\pi^0, 2(\pi^+\pi^-)\eta$	(7-10)%
$3\pi^+3\pi^-, 2\pi^+2\pi^-2\pi^0$	(6-11)%
$KK\pi, KK\eta$	(5-6)%
$K^+K^-\pi\pi$	(8-11)%
$K^+K^-\pi^+\pi^-\pi^0$, $K^+K^-\pi^+\pi^-\eta$	(5-10)%
$2(K^+K^-)$	(9-13)%

the energy region [0.6–0.97] GeV dominated by systematics. BaBar has performed a $\pi^+\pi^-(\gamma)$ cross section measurement based on half a million selected events. The pion form factor is obtained by the ratio $\pi^+\pi^-(\gamma)$ to $\mu^+\mu^-(\gamma)$ which allows a systematic error of 0.5% in the ρ region increasing to 1% outside. The threshold region [$2m_\pi$–0.5 GeV] provides 13% of the total $\pi^+\pi^-$ contribution to the muon anomaly: $a_\mu^{\rm HLO}$ [$2m_\pi$–0.5 GeV] = $(58.0 \pm 2.1)\times 10^{-10}$. To overcome the lack of precision data at threshold energies, the pion form factor is extracted from a parametrization based on ChPT, constrained from spacelike data [21]. The most effective way to measure the cross section near the threshold in the timelike region is provided by ISR events,

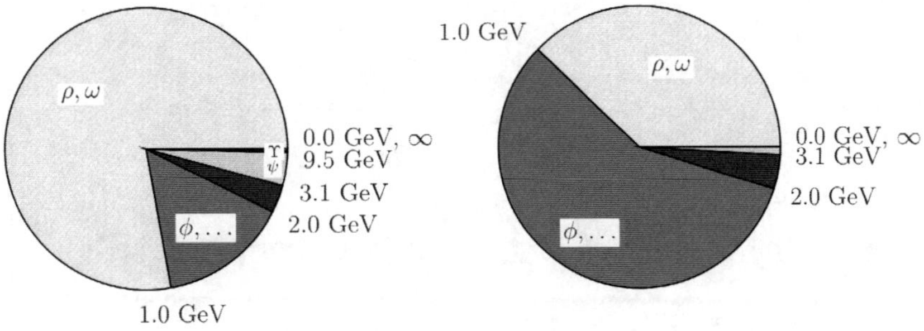

Figure 3. The distribution of contributions (left) and errors (right) in % for $a_\mu^{\rm HLO}$ from different energy regions. The error of a contribution i shown is $\delta^2_{i\,\rm tot}/\sum_i \delta^2_{i\,\rm tot}$ in %. The total error combines statistical and systematic errors in quadrature. From Ref. [5].

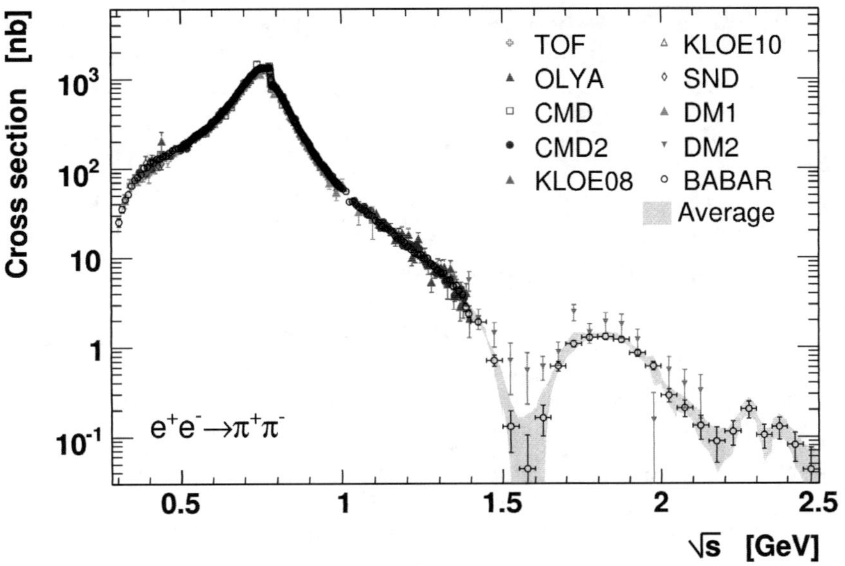

Figure 4. Cross section of $e^+e^- \to \pi^+\pi^-$ as measured by different experiments. (From Ref. [6]).

where the emission of an energetic photon allows the study of the two pions at rest. BaBar has achieved an error between 0.8 and 1.4% in this region, while KLOE has achieved a larger error (up to 7%) due to the point-like approximation for FSR.

There is a fair agreement between the 4 experiments in the region below 1 GeV, with a discrepancy of about 2-3% between KLOE (lowest cross section) and BaBar (highest cross section) at the ρ peak, and CMD2 and SND somehow in the middle. Although small, this difference is larger than the claimed error (0.5% and 1.1% respectively) and can be a limitation for further improvements of $a_\mu^{\rm SM}$. As BaBar and KLOE (published) data use a different normalization (to muon pair and to Bhabha events respectively) it maybe that part of this difference can come in the normalization procedure itself. In order to check this possibility, the KLOE experiment has recently presented a new measurement of the pion form factor derived from the bin-by-bin $\pi^+\pi^-\gamma/\mu^+\mu^-\gamma$ ratio [22] as done in BaBar. As can be shown in Fig. 5, a good agreement is found between the two spectra, which excludes possible problems in the normalization procedure used in KLOE.

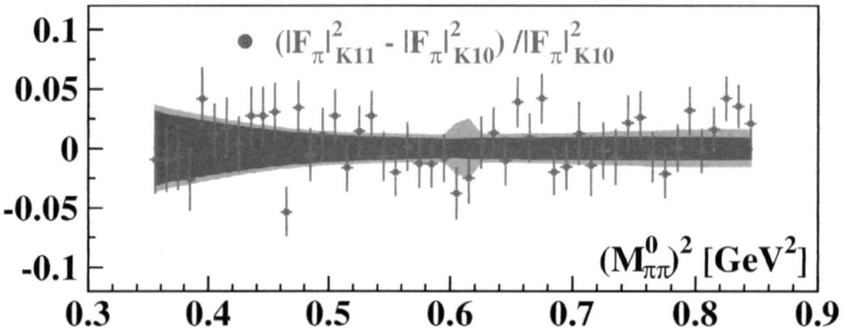

Figure 5. Fractional difference between the published KLOE measurement normalized to Bhabha events and the new preliminary one derived from the bin-by-bin $\pi^+\pi^-\gamma/\mu^+\mu^-\gamma$ ratio [22].

3.2. Measurement of σ_{had} above 1 GeV

The region [1–2.5] GeV, with an uncertainty on σ_{had} between 6 and 10%, is the most poorly known, and contributes about 55% of the uncertainty on $a_\mu^{\rm HLO}$ (see Fig. 3). In this region BaBar using ISR has published results on e^+e^- into three, four, five and six hadrons, with a general improvement respect to the much less precise measurements from M3N, DM1 and DM2. For several channels, BaBar data measured a lower cross sections respect to older experiments, resulting in a reduced contribution from this energy region to $a_\mu^{\rm HLO}$. Recently CMD-3 and SND experiments at the upgraded VEPP-2000 collider in Novosibirsk have presented new measurements on multihadron channels [23]. With about 20 pb^{-1} of collected data, they have already achieved a statistical error comparable to ISR data from B-factories. VEPP-2000 plans to collect an integrated luminosity of 1 fb^{-}1, which would allow a significant improvement for many channels in the region below 2 GeV.
With a specific luminosity of 10^{32}cm^{-2}s^{-1} luminosity, DAΦNE upgraded in energy, can perform a scan in the region from 1 to 2.5 GeV , collecting an integrated luminosity of 20 pb^{-1} per point corresponding to few days of data taking for each energy bin [24]. By assuming an energy step of 25 MeV, the whole region would be thus scanned in one year of data taking. The statistical yield would be one order of magnitude higher than what achieved with 1 ab^{-1} at BaBar, and better than what expected at BESIII with 10 fb^{-1} at 3 GeV.

Finally prospects of reaching an integrated luminosity by a factor of 30-100 exceeding that today appear at Super B-Factories. Such machines will improve accuracy for many processes which studies are now statistically limited.

4. Measuring a_μ

The muon anomaly a_μ has been measured to a always better accuracy during the last 50 years. E821 experiment at Brookhaven has reached an impressive 14-fold improvement in precision respect to the pioneering measurements performed at CERN. Two new proposals, with a fourfold improvement in accuracy, have been proposed at Fermilab [8] (E989) and J-PARC [9]. E989 is based on the well known 'magic-momentum" concept and uses the BNL storage ring as key element. The proposal at J-PARC uses a new approach with ultra-slow muons at off-magic momentum. We will now discuss how the measurement of a_μ is done, having in mind E821 experiment, and its upgrade E989.

The measurement of a_μ uses the spin precession resulting from the torque experienced by the magnetic moment when placed in a magnetic field. An ensemble of polarized muons is introduced into a magnetic field, where they are stored for the measurement period. Assuming that the muon velocity is transverse to the magnetic field ($\vec{\beta} \cdot \vec{B} = 0$), the rate at which the spin turns relative to the momentum vector is given by the difference frequency between the spin precession and cyclotron frequencies. Because electric quadrupoles are used to provide vertical focusing in the storage ring, their electric field is seen in the muon rest frame as a motional magnetic field that can affect the spin precession frequency. In the presence of both $\vec{E}$ and $\vec{B}$ fields, and in the case that $\vec{\beta}$ is perpendicular to both, the anomalous precession frequency (*i.e.* the frequancy at which the muons spin advances relative to its momentum) is

$$\begin{aligned} \vec{\omega_a} &= \omega_S - \omega_C \\ &= -\frac{q}{m}\Big[a_\mu \vec{B} - \Big(a_\mu - \frac{1}{\gamma^2 - 1}\Big)\frac{\vec{\beta} \times \vec{E}}{c}\Big] \end{aligned} \tag{1}$$

The experimentally measured numbers are the muon spin frequency ω_a and the magnetic field, which is measured with proton NMR, calibrated to the Larmor precession frequency, ω_p, of a free proton. The anomaly is related to these two frequencies by

$$a_\mu = \frac{\tilde{\omega_a}/\omega_p}{\lambda - \tilde{\omega_a}/\omega_p} = \frac{R}{\lambda R}, \tag{2}$$

where $\lambda = \mu_\mu/\mu_p = 3.183345137(85)$ (determined experimentally from the hyperfine structure of muonium), and $R = \tilde{\omega_a}/\omega_p$. The tilde over ω_a means it has been corrected for the electric-field and pitch ($\vec{\beta} \cdot \vec{B} \neq 0$) corrections [3]. The magnetic field in Eq. 1 is an average that can be expressed as an integral of the product of the muon distribution times the magnetic field distribution over the storage region. Since the moments of the muon distribution couple to the respective multipoles of the magnetic field, either one needs an exceedingly uniform magnetic field, or exceptionally good information on the muon orbits in the storage ring, to determine $< B_\mu >$-dist to sub-ppm precision. This was possible in E821 where the uncertainty on the magnetic field averaged over the muon distribution was 30 ppb (parts per billion). The coefficient of the $\vec{\beta} \times \vec{E}$ term in Eq. 1 vanishes at the "magic" momentum of 3.094 GeV/c; where $\gamma = 29.3$. Thus a_μ can be determined by a precision measurement of ω_a and B. At this magic momentum, the electric field is used only for muon storage and the magnetic field alone determines the precession frequency. The finite spread in beam momentum and vertical betatron oscillations introduce small (sub ppm) corrections to the precession frequency. These are the only corrections made to the measurement.

The experiment consists of repeated fills of the storage ring, each time introducing an ensemble of muons into a magnetic storage ring, and then measuring the two frequencies ω_a and ω_p. The muon lifetime is 64.4 μs, and the data collection period is typically 700 μs. The g-2 precession period is 4.37 μs, and the cyclotron period ω_C is 149 ns.

Because of parity violation in the weak decay of the muon, a correlation exists between the muon spin and the direction of the high-energy decay electrons. Thus as the spin turns relative to the momentum, the number of high-energy decay electrons is modulated by the frequency ω_a, as shown in Fig. 6. The E821 storage ring was constructed as a super-ferric magnet, meaning that the iron determined the shape of the magnetic field. Thus B0 needed to be well below saturation and was chosen to be 1.45 T. The resulting ring had a central orbit radius of 7.112 m, and 24 detector stations were placed symmetrically around the inner radius of the storage ring. The detectors were made of Pb/SciFi electromagnetic calorimeters which measured the decay electron energy and time of arrival. The detector geometry and number were optimized to detect

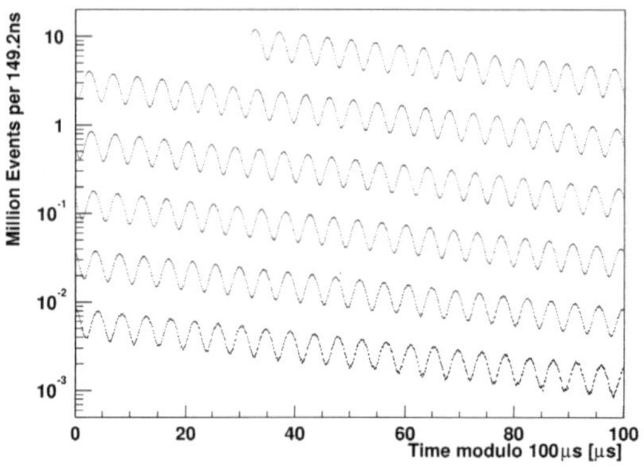

Figure 6. Distribution of electron counts versus time for the 3.6 billion muon decays. The data are wrapped around modulo 100 μs [7].

the high energy decay electrons, which carry the largest asymmetry, and thus information on the muon spin direction at the time of decay. In this design, many of the lower-energy electrons miss the detectors, reducing background and pileup.

5. The Fermilab proposal: E989

The E989 experiment at Fermilab plans to measure a_μ to an uncertainty of 16×10^{11} (0.14 ppm), derived from a 0.10 ppm statistical error and roughly equal 0.07 ppm systematic uncertainties on ω_a and ω_p.

The proposal efficiently uses the unique properties of the Fermilab beam complex to produce the necessary flux of muons, which will be injected and stored in the (relocated) muon storage ring. To achieve a statistical uncertainty of 0.1 ppm, the total data set must contain more than 1.8×10^{11} detected positrons with energy greater than 1.8 GeV, and arrival time greater than 30 μs after injection into the storage ring. The plan uses 6 out of 20 of the 8-GeV Booster proton batches in 15 Hz operational mode, each subdivided into four bunches of intensity 10^{12} p/bunch. The proton bunches fill the muon storage ring at a repetition rate of 18 Hz, to be compared to the 4.4 Hz at BNL. The proton bunch hits a target in the antiproton area, producing a 3.1 GeV/c pion beam that is directed along a 900 m decay line. The resulting pure muon beam is injected into the storage ring. The muons will enter the ring through a new superconducting injector magnet, which will replace the existing one, which is wound in such a manner that the coils intercept the beam on both ends of the magnet. The new injector will result in a higher muon storage efficiency. Once entering the ring, a better optimized pulse-forming network will energize the storage ring kicker to place the beam on a stable orbit. The pion flash (caused by pions entering the ring at injection) will be decreased by a factor of 20 from the BNL level, and the muon flux will be significantly increased because of the ability to take zero-degree muons. The stored muon-per-proton ratio will be increased by a factor of 5 to 10 over BNL.

The E821 muon storage will be relocated to Fermilab, in a new building with a stable floor and good temperature control, neither of which were available at Brookhaven.

The new experiment will require upgrades of detectors, electronics and data acquisition equipment to handle the much higher data volumes and slightly higher instantaneous rates. We have developed high-density, segmented tungsten/scintillating-fiber calorimeters [25] and in-vacuum straw drift tubes. The latter will be used to determine the stored muon distribution from decay positron tracks and to provide data for a greatly improved muon electric dipole moment measurement, which can be obtained in parallel [26]. A modern data acquisition system will be used to read out waveform digitizer data and store it so that both the traditional event mode and a new integrating mode of data analysis can both be used in parallel. The systematic uncertainty on the precession frequency is expected to improve by a factor 3 thanks to the reduced pion contamination, the segmented detectors, and an improved storage ring kick of the muons onto orbit. The storage ring magnetic field will be shimmed to an even more impressive uniformity, and improvements in the field-measuring system will be implemented. The systematic error on the magnetic field is halved by better shimming, relocations of critical NMR probes, and other incremental changes.

In less than two years of running, the statistical goal of 4×10^{20} protons on target can be achieved for positive muons. A follow-up run using negative muons is possible, depending on future scientific motivation. Two additional physics results will be obtained from the same data: a new limit on the muon's electric dipole moment (up to 100 times better); and, a more stringent limit on possible CPT or Lorentz violation in muon spin precession. A technically driven schedule permits data taking to begin in 2016.

6. Prospects on a_μ

With the new experiments planned at Fermilab and J-PARC the uncertainty of the difference Δa_μ between the experimental and the theoretical value of a_μ will be dominated by the uncertainty of the hadronic cross sections at low energies, unless new experimental efforts at low energy are undertaken. The last column of Table 1 shows a future scenario based on realistic improvements in the $e^+e^- \to hadrons$ cross sections measurements. Such improvements could be obtained by reducing the uncertainties of the hadronic cross-sections from 0.7% to 0.4% in the region below 1 GeV and from 6% to 2% in the region between 1 and 2 GeV as shown in Table 3.

	$\delta(\sigma)/\sigma$ present	δa_μpresent	$\delta(\sigma)/\sigma$ prospect	δa_μprospect
$\sqrt{s} < 1$ GeV	0.7%	33	0.4%	19
$1 < \sqrt{s} < 2$ GeV	6%	39	2%	13
$\sqrt{s} > 2$ GeV		12		12
total		53		26

Table 3. Overall uncertainty of the cross-section measurement required to get the reduction of uncertainty on a_μ in units 10^{-11} for three regions of $\sqrt{s}$ (from Ref. [27]).

In this scenario the overall uncertainty on Δa_μ could be reduced by a factor 2. In case the central value would remain the same, the statistical significance would become 7-8 standard deviations, as it can be seen in Fig. 7.

The effort needed to reduce the uncertainties of the $e^+e^- \to hadrons$ cross-sections according to Table 3 is challenging but possible, and certainly well motivated by the excellent opportunity the muon $g-2$ is providing us to unveil (or constrain) "new-physics" effects. A long-term program of hadronic cross section measurements at low energies is clearly warranted and fortunately it has been already started at VEPP-2000. In addition recent theoretical activities focused on lattice calculation have already reached a mature stage and have real prospects to match the future experimental precision.

With the expected reduction of the error on $a_\mu^{\rm HLO}$, and the planned improved precision of the new g-2 experiments, the hadornic Light-by-Light contribution could become the main limitation for further progress on $a_\mu^{\rm SM}$. Although there isn't a direct connection with data, $\gamma-\gamma$ measurements

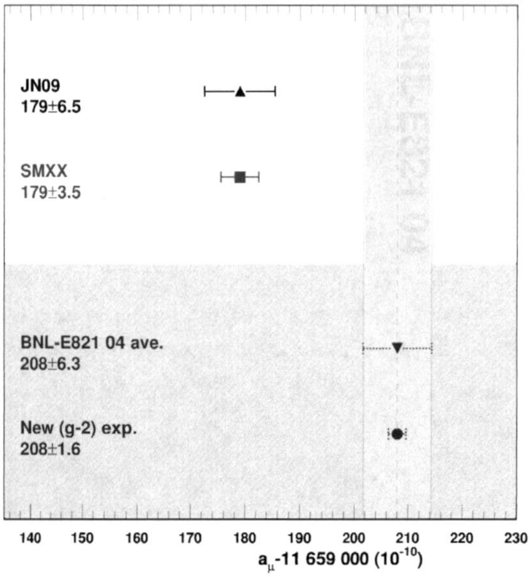

Figure 7. Comparison between a_μ^{SM} and a_μ^{EXP}. "JN09" is the current evalution of a_μ^{SM} using Ref. [5]; "SMXX" is the same central value with a reduced error as expected by the improvement on the hadronic cross section measurement (see text); "BNL-E821 04 ave." is the current experimental value of a_μ; "New (g-2) exp." is the same central value with a fourfold improved accuracy as planned by the future (g-2) experiments [8, 9].

perfomed at e^+e^- colliders will help us to constrain (at least) on-shell form factors [28]. Lattice calculation could help as well.

7. Conclusion

The measurement of the muon anomaly a_μ have been a important benchmark for the development of QED and the Standard Model. In the recent years, following the impressed accuracy (0.54 ppm) reached by E821 experiment at BNL on the measurement of a_μ a worldwide effort from different theoretical and experimental groups have significant improved it SM prediction. At present there appears to be a 3σ difference between the experimental value and the SM prediction of a_μ. This discrepancy, which would fit well with SUSY expectations, is a valuable constraint in restricting physics beyond the standard model, guiding the interpretation of LHC results. In order to clarify the nature of the observed discrepancy between theory and experiment, and eventually firmly establish (or constrain) new physics effects new direct measurements of the muon g-2 with a fourfold improvement in accuracy have been proposed at Fermilab by E989 and J-PARC. First results from E989 could be available around 2017/18.

8. Acknowledgment

I thank Giorgio Giardina and Giuseppe Mandaglio for the warm hospitality and the stimulating atmosphere during the workshop.

[1] F. Jegerlehner, "The anomalous magnetic moment of the muon," Berlin, Germany: Springer (2008) 426 p (Springer tracts in modern physics. 226)
[2] D. Stockinger, "Muon (g-2) and physics beyond the standard model," In Roberts, Lee B., Marciano, William J. (eds.): Lepton dipole moments* 393-438 (Advanced series on directions in high energy physics. 20)
[3] H. N. Brown *et al.* [Muon g-2 Collaboration], Phys. Rev. Lett. **86** (2001) 2227
[4] K. Hagiwara, R. Liao, A. D. Martin, D. Nomura and T. Teubner, J. Phys. G G **38** (2011) 085003
[5] F. Jegerlehner and A. Nyffeler, Phys. Rept. **477** (2009) 1
[6] M. Davier, A. Hoecker, B. Malaescu and Z. Zhang, Eur. Phys. J. C **71** (2011) 1515
[7] G. W. Bennett *et al.* [Muon G-2 Collaboration], Phys. Rev. D **73** (2006) 072003
[8] New Muon (g 2) Collaboration, R.M. Carey *et. al.*, see http://lss.fnal.gov/archive/testproposal/0000/fermilab-proposal-0989.shtml
[9] J. Imazato, Nucl. Phys. Proc. Suppl. **129** (2004) 81.
[10] S. Actis *et al.* [Working Group on Radiative Corrections and Monte Carlo Generators for Low Energies Collaboration], Eur. Phys. J. C **66** (2010) 585
[11] M. S. Chen and P. M. Zerwas, Phys. Rev. D **11** (1975) 58
[12] S. Binner, J. H. Kuhn and K. Melnikov, Phys. Lett. B **459** (1999) 279
[13] M. Benayoun, S. I. Eidelman, V. N. Ivanchenko and Z. K. Silagadze, Mod. Phys. Lett. A **14** (1999) 2605
[14] W. Kluge, Nucl. Phys. Proc. Suppl. **181-182** (2008) 280
[15] V. P. Druzhinin, S. I. Eidelman, S. I. Serednyakov and E. P. Solodov, arXiv:1105.4975 [hep-ex]
[16] J. Prades, E. de Rafael and A. Vainshtein, (Advanced series on directions in high energy physics. 20)
[17] K. Jansen, these proceedings
[18] M. Benayoun, these proceedings
[19] F. Jegerlehner and R. Szafron, Eur. Phys. J. C **71** (2011) 1632
[20] J. Z. Bai *et al.* [BES Collaboration], Phys. Rev. Lett. **88** (2002) 101802
[21] S. R. Amendolia *et al.* [NA7 Collaboration], Nucl. Phys. B **277** (1986) 168
[22] G. Mandaglio, these proceedings.
[23] E. P. Solodov, presentation at the International Workshop on e+e- collisions from Phi to Psi (PHIPSI11), September 19-22 2011, Novosibirsk; S. I. Serednyakov presentation at the International Workshop on e+e- collisions from Phi to Psi (PHIPSI11), September 19-22 2011, Novosibirsk
[24] D. Babusci, C. Bini, F. Bossi, G. Isidori, D. Moricciani, F. Nguyen, P. Raimondi and G. Venanzoni *et al.*, arXiv:1007.5219 [hep-ex].
[25] R. McNabb, J. Blackburn, J. D. Crnkovic, D. W. Hertzog, B. Kiburg, J. Kunkle, E. Thorsland and D. M. Webber *et al.*, Nucl. Instrum. Meth. A **602** (2009) 396
[26] G. W. Bennett *et al.* [Muon (g-2) Collaboration], Phys. Rev. D **80** (2009) 052008
[27] F. Jegerlehner, Nucl. Phys. Proc. Suppl. **181-182** (2008) 26.
[28] D. Babusci, *et al.*, arXiv:1109.2461 [hep-ph]

Journal of Physics: Conference Series **349** (2012) 012009 doi:10.1088/1742-6596/349/1/012009

Study of N* properties at CLAS

D.P. Watts for the CLAS collaboration at Jefferson Lab
SUPA, University of Edinburgh, James Clerk Maxwell Building, King's Buildings, Mayfield Road, Edinburgh, EH9 3JZ

E-mail: daniel.watts@ed.ac.uk

Abstract. The excitation spectrum of the nucleon provides a stringent constraint on the dynamics and interactions of its internal constituents and therefore probes the mechanism of confinement in the light quark sector. Our detailed knowlege of this excitation spectrum is poor, with many predicted states not yet observed in experiment and many "established" states having poorly known properties. To address these shortcomings a worldwide effort is currently underway exploiting the latest generation of electron and photon beams in detailed studies of meson photoproduction from nucleon targets. A major contribution to this effort will come from the experimental programme at Jefferson Lab exploiting the frozen spin target (FROST) with the CLAS spectrometer. The status of this project will be presented along with preliminary results and analyses.

1. Introduction

Obtaining an accurate determination of the excitation spectrum of the nucleon would provide stringent constraints on our understanding of the dynamics and degrees of freedom of its constituents. Significant advances in theory aiming to understand this spectrum have taken place in recent years. One recent highlight is the emergence and expected progression of Lattice QCD predictions, which provide a more direct link between the properties of the excitation spectrum and the underlying non perturbative QCD processes in the light quark sector [1]. Better establishing the excitation spectrum would also guide refinements to QCD-based phenomenological models such as constituent quark models. The recent lattice results tend to support the finding of many constituent quark models that there should be many more excited states of the nucleon than have currently been observed, particularly in the mass region above 1.8 GeV/c^2.

It remains a major challenge to current hadron physics to establish whether this "missing resonance" problem is due to deficiencies in the coverage, quality, or completeness of the world data set to date or whether these states are absent in nature, with significant consequences to our understanding of nucleon dynamics. As well as hunting for unseen resonances, it is important to better establish the properties of "known" resonances, reaching a situation where they are consistently evident in different analyses and have consistent electromagnetic couplings, masses and widths.

Meson photoproduction is an excellent tool to study the nucleon resonance spectrum. The photon offers a clean well understood probe, with the advantage over previous studies with spinless pion beams that the response of the nucleon to polarisation degrees of freedom can be utilised. The new generation of measurements of meson photoproduction carried out or

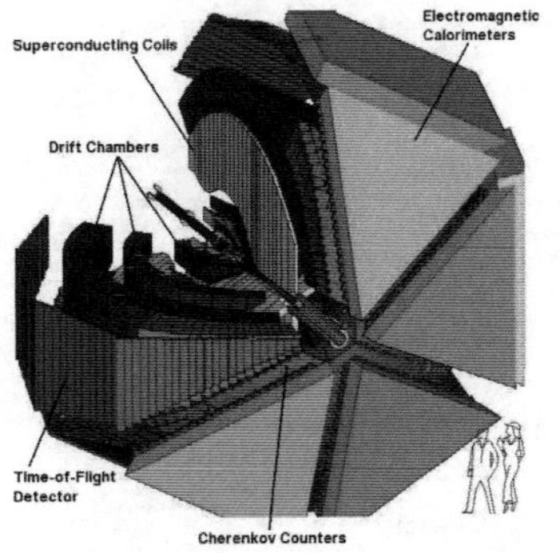

Figure 1. Schematic diagram of the CLAS detector.

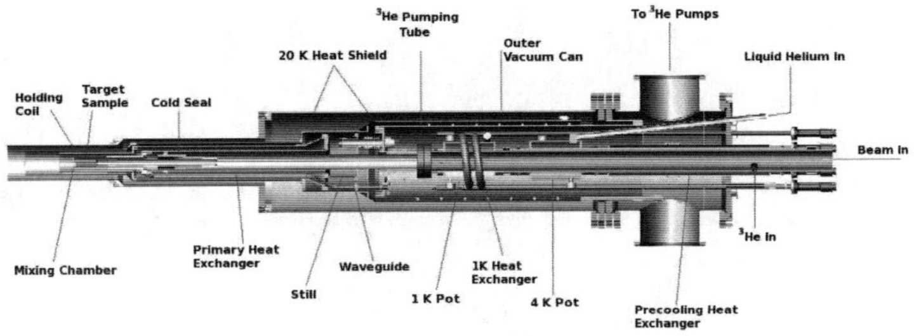

Figure 2. Schematic diagram of the FROST target.

planned at many of the world's electromagnetic beam facilities will provide a step change in the quality of available experimental data. The use of photon beams and targets with high degrees of polarisation, coupled with large acceptance particle detectors is essential for disentangling the spectrum of excited states, which can only be revealed from the data by a detailed partial wave analysis (PWA). The inadequacies of the current world data set are highlighted by the differences in the composition of the resonance spectrum by different partial wave analyses of the same data (for a review see [2]). Significant advances in PWA analyses themselves are also currently underway, including the SAID [3], MAID [4], Bonn-Gatchina [5] and EBAC [6]. One recent highlight from these developments comes from the coupled channel models developed by the EBAC group at Jefferson Lab, which recently led to the suggestion that the low mass of the Roper resonance may arise from strong coupled channel effects [6]. A common problem for all the PWA's in meson photoproduction is that they are currently under-constrained and require more comprehensive experimental data to permit more consistent interpretations of the contributing resonances and reaction mechanisms.

The CLAS measurements presented in this contribution will form a crucial part of this world effort to improve the world data base available for PWA studies. The FROST experiment will provide complete or nearly complete measurements for a whole range of meson photoproduction reactions such as $\gamma p \rightarrow N\pi, p\eta, p\omega, K^+Y$ and $p\pi^+\pi^-$. A "complete" measurement corresponds to the determination of sufficient experimental quantities to eliminate experimental ambiguities

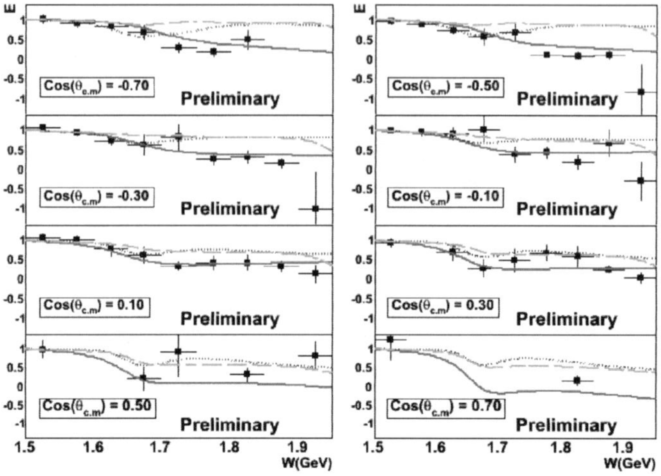

Figure 3. Preliminary results for E in $\vec{\gamma}\vec{p} \to p\eta$ for energies $W = 1.525 - 1.925$ GeV . Curves: η-MAID (dotted line), Bonn-Gatchina PWA (dashed line), and SAID (solid line).

in extracting the scattering amplitude for the reaction and thereby putting as tight constraints on the various PWA's as possible. Achieving a complete measurement requires the precise determination of at least eight well chosen experimental observables (for a recent discussion see [7]). One combination is the cross section, three single-spin and four double-spin observables. For hyperon production channels the ability to determine the polarisation of the recoiling hyperon from the angular distribution of its weak decay allows this complete determination to be achieved entirely from observables measured in the CLAS programme. For non-strange production channels a close to complete set can be obtained at CLAS.

The experimental facilities will be discussed in section 2. A selection of preliminary results for beam-target observales will be presented in sections 3-6.

2. Experimental Setup

The Continuous Electron Beam Accelerator Facility (CEBAF) provides high-intensity electron beams of up to 6 GeV with approximately 100% duty cycle simultaneously to three experimental halls: A, B and C. The upgrade to 12 GeV will be completed in 2014, with associated upgrades to the equipment in the three current experimental halls and a fourth hall, D, currently under construction. The CEBAF accelerator consists of two anti-parallel superconducting RF-linacs connected by recirculation arcs to create a racetrack with a combined length of 1.4 km. The accelerator produces a continuous-wave electron beam which can circulate around the linacs up to 5 times, gaining 1.2 GeV of energy with each pass.

Hall B houses the CEBAF Large Acceptance Spectrometer (CLAS), which can be used with both electron and photon beams. A schematic diagram of CLAS is shown in Fig.1. CLAS is a multi-layered magnetic spectrometer, divided into six independent sectors by six superconducting coils arranged symmetrically around the beamline producing a toroidal magnetic field. There is a magnetic field free region in the centre of the detector, allowing for the use of polarised targets such as FROST. For photon beam experiments the innermost detector is a scintillator start counter surrounding the target providing timing and triggering information on charged reaction products. Outside the start counter, the first three layers of detectors consist of drift chambers to determine the curved trajectories and hence momenta of charged particles. Surrounding the drift chambers, gas Cherenkov Counters identify electron events and allows for the differentiation of electrons from pions. This is then surrounded by a layer of plastic scintillation bars for time-of-flight measurements. The outermost detector system is made up of electromagnetic calorimeters for the detection of photons, electrons and high-energy neutrons.

Near the entrance to Hall B the electrons from CEBAF impinge on a thin radiator in which they undergo bremsstrahlung. The recoiling electrons from this process are momentum analysed in a magnetic spectrometer containing a highly segmented focal plane comprised of thin plastic scintillation detector elements. The device can tag photons in the energy range 0.35 to 5.8 GeV. Circularly polarised photons can be produced by utilising a longitudinally polarised electron beam. The direction of the electron polarisation (and therefore photon circular polarisation) was flipped at a rate of 30 Hz. Linear photon polarisation can be achieved by using a thin diamond radiator. For such cases the lattice structure coherently enhances the momentum transfer direction in the scattering process to directions related to the reciprocal lattice vectors of the crystal, producing linear polarisation for the bremsstrahl photons.

The frozen spin target (FROST) is placed at the centre of the CLAS detector and provides a highly polarised proton target for use in a wide range of experiments. A schematic of the FROST apparatus is shown in Fig. 2. The target material consists of frozen, 1-2 mm beads of butanol (C_4H_9OH) mixed by weight with 5% H_2O and 0.5% TEMPO (2,2,6,6-tetramethylpipendine- 1-oxyl), and is placed inside the target cup which is 50 mm long and 15 mm in diameter. The target material was chosen to have a high maximum polarisation, a high ratio of polarisable nucleons to the total number of nucleons (the quality factor), a high resistance to ionising radiation, and a small number of unwanted polarisable nuclei. In butanol the non-hydrogen components, carbon and oxygen, are spinless particles, so pure butanol has no background polarisation to correct for. To subtract the yield from the unpolarised nuclei the frost target was run simultaneously with a separate carbon and polythene detector downstream of FROST. Events from these targets could be identified from the vertex tracking information of CLAS and their contribution to the butanol yield could be assessed for any variable of interest in the analysis. In operation the FROST target achieved polarisation levels immediately following the polarising process of 85% to 90% and the polarisation degraded at a rate of 0.9% (spin paralell to beam) to 1.5% (anti-parallel to beam axis) per day.

In the following sections some of the preliminary analysis results from the first round of FROST data using a longitudinally polarised proton target will be presented.

3. The beam-target helicty asymmetry, E, in η photoproduction

The measurement of a wide range of final states in meson photoproduction is highly desirable, not only to constrain coupled channel analyses but also as certain channels are expected to filter contributions of nucleon resonances to the scattering amplitude. One example of this is η photoproduction which due to the I=0 isospin of the η can be produced in a single step mechanism only via intermediate nucleon resonances having isospin, $I=\frac{1}{2}$. This restriction simplifies the data interpretation and theoretical efforts to predict the excited states contributing to these reactions.

The polarised cross section for the reaction $\vec{\gamma}\vec{p} \rightarrow p\eta$ of circularly-polarised photons on longitudinally-polarised protons is given by:

$$\frac{d\sigma}{d\Omega} = \frac{d\sigma}{d\Omega_0}\,(1 - P_z\,\delta_\odot\,E)\,, \tag{1}$$

where $d\sigma/d\Omega_0$ is the unpolarised cross section. P_z and $\delta_\odot$ are the degrees of target and beam polarisation, respectively. E denotes the helicity asymmetry.

The asymmetry is extracted from the difference in reaction yields between the photon and target nucleons spins aligned or antialigned. Preliminary results [11] of the helicity asymmetry,

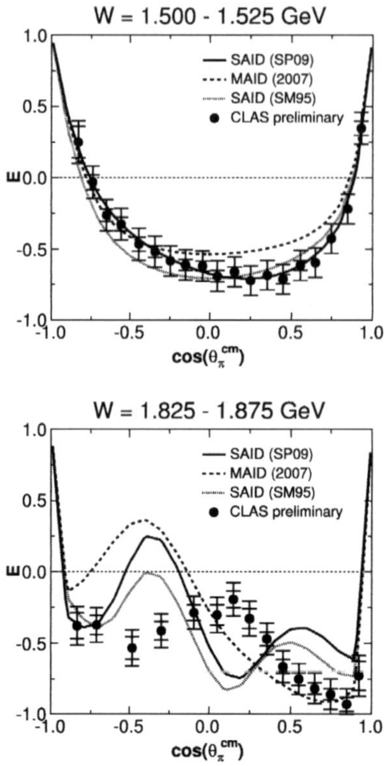

Figure 4. Preliminary results of the double-polarisation observable E (helicity difference) for $\vec{\gamma}\vec{p} \to n\pi^+$[8]. The inner error bars indicate stat. uncertainties; the outer error bars include a 10 % sys. uncertainty, which is expected to be reduced in the final analysis. The curves show solutions of the SAID SP09 [3], MAID [4] and SAID SM95 PWA.

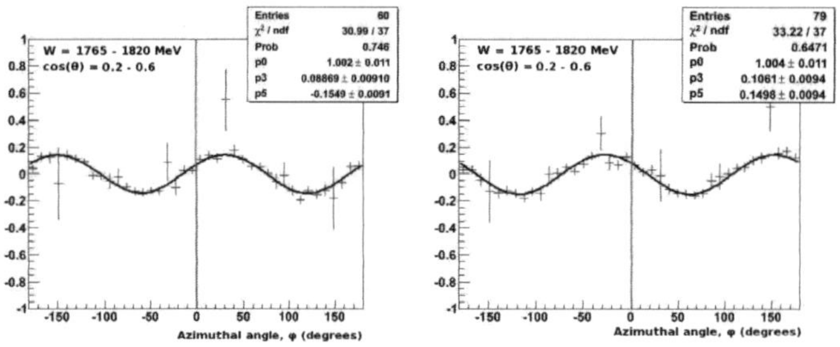

Figure 5. (Upper) phi asymmetry for the positively polarised target setting. (Lower) phi asymmetry for negative target setting. The vertical red line at 0^0 is present to help demonstrate the phase shift in the asymmetry due to the G contribution to the cross section [9]

E, for $\vec{\gamma}\vec{p} \to p\eta$ are shown in Fig. 3. Since the η threshold is dominated by the N(1535)S_{11} resonance, the observable exhibits values close to unity for W $\leq$ 1.6 GeV/c^2 . The preliminary results indicate that the observable remains positive below about W = 2 GeV, providing new information to constrain the partial wave analyses for this channel.

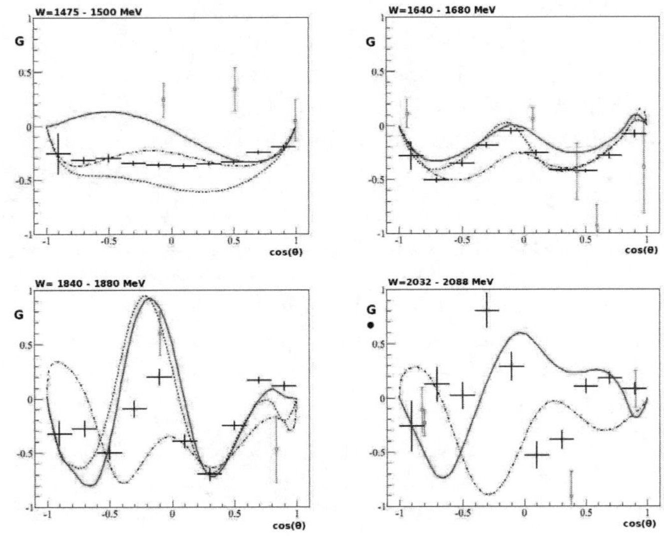

Figure 6. Preliminary FROST values of the double-polarisation observable G (black points) plot as a function of *cos(θ)* for a small fraction of the available mass (W) bins [9]. Overlaid on the plots are the current SAID [3] (solid-red line) and MAID2007 [4] (dashed-blue line) solutions. The open pink squares show the previous experimental data for G.

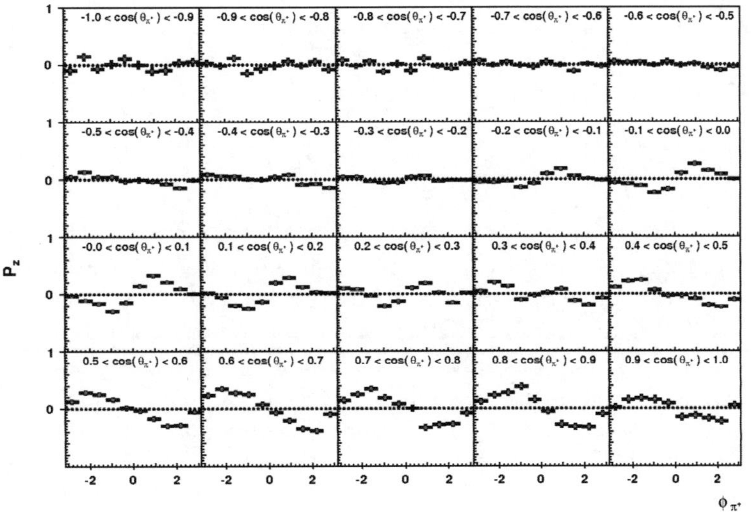

Figure 7. Preliminary target asymmetry, P_z, from FROST for the reaction $\vec{\gamma}\vec{p} \to p\pi^+\pi^-$ for $E_\gamma \in [0.7, 0.8]$ GeV [10]. Error bars are statistical only.

4. The beam-target helicity asymmetry E in π^+ photoproduction

Comprehensive measurement of the pion production reaction is important as, of all the production channels accesible, the pion photoproduction is expected to give sensitivity to the largest set of resonances. The current database for pion photoproduction is mainly populated by unpolarised cross section data and single-spin observables and is therefore significantly under-constrained.

Figure 4 shows preliminary results [8] of the double polarisation E for $\vec{\gamma}\vec{p} \to \pi^+ n$ (Eqn. 1). While the predictions shown in Fig. 4 agree nicely with the new data at low energies, discrepancies emerge at higher energies for W $\geq$1.7 GeV/c^2. Single-pion photoproduction in this region appears less well understood than previously expected. The implications for the resonance spectrum and properties will be explored when the finalised results are included in the various PWAs.

5. The beam-target linear asymmetry G in π^+ photoproduction

The use of a linearly polarised photon beam with a longitudinally polarised proton target gives access to the observable G, for which only sparse data exists in the current world data base. The differential cross section for the case of a linearly polarised photon beam and longitudinally polarised proton target can be written as:

$$\frac{d\sigma}{d\Omega} = \left(\frac{d\sigma}{d\Omega}\right)_0 (1 - P_L\Sigma\cos(2\phi) + P_L\delta_\odot G\sin(2\phi)), \tag{2}$$

where $d\sigma/d\Omega_0$ is the unpolarised cross section. P_L and $\delta_\odot$ are the degrees of linear beam and target polarisation, respectively. G denotes the beam-target linear asymmetry observable and Σ is the single beam polarisation observable.

There were two possible settings for the photon beam polarisation in the first FROST experiment: PARA ($||$) in which the electric field vector was parallel to the floor and PERP ($\perp$) in which the electric field vector was perpendicular to the floor. In addition, the target was polarised either parallel or anti-parallel to the beam direction, defined as positive and negative target polarisation, respectively. The combination of these resulted in a total of four possible combinations of polarised beam-polarised target settings during the run. An asymmetry in the reaction yield as a function of pion azimuthal angle with respect to the photon polarisation plane can be obtained:

$$\frac{N(\phi)_\perp - N(\phi)_{||}}{N(\phi)_\perp + N(\phi)_{||}} = P_L\Sigma\cos(2\phi) - P_L\delta_\odot G\sin(2\phi). \tag{3}$$

The value of G could be extracted by fitting these phi asymmetry spectra. Typical plots for positive and negative target polarisation settings are shown in Fig. 5. The effect of G in equation (3) gives the expectation of a change in sign of the ϕ offset when the target is polarised in the positive or negative direction. This is clearly seen when comparing the 2 plots in the Fig. 5 .

Preliminary values for G for a sample of the available W bins are shown in Fig. 6. It should be noted that the systematics in the degree of linear polarisation of the beam are still being evaluated so the results are at an early stage. Despite the preliminary nature it is clear that the new data show a great improvement in accuracy and kinematic coverage compared to previous measurements. Even in the low lying part of the excitation spectrum near to the Roper resonance it is seen that current PWAs give different expectations for G. These significant differences highlight how the current world data does not constrain this reaction even in the low lying part of the spectrum. The new data will therefore give valuable new constraints on PWAs over the mass range from 1400 up to 2300 MeV.

6. Polarisation observables for $\pi^+\pi^-$ production on the proton

One of the key experiments in the search for yet unobserved states is the investigation of double-pion photoproduction. Quark models predict large couplings of those states to $\Delta\pi$, for instance. The five-dimensional cross section for the photoproduction of two pseudoscalar mesons using longitudinal target polarisation and circularly-polarised (or unpolarised) beam can be written in the form [12]:

$$I = I_0\{(1 + \Lambda_z \cdot P_z) + \delta_\odot(I^\odot + \Lambda_z \cdot P_z^\odot)\},$$

where I_0 denotes the unpolarised cross section and $\delta_\odot$ and Λ_z denote the degree of beam and target polarisation, respectively. The additional polarisation observables, P_z and $I^\odot$, for the two-meson final state arise since the reaction is no longer restricted to a single plane.

Figure 7 shows an example for the observable P_z in $\gamma\vec{p} \rightarrow \pi^+\pi^-$[10]. The variables θ and ϕ denote the polar and azimuthal angle of the π^+ in the rest frame of the two mesons. The observable acquires surprisingly large values for positive cos θ_{π^+} with the statistical errors in some cases smaller than the symbol size. The expected odd behaviour of the distribution is clearly visible.

7. Conclusion

Nucleon spectroscopy is undergoing an exciting period with major advances in theory being complemented by a step change in the quality of experimental data. Experimental programmes of meson photoproduction measurements at the world's electromagnetic beam facilities will provide the long sought extensive data sets to constrain the various partial wave analyses and more reliably extract the excitation spectrum of the nucleon and its properties. The FROST programme at Jefferson Lab will provide a large proportion of this new high quality experimental data. In the coming decade the combined efforts of experimentalists and theorists in this area is expected to allow deeper insights into our understanding of QCD confinement in the light quark sector.

8. Acknowlegements

The work is supported by DOE and NSF grants. The author thanks the Science and Technology Funding Council (UK) for support.

references

[1] J. Bulava, R. Edwards, E. Engelson, B. Joó, H.-W. Lin, C. Morningstar, D. Richards, S. Wallace, Phys. Rev. D 82 (2010) 014507.
[2] E. Klempt, J.-M. Richard, Rev. Mod. Physics 82 (2010) 1095.
[3] M. Dugger, CLAS-Collaboration, Phys. Rev. C 79 (2009) 065206.
[4] D. Drechsel, O. Hanstein, S. S. Kamalov, L. Tiator, Nucl. Phys. A 645 (1999) 145.
[5] E. Klempt, A. Anisovich, V. Nikonov, A. Sarantsev, U. Thoma., Eur. Phys. J. A 24 (2005) 111.
[6] H. Kamano, S. X. Nakamura, T. S. H. Lee, T. Sato, Phys. Rev. C 81 (2010) 065207.
[7] B. Dey, M. McCracken, D. Ireland, C. Meyer, Phys. Rev. C 83 (2011) 055208.
[8] S.Strauch, NSTAR 2011 conference, Jefferson Lab.arXiv:nucl-ex/1108.3050.
[9] J. McAndrew, Ph.D. thesis, Edinburgh University (2011).
[10] S.Park, NSTAR 2011 conference, Jefferson Lab.
[11] B.Morrison, NSTAR 2011 conference, Jefferson Lab.
[12] W.Roberts, T.Oed, Phys. Rev. C 71 (2005) 055201.

Parity violating electron scattering at MAMI

Sebastian Baunack

Institut für Kernphysik, Universität Mainz, J. J. Becherweg 45, 55099 Mainz, Germany

E-mail: baunack@kph.uni-mainz.de

Abstract. The investigation of the structure of the nucleon can help to understand the non-pertubative regime of the QCD. In the viewpoint of QCD, the nucleon is made up of constituent quarks, sea quarks and gluons. The nucleon structure can be described by the electromagnetic form factors. Parity violating electron scattering offers a tool to investigate the strange quark contribution to the nucleon form factors. Such measurements are carried out at the electron accelerator facility MAMI at Mainz. Recent results are presented here.

1. Strange form factors

The proton is in the viewpoint of QCD a highly relativistic strong interacting system, where, beside the constituent quarks, also sea quarks and gluons contribute to its properties. Strange quark contributions are of a specific interest since this is a pure sea quark effect. Various matrix elements may be used for an investigation of strange quark effects [1]. The vector element $< p|\bar{s}\gamma_\mu s|p >$ provides access to the contribution of strange quarks to the vector current of the nucleon expressed by the vector form factors [2, 3]. In order to determine the contribution of individual quark flavors to the nucleon form factors, a flavor decomposition is done omitting the heavier quarks c, b and t:

$$G^p_{E,M} = \sum_{f=u,d,s} q_f G^{f,p}_{E,M} = \frac{2}{3}G^{p,u}_{E,M} - \frac{1}{3}G^{p,d}_{E,M} - \frac{1}{3}G^{p,s}_{E,M} \qquad (1)$$

$$G^n_{E,M} = \sum_{f=u,d,s} q_f G^{f,n}_{E,M} = \frac{2}{3}G^{n,u}_{E,M} - \frac{1}{3}G^{n,d}_{E,M} - \frac{1}{3}G^{n,s}_{E,M} \qquad (2)$$

$Gp/n, x_{E,M}$ denotes the contribution of the quark flavor x to the electric or magnetic form factor of the proton or the neutron. The factors $2/3$ and $-1/3$ reflect the electric charges of the corresponding quarks. Isospin symmetry reduces the number of unknown flavor form factors from 12 to 6. The weak interaction provides additional information:

$$\tilde{G}^p_{E,M} = (\frac{1}{4} - \frac{2}{3}\sin^2\theta_W)G^u_{E,M} - (\frac{1}{4} - \frac{1}{3}\sin^2\theta_W)G^d_{E,M} - (\frac{1}{4} - \frac{1}{3}\sin^2\theta_W)G^s_{E,M} \qquad (3)$$

$\tilde{G}^{p,n}_{E,M}$ denote the weak form factors of proton and neutron and the factors in front of the flavor form factors represent the weak charges of the corresponding quarks. The interference between tree level electromagnetic and weak amplitudes leads to a parity violating asymmetry in the elastic scattering cross section of right- and left-handed electrons σ^+, σ^-: $A_{PV} = (\sigma^+ - \sigma^-)/(\sigma^+ + \sigma^-)$. This asymmetry can be expressed by the known electromagnetic form factors of proton and neutron, by the weak axial form factor of the proton $\tilde{G}_A^p$ and the unknown strange form factors G_E^s and G_M^s as a sum of three terms, $A_{PV} = A_V + A_S + A_A$: A_V represents the vector coupling on the proton vertex where the strangeness contribution has been taken out, A_S contains the strange quark vector contribution, and A_A represents the axial coupling on the proton vertex [2]:

$$A_{PV} = -a\rho'_{eq}\left((1-4\hat{\kappa}'_{eq}\hat{s}_Z^2) - \frac{\epsilon G_E^p G_E^n + \tau G_M^p G_M^n}{\epsilon (G_E^p)^2 + \tau (G_M^p)^2}\right) \tag{4}$$

$$A_S = a\rho'_{eq}\frac{\epsilon G_E^p G_E^s + \tau G_M^p G_M^s}{\epsilon (G_E^p)^2 + \tau (G_M^p)^2} \tag{5}$$

$$A_A = a\frac{(1-4\hat{s}_Z^2)\sqrt{1-\epsilon^2}\sqrt{\tau(1+\tau)}G_M^p \tilde{G}_A^p}{\epsilon (G_E^p)^2 + \tau (G_M^p)^2} \tag{6}$$

(7)

with $a = \frac{G_\mu Q^2}{4\pi\alpha\sqrt{2}}$, G_μ the Fermi coupling constant, α the fine structure constant, $\tau = \frac{Q^2}{4M_p^2}$, Q^2 the negative squared momentum transfer and M_p the proton mass, $\epsilon = [1 + 2(1+\tau)\tan^2\frac{\Theta}{2}]^{-1}$, Θ the scattering angle in the laboratory frame, $\hat{s}_Z^2(M_Z) = 0.23116(13)$[4]. ρ'_{eq} and $\hat{\kappa}'_{eq}$ include the electro-weak radiative corrections evaluated in the $\overline{MS}$ scheme. One measurement of the parity violating asymmetry A_{PV} yields one linear combination of strange electric and magnetic form factor $G_E^s + \alpha G_M^s$ where α depends on the kinematics. Two independent measurements at the same momentum transfer Q^2 but with different kinematics allow to disentangle G_E^s and G_M^s when $\tilde{G}_A^p$ is used as an input parameter. There are numerous theoretical approaches to the strange quark contribution to the vector current of the proton, covering the wide range from nonrelativistic constituent quark models [6] until quenched lattice calculations [7, 8].

2. Measurement of strange form factors

A measurement at forward angles with a proton target yields a linear combination of G_E^s and G_M^s, a measurement at forward angles on a helium target gives a measurement of G_E^s alone, a measurement at backward angles measurements on a proton and a deuterium target yield two different linear combinations of G_M^s and $\tilde{G}_A^p$. Table 1 summarizes the experiments that made such measurements. At the negative squared momentum transfer $Q^2 = 0.1$ GeV^2 there are four collaborations that have published results: the SAMPLE collaboration at MIT MIT-Bates [9], the A4 collaboration at Mainz [10], the G^0 collaboration at TJNAF [11] and the HAPPEX collaboration at TJNAF [12]. A global analysis yields the values of $G_E^s = -0.005 \pm 0.019$ and $G_M^s = 0.18 \pm 0.27$ [12] favoring calculations that predict only small strangeness contributions.
At $Q^2 = 0.23$ GeV^2 there are results for forward and backward angle measurements published by the collaborations A4 [10, 13] and G^0 [11, 16], also excluding large strangeness contributions. The experimental setup of the A4 collaboration and new results will be presented in the following chapters.

3. The A4 experimental setup

The measurement of small parity violating asymmetries of the order 10^{-6} requires high statistics on the one hand since the statistical uncertainty in the measured asymmetry scales with $1/\sqrt{N}$ where N is the number of registered elastically scattered electrons, and control of systematic effects on the other hand.
The concept of the A4 experiment [17] at MAMI is a single arm calorimetric measurement. The electron beam has an intensity of 20 μA and an average polarization of about 80%. The electrons are accelerated up to $E = 1.5$ GeV. Two polarimeters are in the beamline of the A4 experiment, a

Table 1. Experiments that measure parity violating asymmetries, and their kinematical conditions and targets. A wide range of momentum transfers and scattering angles is already covered. The A4 data point shown in italic is scheduled for the measurement in 2012.

	p target forward angle	p target backward angle	^{4}He target forward angle	d target backward angle
SAMPLE		0.10 GeV^2		0.04 GeV^2 0.1 GeV^2
Happex	0.10 GeV^2 0.48 GeV^2 0.62 GeV^2		0.10 GeV^2	
A4	0.11 GeV^2 0.23 GeV^2 0.62 GeV^2	0.11 GeV^2 0.23 GeV^2		*0.11 GeV^2* 0.23 GeV^2
G^0	(0.12...1) GeV^2	0.23 GeV^2 0.62 GeV^2		0.23 GeV^2 0.62 GeV^2

Compton Backscatter polarimeter [18] and a Transmission Compton polarimeter [19]. Additional polarization measurements were made by the Moeller polarimeter of the A1 collaboration and by a Mott polarimeter situated close to the electron source. Altogether a 4% uncertainty in the knowledge of the beam polarization is achieved.

The helicity of the electrons is changed every 20 ms. In order to suppress helicity correlated false asymmetries due to different beam position, angle, intensity or energy for different helicity states, several feedback stabilization systems are installed along the way of the electrons from the accelerator to the target. The high power liquid hydrogen target [20] is 10 cm long for the forward angle measurements and 23.4 cm long for the backward angle measurement. The luminosity is $L \approx 0.5 \cdot 10^{38}$ cm^{-2}s^{-1} or $L \approx 1.2 \cdot 10^{38}$ cm^{-2}s^{-1} respectively.

The scattered electrons are detected in a homogenous electromagnetic calorimeter that consists of 1022 lead fluoride crystals [14] (see fig. 1). The calorimeter can cope with event rates of 100 MHz, the energy resolution is about $\Delta E/E = 3.9\%/\sqrt{E[\mathrm{GeV}]}$. The covered angles are 2π in the azimuthal range and $30° \leq \theta \leq 40°$ (forward angle configuration) or $140° \leq \theta \leq 150°$ (backward angle configuration) in the polar range. The covered solid angle is $\Delta\Omega = 0.62$ sr. 72 plastic scintillators are installed in front of the PbF_2 crystals for a separation of neutral from charged particles. This is necessary at backward angles because photons from π^0 decay cannot be distinguished by the PbF_2 calorimeter itself because the photons produce in the crystals electromagnetic showers very similar to those coming from electrons.

4. Recent results from the A4 experiment

The preliminary results of the two most recent measurements are presented here: A backward angle measurement with a deuterium target at a low beam energy and a forward angle measurement with a proton target at a high beam energy.

In order to obtain the strange form factors G_E^s and G_M^s, the axial form factor $\tilde{G}_A^p$ of the proton must also be known. At backward angles, a measurement with a proton target is mainly sensitive to a linear combination of G_M^s and $\tilde{G}_A^p$. A measurement with a deuterium target at the same momentum transfer yields a different linear combination of these two form factors and allows a determination of $\tilde{G}_A^p$. About 1100 h of data were taken with a deuterium target at the momentum transfer of $Q^2 = 0.23$ GeV^2 which matches to the previous measurement with a hydrogen target [13]. Compared to the analysis of to the hydrogen data the analysis is more challenging: The peak of the quasielastic scattered electrons is broader due to the

Journal of Physics: Conference Series **349** (2012) 012010 doi:10.1088/1742-6596/349/1/012010

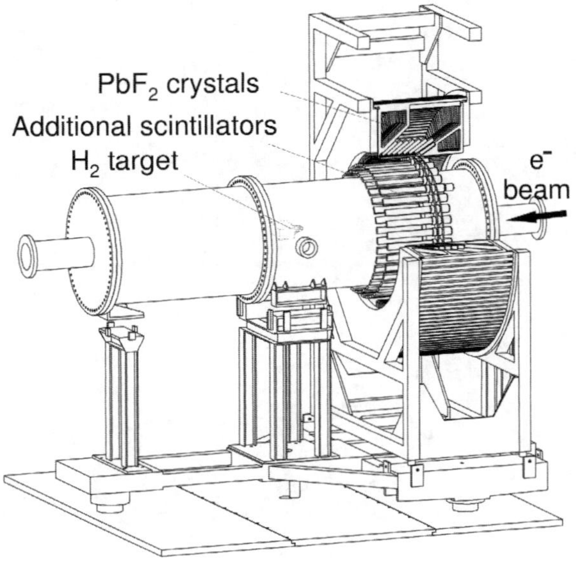

Figure 1. Drawing of the A4 lead fluoride calorimeter and the scattering chamber [13]. The calorimeter is mounted on a rotatable platform so that one can easily change between forward and backward angle configuration. The backward angle configuration is shown here. The lead fluoride crystals cover the 2π azimuthal range and polar angles between 30° and 40° in forward angle configuration and between 140° and 150° in backward angle configuration. Plastic scintillators are installed between the crystals and the scattering chambers for a separation of neutral and charged particles.

Fermi motion of the nucleons within the nucleus and the signal to background ratio is worse because the rate of charged particles increases by a factor of 1.5, while the rate of decay photons increases by a factor of 2. This requires a careful study of the background subtraction procedure. We obtain a preliminary value for the asymmetry in quasielastic *ed* scattering of $A_d = (-20.02 \pm 0.84_{stat} \pm 1.25_{syst.}) \cdot 10^{-6}$. Adding these uncertainty in quadrature, we get for the form factors a linear combination of $\tilde{G}^p_A$+$0.61 \cdot G^s_M$= -0.55 ± 0.35. For $\tilde{G}^p_A$ alone, we obtain the preliminary value of $\tilde{G}^p_A$= -0.47 ± 0.31 which is smaller than the value taken from the calculation [15], but in good agreement with the result from the G0 collaboration at the same momentum transfer [16].

In the year 2009, a measurement with a proton target at high momentum transfer was carried out at the MAMI-C beam energy of 1508 MeV. The resulting momentum transfer is $Q^2 = 0.62$ GeV^2. In this momentum transfer region the results of the G0 forward angle measurements suggested nonvanishing strange quark contributions to the vector form factors of the proton [11]. 600 h of data were taken for this measurement. The average beam polarization was 85%. Fig. 2 shows the preliminary results with respect to the sign of the electron beam polarization (halfwave plate at the source in or out). We obtain a preliminary result for the linear combination of the strange form factors of $G^s_E + 0.628\ G^s_M = 0.067 \pm 0.030$, where all errors have been added in quadrature.

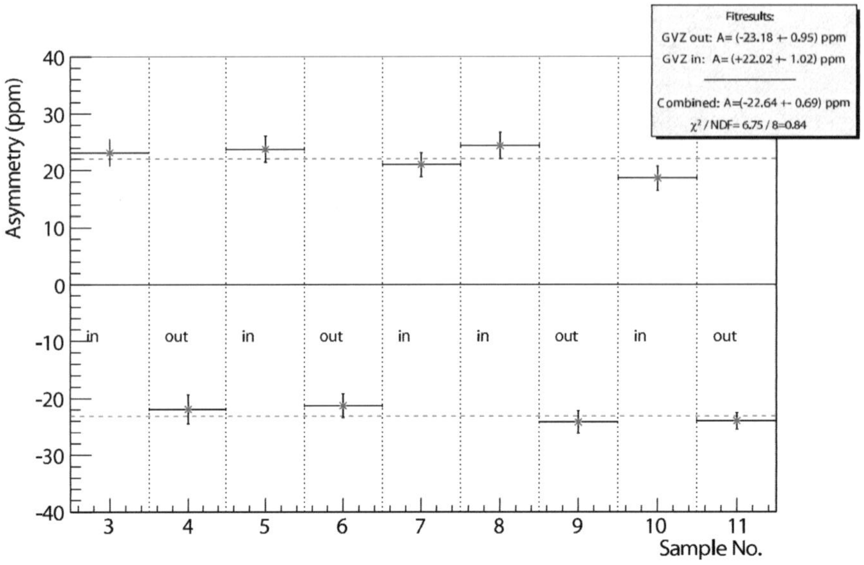

Figure 2. Preliminary result from the forward angle measurement at $Q^2 = 0.62$ GeV^2. The asymmetries that are shown are corrected for false asymmetries and the beam polarization. "Out" and "In" denote subsets of the measurement with either a positive or a negative electron beam polarization (halfwave plate at the source out or in). The sign change in the measured asymmetries can be easily observed.

5. Conclusions and outlook

Parity violating electron scattering can be used to determine the strange quark contributions to the vector form factors of the proton. After some years of operation, the A4 experiment at MAMI measures routinely such asymmetries of the order of 10^{-6}. Our recent measurements are in good agreement with results from other collaboration and confirm that the strangeness contribution to the vector form factors of the protons are small. With the existing techniques it will be difficult to measure non-zero strange form factors significantly. However, strange quark contributions are a source of hadronic uncertainty in precision measurements of the weak charge of the proton like the Qweak experiment at Jefferson Lab is carried out right now [21] or the new planned project P2 at MAMI. Therefore it is necessary to determine the strange form factors as precise as possible especially at low momentum transfer. Currently, the A4 collaboration is performing backward angle measurements with a proton and a deuterium target at $Q^2 = 0.1$ GeV^2 The projected measurement times are 1000 h for each target. The aim is to reduce the existing uncertainties by factor of 2.

References

[1] B Kaplan and A Manohar 1988 *Nucl. Phys.* B **310** 527
[2] M J Musolf et al. 1994 *Phys. Rep.* **239** 1
[3] D H Beck and B R Holstein 2001 *Int. J. Mod. Phys. E* **10** 1-41
[4] K Nakamura et al. (Particle Data Group) 2010 *J. Phys.* G **37** 075021
[5] D O Riska 2007 *Eur. Phys. J.* A **32** 389
[6] A Kiswandhi et al. 2011 *Phys. Lett.* B**704** 373-377
[7] D B Leinweber et al. 2006 *Phys. Rev. Lett.* **97** 022001
[8] P Wang et al. 2009 *Phys.Rev.* C **79** 065202
[9] D T Spayde et al. 2004 *Phys. Lett.* **B583** 79
[10] F E Maas et al. 2004 *Phys. Rev. Lett.* **93** 022002

Journal of Physics: Conference Series **349** (2012) 012010 doi:10.1088/1742-6596/349/1/012010

[11] D Armstrong et al. 25 *Phys. Rev. Lett.* **95** 092001
[12] A Acha et al. 2007 *Phys. Rev. Lett.* **98** 032301
[13] S Baunack et al. 2009 *Phys.Rev.Lett.* **102** 151803
[14] S Baunack et al 2011 *Nucl. Inst. Meth.* A **640** 58-68
[15] S L Zhu et al. 2000 *Phys. Rev.*D **62** 033008
[16] D Androic et al. 2010 *Phys.Rev.Lett.* **104** 012001
[17] F E Maas et al. 2003 *Eur. Phys. J.* A **17** 339
[18] J Diefenbach et al. 2007 *Eur. Phys. J.* A **32** 555-559
[19] C Weinrich et al. *Eur. Phys. J.* A **24S2** 129-130
[20] I Altarev et al. 2006 *Nucl. Instr. Meth.* A **564** 13
[21] R D Carlini 2010 *AIP Conf.Proc.* **1261** 172-178

Journal of Physics: Conference Series **349** (2012) 012011 doi:10.1088/1742-6596/349/1/012011

Measurement of helicity dependent total inclusive γ–^{3}He cross section and the GDH sum rule on the neutron

S Costanza[*1]

[*] INFN, Sezione di Pavia, Via A. Bassi 6, 27100 Pavia, Italy

E-mail: `Susanna.Costanza@pv.infn.it`

Abstract. A first experiment using a polarised ^{3}He target has been carried out in Mainz at the MAMI accelerator, with the aim of investigate the Gerasimov-Drell-Hearn sum rule on the neutron. In fact, due to the spin structure of ^{3}He, the data obtained with the polarised ^{3}He target give a complementary and more direct access to the neutron compared to the existing data on the deuteron. Results of the unpolarised and helicity dependent total photoabsorption cross section $\gamma\,^3He \to X$, of the pion photoproduction channels $\gamma\,^3He \to \pi X$ and of the $\gamma\,^3He \to ppn$ channel, obtained in a photon energy range between 200 MeV and 500 MeV, will be presented.

1. The GDH sum rule

Since the beginning of the 1960s, a central issue of nuclear and particle physics has been the study of the internal structure of the nucleon, in particular of its spin structure, which is not so well understood as other nucleon properties.

For this study, of particular interest are sum rules, which connect information from all energies to fundamental parameters of the present interaction models.

The Gerasimov-Drell-Hearn sum rule [1, 2] is a good example of these rules: it relates the nucleon anomalous magnetic moment (AMM) κ, the spin S and the mass M of a nucleon to the integral over the weighted helicity difference of the total absorption cross section for circularly polarised photons on a longitudinally polarised nucleon target.

$$I_{GDH} = \int_{\nu_{th}}^{\infty} \frac{\sigma_p - \sigma_a}{\nu} d\nu = 4\pi^2\kappa^2 \frac{e^2}{M^2} S, \qquad (1)$$

where ν is the photon energy, and σ_p and σ_a denote the total absorption cross section for parallel and antiparallel orientation of photon and particle spins, respectively. The lower limit of the integral, ν_{th}, corresponds to pion production and photodisintegration threshold for a nucleonic and nuclear target, respectively.

Table 1 reports the magnetic moment μ and the AMM κ values for protons, neutrons, deuterons

[1] on behalf of the A2 Collaboration

Table 1. Magnetic moment μ and AMM κ (units: μ_N, nuclear magneton) for protons, neutrons, deuterons and ^{3}He nuclei.

	p	n	d	^{3}He
μ	2.79	-1.91	0.86	-2.13
κ	1.79	-1.92	0.65	498

and ^{3}He nuclei.

The GDH sum rule is derived from very general fundamental physical principles, in particular from the forward Compton scattering, the optical theorem and the low energy one.
In the past, there have been several attempts to find causes for a failure of the GDH sum rule. The only "weak" hypothesis is the assumption that the Compton scattering becomes spin-independent as the energy tends to infinity.
Possible explanations for this violation could be the exchange of a a_1-like meson between the photon and the nucleon or the non-pointlike quark structure.
The test of this relation provides a fundamental check of our knowledge of the γ-nucleon interaction, as well as of the physics of strongly interacting systems; it is also a check of the existing photo-reaction models. In addition, through the helicity dependent partial channels, it will be possible to access new observables and to study the baryonic resonances.

2. Theoretical predictions for the GDH on the nucleon

It is possible to estimate the GDH sum rule value for the nucleon by using a combination of multipole analyses of the available single pion photoproduction data (mostly from unpolarised experiments) [3, 4] and phenomenological models of multipion and heavy meson photoproduction reactions [5, 6, 7] up to $E_\gamma \simeq 2$ GeV. Above this energy, the contribution can be estimated from Regge-type approaches [8].
Table 2 reports the current theoretical estimates of the GDH sum rule values both for the proton and the neutron. Predictions for the $N\pi$ channel are from the SAID [3] and MAID [4] (within brackets) multipole analyses; the contribution for the $N\pi\pi$ channel are from [5]; the estimates for $N\eta$ are from [4], the kaon contributions are from [6] and the vector meson ones are from [7]; finally, the Regge estimates are from [8].
There is a disagreement between these theoretical predictions and the GDH sum rule value for the proton, while the neutron GDH value is roughly reproduced.
In addition, it is worth noting that according to the models, I_{GDH} for the proton and the neutron are roughly the same, while this is not the case for the GDH sum rule values.
In order to find out the reasons for this discrepancy, a precise measurement of the GDH integral for both the proton and the neutron is needed, as well as a systematic study of the partial channels, in particular of the N$\pi(\pi)$ ones, which give the dominant contribution to the GDH integral.

3. The GDH sum rule on the proton

The first experimental check of the GDH sum rule for the proton was performed by the GDH collaboration jointly at the Mainz and Bonn tagged photon facilities, where the I^p_{GDH} was experimentally evaluated in the photon energy range between 200 MeV and 2.9 GeV [9, 12].
The estimated value of the GDH integral is $211 \pm 5\,(stat) \pm 12\,(sys)$ μb. This value comes out from the combination of the experimental results, shown in figure 1, with the theoretical

predictions for the unmeasured energy ranges, as reported in table 3.
The obtained result supports the validity of the GDH sum rule for the proton, contrarily to the estimates given in table 2.
The discrepancy is mainly due to the oscillating photon energy dependence of the GDH integrand, because of the alternating sign of the multipole contributions.
In order to have a reliable prediction of the GDH integral, it is then needed a very high accurate model that has not been reached yet.

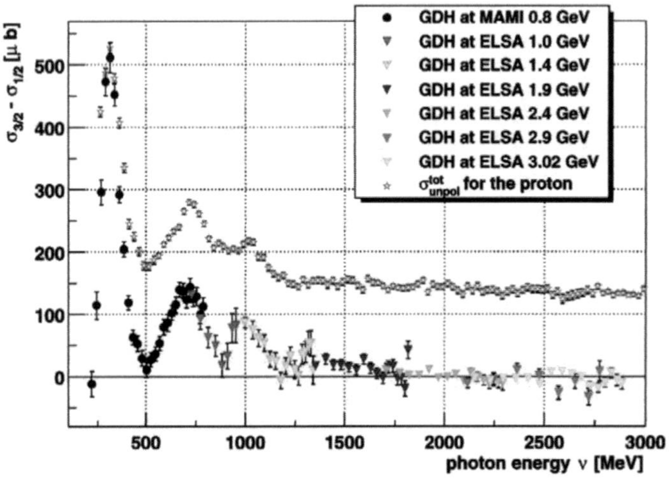

Figure 1. Helicity dependent cross section on the proton.

4. The GDH sum rule for the neutron

In the neutron case, the experimental verification of the GDH sum rule is complicated by the lack of free neutron targets.
In order to compensate for this lack, deuteron and ^{3}He targets can be used. In both cases, nuclear structure effects and final state interactions will prevent the direct access to the cross section of the "free nucleon" and the evaluation of the free neutron contribution will be model

Table 2. Theoretical estimates of the GDH sum rule values for the proton and the neutron.

	I_{GDH} proton	I_{GDH} neutron
$\gamma N \to N\pi$	172 [164]	147 [131]
$\gamma N \to N\pi\pi$	94	82
$\gamma N \to N\rho$	-8	-6
$\gamma N \to K\Lambda(\Sigma)$	-4	2
$\gamma N \to N\rho(\omega)$	0	2
Regge contribution ($E_\gamma > 2$ GeV)	-14	20
TOTAL	$\sim$ 239 [231]	$\sim$ 247 [231]
GDH sum rule	204	233

dependent. On the other hand, the comparison between the two different free neutron values extracted from both targets and the usage of different nuclear models will play a crucial role in constraining the theoretical analyses and will give a fundamental cross check of the reliability of the extraction procedures.

4.1. Deuteron targets as free neutron substitutes

Deuterons are systems consisting of one proton and one neutron with paired spins, with the 96% of probability to be in relative S state. Hence the magnetic moment of the deuteron is the sum of the magnetic moment of the proton and of the neutron:

$$\mu_d \approx \mu_p + \mu_n.$$

The results of the helicity dependent total inclusive photoabsorption cross section obtained by the GDH collaboration at Bonn [15] and at Mainz [14] are shown in figure 2 and compared to a nuclear model proposed by Arenhövel, Fix and Schwamb (AFS) [16].
Above 700 MeV the AFS model overestimated the data: this could be due to a poor treatment of the double pion photoproduction channels, whose contribution becomes relevant at high energies. The experimental value of I^d_{GDH} between $E_\gamma = 200$ MeV up to $E_\gamma = 1.8$ GeV is $452 \pm 9\,(stat) \pm 24\,(sys)$ μb.
In a partial wave analysis approach, assuming that the incoherent, quasi-free meson production processes dominate in the measured energy region, the GDH integral above the pion production threshold can be approximated by the sum of the proton and neutron contribution, weighted on effective degrees of proton and neutron polarisation p^d_p and p^d_n, respectively:

$$I^d_{GDH}|_{\nu > m_\pi} \sim p^d_p \cdot I^p_{GDH} + p^d_n \cdot I^n_{GDH} \tag{2}$$

where $p^d_p = p^d_n = 0.93$ as evaluated by [13] when taking into account all nuclear wave function components.
This formula was used to extract a rough value of the GDH integral for the free neutron from the measured values of the GDH integral for the deuteron and for the proton in the energy region between 0.2 and 1.8 GeV. In this way, a value $I^n_{GDH}\,[0.2 - 1.8\,\mathrm{GeV}]$ comparable with the GDH sum rule prediction for the neutron ($\sim 233\,\mu$b) was found.
This estimation is quite crude, since the low ($E_\gamma < 200$ MeV) and very high ($E_\gamma > 1.8$ GeV) are missing in the available data. In addition, there are also other channels ($\gamma d \to \pi^0 d$ and $\gamma d \to pn$) which contribute to I^d_{GDH}. Unfortunately, these channels cannot be treated in the quasi-free approximation; on the other hand, their contribution is quite small compared to the

Table 3. Contribution of the different energy regions to the GDH integral I^p_{GDH} on the proton.

E_γ (MeV)	I^p_{GDH} (μb)
≤ 0.2 (theoretical)	-28.5 ± 2
$0.2 - 0.8$ (measured)	$226 \pm 5 \pm 12$
$0.8 - 2.9$ (measured)	$27.5 \pm 2 \pm 1.2$
≥ 2.9 (theoretical)	-14 ± 2
Total	$211 \pm 5 \pm 12$
GDH sum rule	204

incoherent pion production.
Furthermore, the reactions on the deuteron are affected by its internal structure, so these experimental results represent a good test of our knowledge and understanding of nuclear dynamics.

4.2. ^3He targets as free neutron substitutes
^{3}He is a system consisting of two protons with spins paired off and an "active" unpaired neutron, in relative s state with 90% probability. As a result, the contribution of the two protons to the magnetic moment of ^{3}He cancels off and it can be approximated with the magnetic moment of the neutron:

$$\mu_{^3He} \approx \mu_n.$$

So it is clear that the spin structure function of ^{3}He is much closer to the free neutron than the deuteron one. Therefore, it is expected that the measured GDH integral function for ^{3}He above the pion production threshold will be a good approximation of the GDH integral value for the neutron. In a PWA approach, the following formula can be used:

$$I^{^3He}_{GDH}|_{\nu > m_\pi} \approx -2 \cdot p_p^{^3He} \cdot I^p_{GDH} + p_n^{^3He} \cdot I^n_{GDH}, \tag{3}$$

where $p_p^{^3He} = -0.026$ and $p_n^{^3He} = 0.87$ are the effective degrees of proton and neutron polarisation, respectively.
Since the proton contribution to the measured helicity dependent yields is much smaller for ^{3}He than for the deuteron, it can be clearly seen that the most accurate evaluation of I^n_{GDH} will come from ^{3}He, which has been used at MAMI as the substitute for polarised neutron target.
While the GDH integral for ^{3}He is expected to have approximately the same value of the GDH integral for the neutron above the pion production threshold, this is not true if the photodisintegration energy region is taken into account in the integral, as comes out from the generalized formula of the GDH sum rule for nucluei:

$$I_{GDH}\left(^3He\right) = 4\pi^2 S \left(\frac{1}{S}\mu - \frac{Q}{M}\right)^2 = \int_{\nu_{th}}^{\infty} \frac{\sigma_{3/2}(\nu) - \sigma_{1/2}(\nu)}{\nu} d\nu, \tag{4}$$

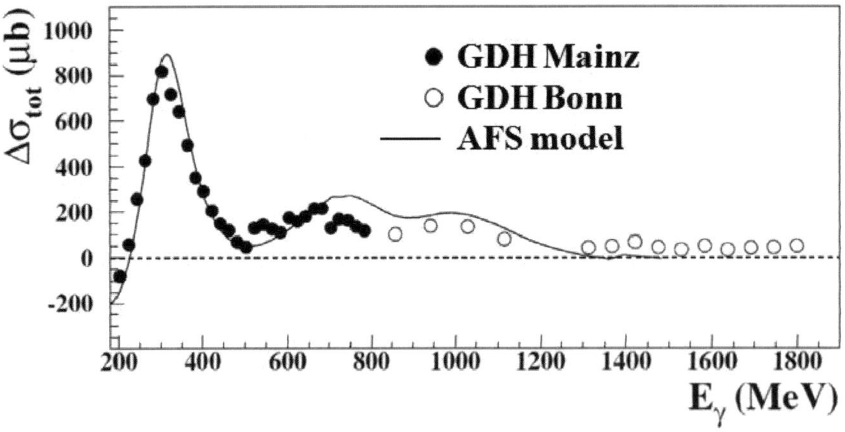

Figure 2. The helicity dependent total inclusive photoabsorption cross section obtained in [14] (full circles) and in [15] (open circles), compared to the theoretical predictions of AFS (Arenhövel, Fix and Schwamb) [16].

which is expected to be of the order of 498 μb, $\gg I^n_{GDH}$. This means that the contribution from the two-body break up to the pion production threshold is quite large, in order to satisfy the predicted value of 498 μb. Precise experimental data are clearly required from the photodisintegration threshold upwards.
The Mainz tagged facility cannot access such low energies: MAMI can perform investigations beyond the π production threshold. The direct measurement of $I_{GDH}(d)$ and $I_{GDH}(^3He)$ from the break-up threshold region up to about 60 MeV is planned at the new upgraded HIγS facility, of the TUNL laboratory (Durham NC, USA) [17].
The results obtained from both measurements will provide useful information to better understand the GDH sum rule integrand for ^{3}He and to extract from this the neutron information.

5. Helicity dependent results on ^{3}He
A first experiment for the measurement of the double polarised photoabsorption cross section on ^{3}He in the $\Delta(1232)$ baryon resonance region has been performed at the tagged photon facility of the MAMI accelerator in Mainz, during July 2009.
This experiment was carried out with a circularly polarised photon beam produced via bremmstrahlung process of a longitudinally polarised electron beam.
In the following, some preliminary results from runs limited to 525 MeV will be shown, as a function of the incoming photon energy up to 500 MeV.

5.1. The MAMI experimental setup at Mainz
The experiment was carried out at the tagged photon facility of the MAMI accelerator in Mainz. Circularly polarized photons were obtained by bremsstrahlung of longitudinally polarized electrons with an average polarization of 75% [18]. The bremsstrahlung photons were tagged using the Glasgow-Mainz magnetic spectrometer with an energy resolution of about 2 MeV [19, 20]. The relative tagging efficiency was monitored throughout the experiment by a CCD photon camera and absolute measurements were made regularly using a lead-glass detector.
The reaction products were detected by the central detector system, consisting of the Crystal Ball (CB) NaI spectrometer, complemented by the Multi-Wire Proportional Chambers (MWPCs), used to identify and track the charged particles, and the cylindrical Particle Identification Detector (PID), used to distinguish the charged from the neutral particles detected by the CB. The combined information provided by these three detectors provides accurate energy, angle and particle identification in the azimuthal (ϕ) and polar (θ) angular regions from 0° to 360° and from 21° to 159°, respectively.
In addition, the TAPS spectrometer provides information about the particles flying in the forward direction, outside the acceptance region of CB, but it has not been included in the data analysis that will be presented in the following.
The polarised ^{3}He gas target has been produced at the Physics Institute of the Mainz University as a substitute for a polarised neutron target. The gas has been polarised via the Metastability Exchange Optical Pumping (MEOP) method [21].
The polarisation process starts by applying a weak gas discharge to the gas sample maintained at a pressure of 0.8-1 mbar: this induces a small fraction of the atoms ($\sim 10^{-6}$%) to reach the metastable state $2\,^3S_1$.
In the presence of a non-null magnetic field, the metastable state is split by the fine and hyperfine interactions in two sublevels with quantum numbers $m_F = \pm 1/2$.
The absorption of right-handed circularly polarised laser light with $\lambda = 1083$ nm causes then the transition to the $2\,^3P_{0,1,2}$ states. According to the angular momentum conservation laws, only transitions with $\Delta m_F = 1$ are allowed.
Then, after excitation, a spontaneous remission to both sublevels of the metastable occurs.

Journal of Physics: Conference Series **349** (2012) 012011 doi:10.1088/1742-6596/349/1/012011

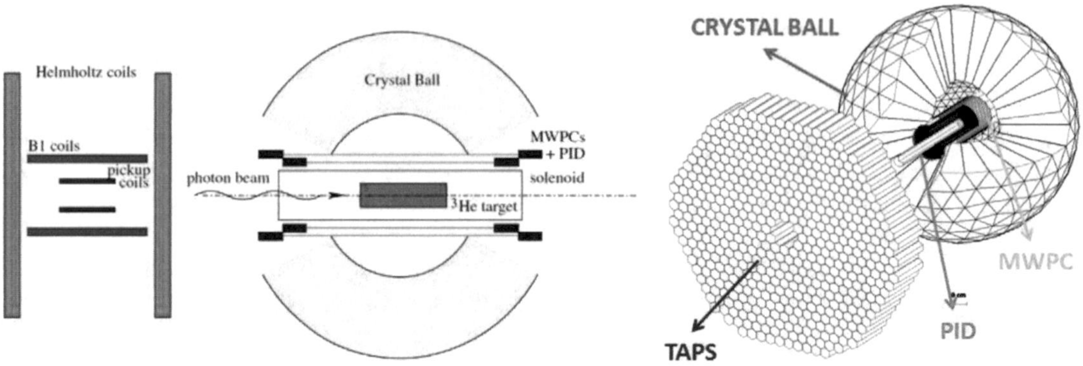

Figure 3. Side view of the MAMI experimental setup used for the ^{3}He measurements (left); sketch of the detectors (right): the Crystal Ball (CB), MultiWire Proportional Chambers (MWPCs), Particle Identification Detector (PID) and TAPS subdetectors are shown.

However, the continue depopulation of the $2\,^3\mathrm{S}_1(m_F = -1/2)$ state results in a higher population of the $2\,^3\mathrm{S}_1(m_F = 1/2)$ sublevel.
Finally, the nuclear polarisation of the metastable state is transferred to the unpolarised ground state atoms via atomic collisions, during which they exchange the excitation of their atomic electrons without altering the nuclear spins of the atoms involved:

$$^3\vec{\mathrm{He}}(2^3\mathrm{S}_1) + {}^3\mathrm{He}(1^1\mathrm{S}_0) \rightarrow {}^3\mathrm{He}(2^3\mathrm{S}_1) + {}^3\vec{\mathrm{He}}(1^1\mathrm{S}_0).$$

As a result, one obtains a ground state with polarised nucleus and a new metastable atom can be further optically pumped by the laser.
After the polarisation process, the gas is compressed with minimal polarisation losses to the desired pressure by a non-magnetic piston.
The target cell (figure 4) is cylindrical, with an outer diameter of 6 cm and a total length of 20 cm; it is made from quartz glass with two 50 μm thick titanium foils as entry and exit windows for the photon beam. This material was chosen since it provides the necessary gas tightness and give an acceptably long relaxation time of the gas polarisation.
The target cell is then inserted inside the CB detector, where the polarisation alignment is maintained by a solenoid inside a region with a very low magnetic field gradient. The measurement of the polarisation was carried out via a NMR technique, with an external magnetic field produced in the z-direction by Helmholtz coils, providing a quantization axis along which the ^{3}He spins are oriented. The setup used is the one shown in figure 3.

Under these conditions, with a gas pressure of 5 bar, the ^{3}He gas target density is relatively low ($N_T \sim 2.5 \cdot 10^{21}$ cm^{-2}) compared to that of a solid or liquid target (about 100 times less). Despite this, the ^{3}He gas target is pure, so it has a greater fraction of polarised neutron with respect to the deuterated butanol case.
The plot shown in figure 5a) shows the z-vertices of the emitted charged hadrons, as reconstructed by using the MWPCs information. Around -100 and +100 mm the effect of the target cell windows is clearly visible and much bigger than the ^{3}He contribution. The influence of the windows cancels exactly when taking into account the $(N_P - N_A)$ difference, where $N_{P(A)}$ is the number of experimental events with parallel and antiparallel γ-^{3}He spin configuration. As shown in figure 5b), in this case the only contribution from the polarised ^{3}He gas gives rise to a non-zero count rate in the target region.

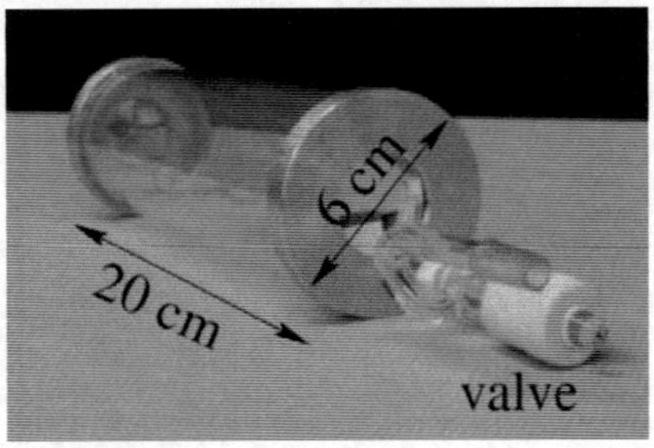

Figure 4. The ^{3}He target cell used at MAMI.

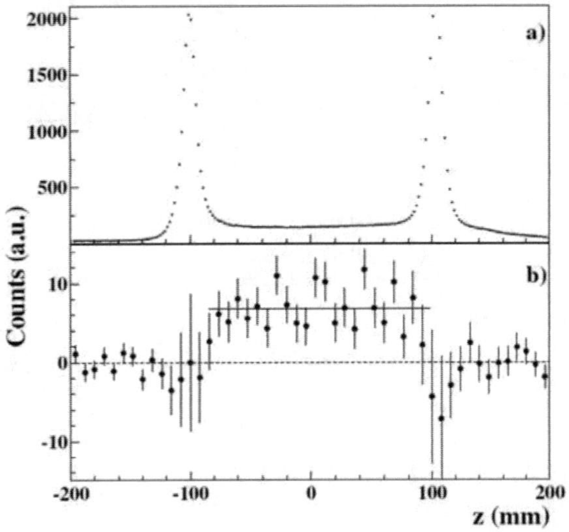

Figure 5. a) Distribution of the z-vertices of all charged particle events, with tracks reconstructed with the MWPCs; b) same as in a) but for the difference between the number of events emitted with parallel or antiparallel $\gamma-{}^3$He spin configuration. The solid line is the best fit to a constraint constant function.

6. Data Analysis

In the following, the obtained unpolarised and helicity dependent photoabsorption cross sections on ^{3}He will be showed. In particular, the total inclusive cross sections will be presented, as well as the results for the partial channels $\gamma\,^3He \to \pi^{\pm}X$, $\gamma\,^3He \to \pi^0 X$ and $\gamma\,^3He \to ppn$. In addition, a comparison between the total inclusive and the sum of partial channels will be shown.

Concerning the total inclusive cross section, the analysis has been performed without partial channels separation; concerning the semi-inclusive channels, the π/p separation has been performed through the dE/dx vs E technique, using PID and MWPCs information, while the π^0 have been identified by the $\gamma\gamma$ invariant mass plot, using only CB information.

Finally, concerning the *ppn* partial channel, these events have been separated by the $ppn\pi^0$ events using missing mass and missing energy plots.

6.1. Unpolarised cross section

Figure 6 shows the preliminary results obtained for the unpolarised total inclusive photoabsorption cross section (blue circles) as a function of the incoming photon energy. These results are compared to the experimental data previously published by the DAPHNE Collaboration. The good agreement between the two different data sets gives confidence in the applied data analysis procedure.

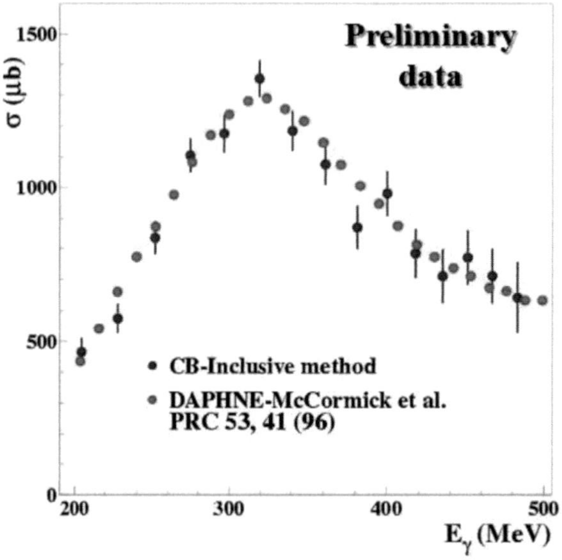

Figure 6. The unpolarised total inclusive photoabsorption cross section on ^{3}He (blue circles) is compared to the unpolarised results obtained by the DAPHNE collaboration (red circles) [22].

Concerning the unpolarised cross sections for the partial channels, the preliminary results for the $\pi^{\pm}X$ and π^0X channels are shown in figure 7, where the data are compared to the predictions from the Fix model (full line) and from the MAID multipole analysis (dotted line) [23].
In the MAID-based empirical model, the ^{3}He prediction is evaluated simply assuming that the protons and the neutron inside ^{3}He behave as free nucleons.
On the contrary, the Fix semi-empirical model [24] accounts also for the Fermi motion of the interacting nucleons inside the nuclei and the nuclear effects, like the Final State Interactions (FSI). These effects result in a damp and broadening of the peak corresponding to the Δ-resonance, with respect to the free nucleons MAID estimation.
The data are in a good agreement with the Fix model in the case of the charged partial channels, whereas there is a sizeable discrepancy for the neutral channel, where a bigger contribution is due to nuclear structure effects, which are more difficult to be reproduced.
The first preliminary results for the $\gamma\,^3He \rightarrow ppn$ partial channel are shown in figure 8.
In this case, the model predictions come from the so-called "quasi-deuteron" approximation, as if the photon interacts only with a proton and a neutron paired in a virtual deuteron inside the ^{3}He nucleus, with the remaining spectator proton.
In this approximation, the cross section for ^{3}He can be obtained by multiplying the cross section for the deuteron, estimated with the Schwamb model [25], by the factor 1.68 [26], that takes into account the number of "quasi-deuterons" contained in the ^{3}He nucleus.
From the plot, there is a clear discrepancy between the experimental data and the "quasi-deuteron" model; this is mainly due to the fact that this model does not take into account the

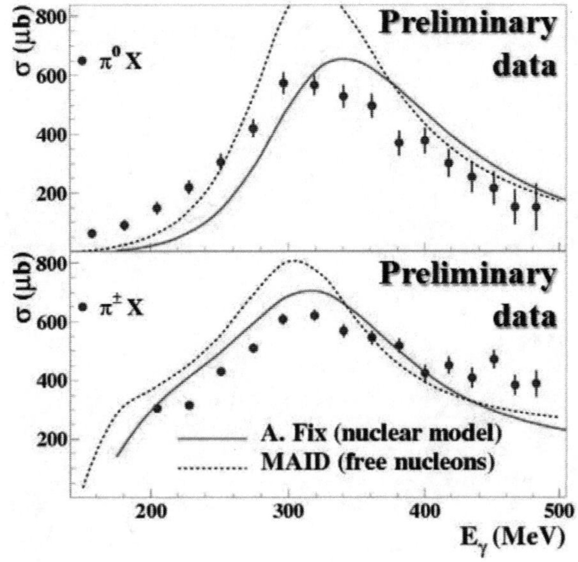

Figure 7. Preliminary results of the unpolarised total cross sections for the semi-inclusive channels $\gamma\,^3\mathrm{He} \rightarrow \pi^0 X$ (upper plot) and $\gamma\,^3\mathrm{He} \rightarrow \pi^\pm X$ (lower plot). The experimental data (blue circles) are compared to the Fix (full line) and MAID (dotted line) models.

three body effects, which are to contribute to this channel for about 30% [27].

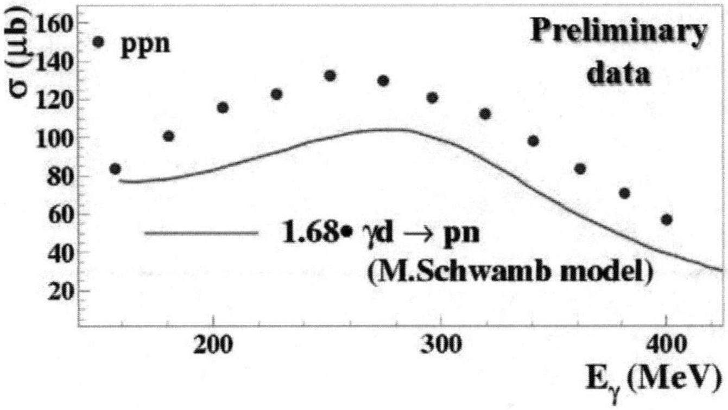

Figure 8. Preliminary unpolarised total cross section for the partial $\gamma\,^3\mathrm{He} \rightarrow ppn$ channel. The experimental data (blue circles) are compared to the "quasi-deuteron" approximation (full line) evaluated from the Schwamb model for the deuteron.

Finally, if we compare the total inclusive cross section to the sum of the partial channels (figure 9), a good agreement is obtained, demonstrating that all different analyses have been performed in a proper and consistent way.

6.2. Helicity dependent cross section on ^3He

In the polarised cross section case, the same channels as before (both the total inclusive and the partial ones) will be presented in the following.

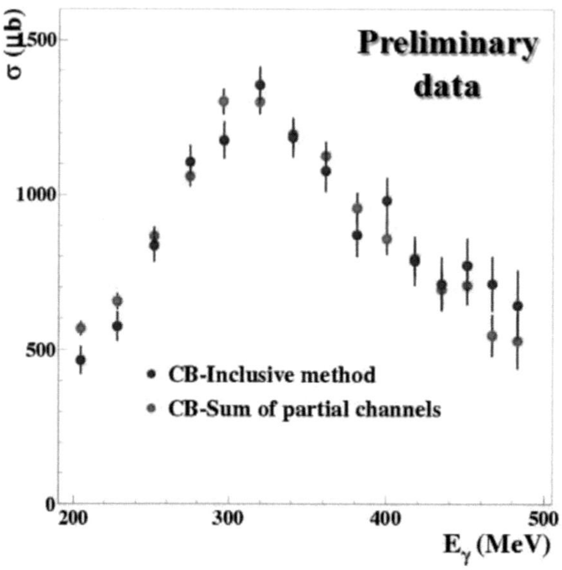

Figure 9. Comparison between the unpolarised $\gamma\,^3$He total cross sections obtained with the inclusive analysis (blue circles) and as sum of the partial/semi-inclusive channels (red circles).

The plot in figure 10 shows the preliminary results for the helicity dependent cross section difference $\Delta\sigma$ on ^{3}He as a function of the incoming photon energy. The data are compared to the prediction of two models: the red one is a combination of the Fix and Schwamb contributions, in order to take into account both the nuclear effects and the FSI and the "quasi-deuteron" description of the *ppn* partial channel, respectively. The other prediction in the plot (green line) is that of the MAID-based model, where the nucleons inside the nucleus are assumed to be free and only the effect of the spin alignments are taken into account. In this framework, the prediction is evaluated as:

$$\Delta\sigma_{^3He} = 0.87 \cdot \Delta\sigma_n - 2 \cdot 0.026 \cdot \Delta\sigma_p, \quad (5)$$

where the numbers are the effective nucleon polarisations and $\Delta\sigma_{n(p)}$ are the free neutron and proton helicity dependent $N\pi$ cross sections obtained from the MAID model.
The agreement between the experimental data and these empirical models is reasonable, taking into account the non-negligible statistical experimental errors.
In figure 11, the total helicity dependent cross section difference $\Delta\sigma_\pi = \sigma_{\pi,a} - \sigma_{\pi,p}$ for the $\gamma\,^3He \rightarrow \pi^0 X$ (upper plot) and $\gamma\,^3He \rightarrow \pi^\pm X$ channels (lower plot) are shown and compared to the corresponding model predictions from Fix and MAID.
It seems that in these cases the nuclear effects are not so important as for the unpolarised case. Both models fail to predict the measured shape of the $\pi^0 X$ channel, while they better reproduce the $\pi^\pm X$ channels, especially in the lower measured energy range.
As for the unpolarised case, the *ppn* partial channel described in the "quasi-deuteron" approach is shown in figure 12. For this channel, the same considerations as for the corresponding unpolarised channel hold.

6.3. Estimation of the GDH integral for the neutron

From the results presented in the previous sections, it is clear that the present models cannot be used for a reliable extraction of the free neutron information. Nevertheless they can provide

Journal of Physics: Conference Series **349** (2012) 012011 doi:10.1088/1742-6596/349/1/012011

a very rough neutron extraction method, as shown in figure 13.
The plot in the upper part is the helicity dependent total cross section difference as a function of the photon energy obtained with the ^{3}He measurement, compared to the results obtained by the GDH collaboration for the proton and the deuteron.
From these results, the "free neutron" contribution can be extracted in a very simple way, both from the ^{3}He and from the deuteron data, disregarding nuclear effects and taking into account only the effects of the different spin alignment of the nucleons.
According to equations (2) and (3), the "free neutron" contribution can be obtained with the following approximations for the deuteron:

$$\Delta\sigma_{neutron} = \frac{\Delta\sigma_{deuteron}}{0.92} - \Delta\sigma_{proton} \tag{6}$$

and for ^{3}He:

$$\Delta\sigma_{neutron} = \frac{\Delta\sigma_{^3He} + 2 \cdot 0.026 \cdot \Delta\sigma_{proton}}{0.87}, \tag{7}$$

where the numerical factors in the previous equations are the effective nucleon polarisation coefficients, as explained in section 4.
The comparison of the "free neutron" extracted contributions is shown in the lower plot of figure 13. The results obtained with this simple model are then compared to the MAID model for the sum of all the single pion photoproduction channels on the neutron ($\gamma n \to N\pi$).
Given the very rough nature of the extraction method, we can say that there is a good agreement between the two "free neutron" extracted data.
Comparing the upper and lower plots, it is easy to see that the difference between the helicity dependent total photoabsorption cross section difference for the deuteron and the "free neutron" contribution extracted from this cross section is bigger than the corresponding difference for ^{3}He. This is due to the fact that the proton contribution to the helicity dependent cross section

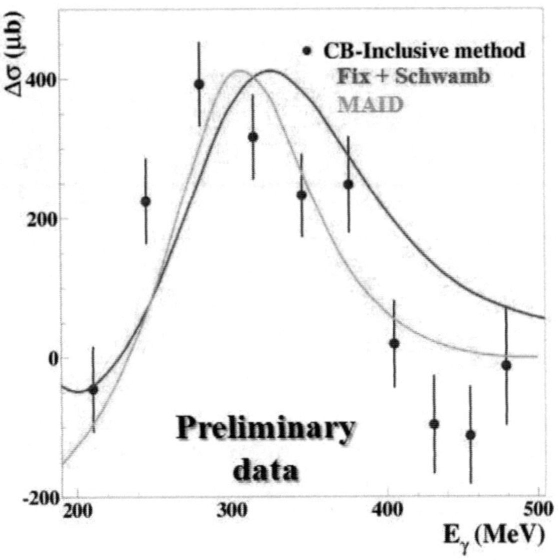

Figure 10. Preliminary results of the helicity dependent total inclusive cross section $\Delta\sigma_{tot}$ on ^{3}He, compared to the theoretical predictions of the Fix and Schwamb combined model (red line) and of the MAID model (green line).

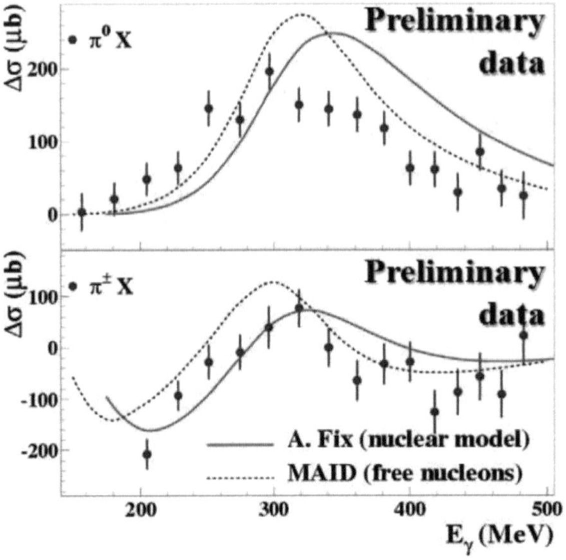

Figure 11. Preliminary results for the helicity dependent total cross section $\Delta\sigma$ for the semi-inclusive channels $\gamma\,^3\mathrm{He} \to \pi^0 X$ (upper plot) and $\gamma\,^3\mathrm{He} \to \pi^{\pm} X$ (lower plot). The experimental data (blue circles) are compared to the Fix (full line) and MAID (dotted line) models.

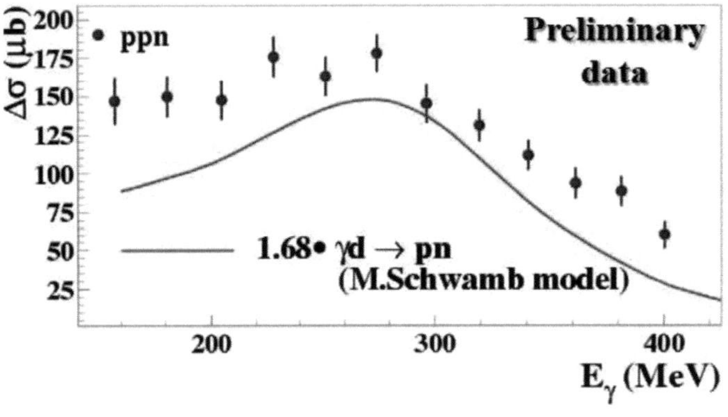

Figure 12. Preliminary results for the helicity dependent total for the partial $\gamma\,^3\mathrm{He} \to ppn$ channel. The experimental data (blue circles) are compared to the "quasi-deuteron" approximation (full line) evaluated from the Schwamb model for the deuteron.

difference is much smaller for ^{3}He than for the deuteron, because of the spin structure of ^{3}He. This means that photoabsorption on polarised ^{3}He targets is a very promising alternative method to investigate the GDH integral for the neutron.

7. Conclusions

Preliminary results on the unpolarised and helicity dependent total photoabsorption cross section on ^{3}He have been obtained from the analysis of the data collected at MAMI (Mainz) in the energy region $200 < E_\gamma < 500$ MeV.

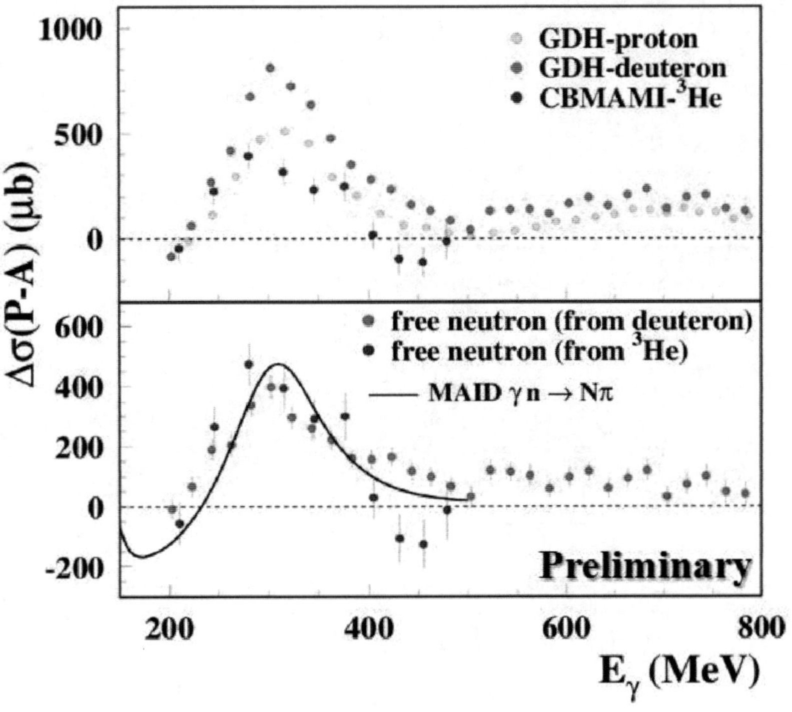

Figure 13. The upper plot shows the helicity dependent total cross section difference on ^{3}He as a function of the energy of the incoming photon (blue circles); the data are compared to the experimental results obtained by the GDH collaboration on the proton (green circles) and the deuteron (red circles). The lower plot is a comparison between the "free neutron" contribution extracted from ^{3}He (blue circles) and from the deuteron (red circles).

This first experimental test has provided unprecedented data of this kind and preliminary results proving the feasibility of the use of polarised ^{3}He gas target to check the GDH sum rule on the neutron.
The unpolarised total inclusive photoabsorption cross section on ^{3}He ($\gamma\,^3He \to X$) has been compared to previous data obtained with the DAPHNE detector. The good agreement found demonstrates the good quality of the data and of the analysis method.
The unpolarised and polarised cross sections for the partial channels $\gamma\,^3He \to \pi^{\pm}X$, $\gamma\,^3He \to \pi^0X$ and $\gamma\,^3He \to ppn$ in the Δ resonance region have also been measured for the first time.
The present theoretical models are not able to describe in a satisfactory manner the experimental results. They provide additional constraints for nuclear and subnuclear models, as well as for improved theoretical calculations, which are needed for a better description of the considered processes.
Further measurements to improve statistics and investigate a wider energy range are needed.

References

[1] Gerasimov S B 1966 *Sov. J. Nucl. Phys.* **2** 430
[2] Drell S D and Hearn A C 1966 *Phys. Rev.* **96** 1428
[3] Arndt R A et al 2002 *Phys. Rev.* C **66** 055213
[4] Drechsel D et al 2001 *Phys. Rev.* D **63** 114010
[5] Fix A and Arenhövel H 2005 *Eur. Phys. J.* A **25** 115

[6] Sumowigado S and Mart T 1999 *Phys. Rev.* C **60** 028201
[7] Zhao Q, Al-Khalili J S and Bennhold C 2002 *Phys. Rev.* C **65** 032201
[8] Bianchi N and Thomas E 1999 *Phys. Lett.* B **450** 439
[9] Ahrens J et al 2000 *Phys. Rev. Lett.* **84** 5950
[10] Ahrens J et al 2001 *Phys. Rev. Lett.* **87** 022003
[11] Dutz H et al 2003 *Phys. Rev. Lett.* **91** 192001
[12] Dutz H et al 2004 *Phys. Rev. Lett.* **93** 032003
[13] Rondon O A 1999 *Phys. Rev.* C **60** 035201
[14] Ahrens J et al 2009 *Phys. Lett.* B **672** 328
[15] Dutz H et al 2005 *Phys. Rev. Lett.* **94** 162001
[16] Arenhövel H, Fix A and Schwamb M, 2004 *Phys. Rev. Lett.* **93** 202301
[17] Weller H et al 2009 *Prog. Part. Nucl. Phys.* **62** 257
[18] Aulenbacher K et al 1997 *Nucl. Instrum. Meth.* A **391** 498
[19] Anthony I et al 1991 *Nucl. Instrum. Meth.* A **301** 230
[20] Hall S J et al 1996 *Nucl. Instrum. Meth.* A **368** 698
[21] Colegrove F et al 1963 *Phys. Rev.* **132** 2561
[22] Mc-Cormick et al 1996 *Phys. Rev.* C **53** 41
[23] Drechsel D et al 2007 *Eur. Phys. J.* A **34** 69
[24] Fix A Private Communication
[25] Schwamb M and Arenhövel H 2001 *Nucl. Phys.* A **690** 682
[26] d'Hose N et al 1989 *Phys. Rev. Lett.* **63**
[27] Audit G et al 1997 *Nucl. Phys.* A **614** 461

Journal of Physics: Conference Series **349** (2012) 012012 doi:10.1088/1742-6596/349/1/012012

Status and perspectives in strangeness photoproduction

T C Jude
Physikalisches Institut, Universität Bonn
E-mail: jude@physik.uni-bonn.de

Abstract.
A review of recent results in strangeness photoproduction is presented, in preparation for experimental proposals for the new BGO-OD experiment at the ELSA facilty, Bonn. Cross section measurements and polarisation observables in $\gamma(p, K^+)\Lambda$ are discussed, and the relation to these measurements and what can be gleaned from neutral kaon photoproduction. Recent cross section data for $\gamma(p, K^0)\Sigma^+$ and $\gamma(n, K^0)\Lambda$ is presented, and measurement of the $\Lambda(1405)$ lineshape from $\Sigma\pi$ decays and comparison to theory. Hypernuclei are introduced, and what can be acheived using a real photon beam.

1. Introduction

Over the last 20 years, increases in electron beam energy and intensity at accelerator facilities has markedly increased statistics for the photoproduction of mesons and baryons with non-zero strangeness. The intention of this paper is to highlight recent progess and areas where further research is required. This is expected to form the basis of experimental proposals in open strangeness for the newly commissioned BGO-OD experiment [1] at the ELSA accelerator facility [2], Bonn. The experiment uses a real, bremstrahlung photon beam to an energy of 3.5 GeV, which impinges upon a static target at the centre of the BGO Ball. The BGO Ball, a highly segmented calorimeter, covers polar angles from 155^0 to 25^0 degrees. A momentum spectrometer, consisiting of a series of tracking detectors and a large magnetic field covers more forward angles. The excellent momentum reconstruction makes the experiment ideally suited to strangeness photoproduction, where for example, t-channel exchange is expected to be dominant, and so acceptance at forward centre of mass angles is crucial in understanding the photoproduction mechanism. For further details see [3].

The reviews in this paper are not exhaustive, rather it is intended to provide a platform for further discussion. The paper is organised as follows. Progress in charged strangeness photoproduction is presented in section 2. Comparison and relation to neutral kaon photoproduction and discussion of recent results is in section 3. Section 4 describes theory and measurement of the $\Lambda(1405)$. Section 5 briefly describes progress in hypernuclei research and what can could be achieved using a real photon beam as a production mechanism.

2. $K^+\Lambda$ photoproduction

Due to the absence of analytical solutions to QCD, constituent quark models have often been used to describe the photoproduction mechanism. Many models have predicted resonance states however which have not been observed in the excitation spectrum of non-strange photoproduction channels. It has been argued that this may be that the degrees of freedom that have been used are incorrect (for example, di-quark models have been proposed, where there are two tightly bound constituent quarks with antisymmetric wave functions [4]), or alternatively that experiments have been insensitive to these resonance contributions. Capstick and Roberts [5] suggested that many so called "missing resonances" may couple strongly to photoproduction channels with open strangeness.

The photoproduction mechanism can be described as a series of transversity or helicity amplitudes. An understanding of the relative phases and magnitudes allows a complete, model independent measurement. These amplitudes can be extracted from combinations of the polarisation of the beam, target and recoil nucleon. A judiciously selected set of these polarisation observables allows a complete measurement of all amplitudes and phases [7].

There is currently a global effort for complete measurements of photoproduction channels, and $\gamma(p, K^+)\Lambda$ offers one of the best possibilities. The pseudoscaler K^+ with zero spin limits the required number of polarisation observables (compared to vector mesons), and the self analysing weak decay of the Λ allows access to the recoil nucleon polarisation via the distribution of decay particles (to determine the polarisation of a recoil proton, for example, requires the secondary scattering of the proton from material surrounding the reaction vertex).

The search for missing resonances and the goal of a complete model independent measurement have provided motivation for the most extensively measured strangeness channel, $\gamma(p, K^+)\Lambda$. Differential cross section data with high statistics have been measured with the SAPHIR detector [8, 9] at ELSA and CLAS detector [10, 12] at Jefferson Lab. Discrepancies between the SAPHIR and CLAS data (fig. 1) however leave ambiguities to the contributing s-channel resonances. The most notable difference is around the structure at approximately $\sqrt{s} = 1.9$ GeV. Different partial wave analyses, constituent quark models and isobar models cannot agree which resonances contribute. This is despite the extensive data set of polarisation observables (see below). For example, an isobar model of Mart and Bennhold predicted a contribution from $D_{13}(1900)$ [13]. A model of Saghai suggested that $D_{13}(1900)$ was not required to fit to data if $P_{01}(1810)$ and $P_{03}(1890)$ hyperonic resonances were included. A coupled channel approach by Julia-Diaz *et al.* (where intermediate states, $\gamma p \to \pi N \to KY$ are also included), suggested large contributions from S_{11} states.

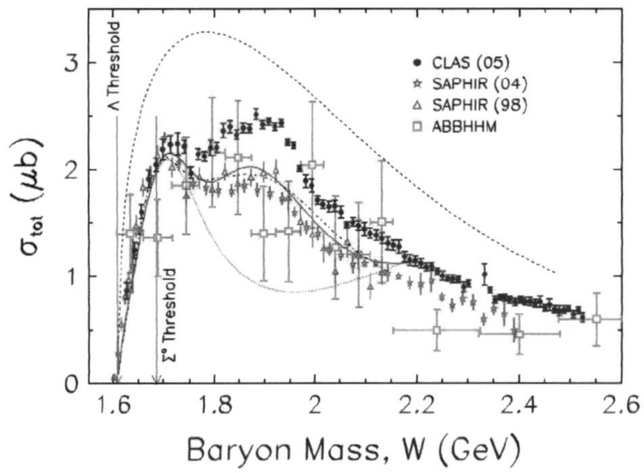

Figure 1. $\gamma(p, K^+)\Lambda$ cross section data from CLAS (blue points) [10] and SAPHIR (red points) [8, 9] detectors. Fitted models are referenced in [10].

Data will soon be published from the Crystal Ball experiment at MAMI [6], which will add a third data set with high statistics and beam energy resolution, and will constrain the excitation spectrum further (fig. 2). This will increase statistical accuracy particularly near threshold, which is important for the testing of models based upon chiral effective Lagrangians ([18], for example). Due to the large strange quark mass, chiral symmetry breaking limits these models further than when applying them to non-strange final states. It will also allow the search for narrow structure, which previous data with beam energy resolutions of the order of 20 MeV would have been unable to resolve.

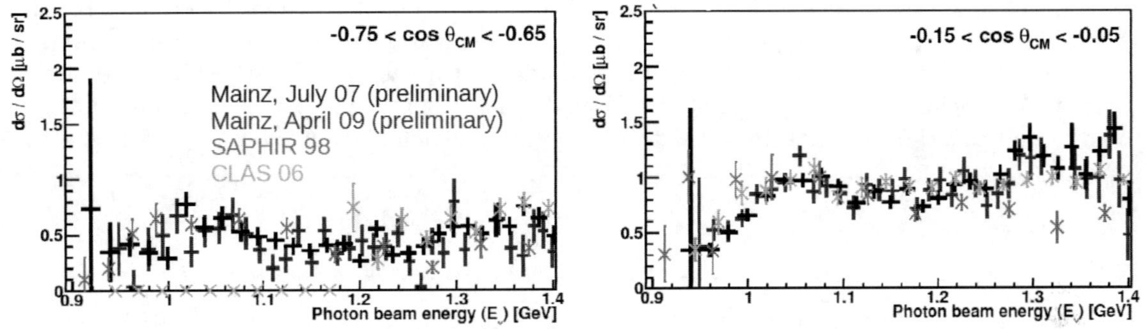

Figure 2. Preliminary differential cross section data for $\gamma(p, K^+)\Lambda$ measured at the Crystal Ball detector [6]. Dark blue and black points are data measured from two different data taking periods (inset). Red and light blue data were measured at SAPHIR [9] and CLAS [10] respectively. Angular range, θ_{CM}, is the centre of mass polar angle of K^+ detection.

Polarisation observables have been extensively measured for this channel. This includes, for example, beam asymmetry, Σ [15, 16], beam-recoil polarisation (C_Z and C_X) [17] (fig. 3(a)) and recoil polarisation, P_Λ [15, 16] (fig. 3(b)).

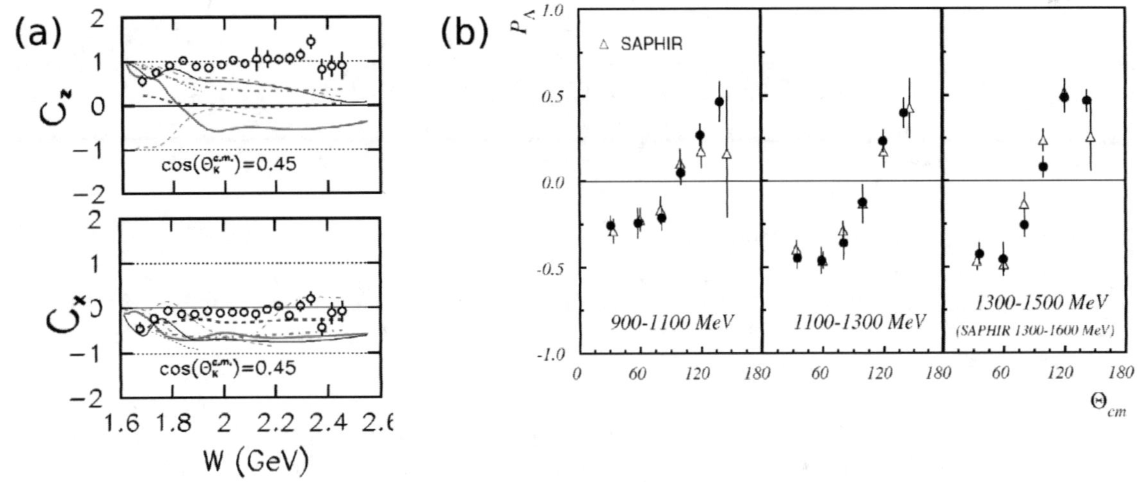

Figure 3. (a) Beam-recoil polarisation, C_Z (top) and C_X (bottom) for $\gamma(p, K^+)\Lambda$. Taken from, and model fits described in [17]. (b) Recoil polarisation, P for $\gamma(p, K^+)\Lambda$. Taken from [15].

Data from recoil nucleon polarisation exhibit unexpected behaviour. The magnitude of the hyperon polarisation, R_Λ, can be described by the recoil polarisation, P, and the beam-recoil polarisations, C_X and C_Z (eq. 1).

$$|R_\Lambda| \equiv \sqrt{P^2 + C_X^2 + C_Y^2} \tag{1}$$

R_Λ must be equal or less than one, with the polarisation shared between R_Λ and the relative orbital angular momentum of the K^+ and Λ. Fig. 4(a) [11] shows the surprising result that R_Λ is nearly one over all energies and angles. This would occur if K^+ and Λ were in a relative s-wave, however this was not predicted by any partial wave analysis before the measurement. It was suggested that this may hint at more fundamental dynamics of the photoproduction reaction [11]. Schumacher [11] suggested a "toy" model, where the incoming photon fluctuates to an $s\bar{s}$ pair and the s quark retains the full polarisation which is transfered to the Λ (fig. 4(b)).

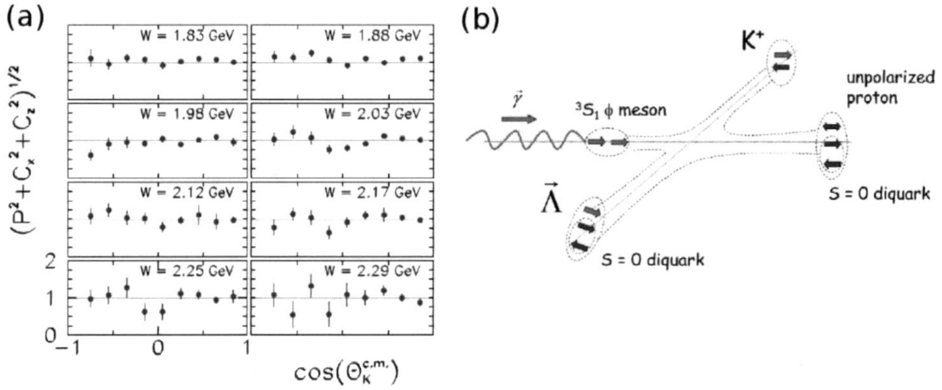

Figure 4. (a) Hyperon polarisation, R_Λ versus the cosine of the K^+ centre of mass polar angle for different centre of mass energies, W. Taken from [17]. (b) "Toy" model of polarisation transfer to the recoiling hyperon [11]. Description in the text.

3. Neutral kaon photoproduction

Determining s-channel resonances which contribute to the $\gamma(p, K^+)\Lambda$ spectrum is further complicated by large t-channel contributions. These contributions must be understood before a partial wave analysis of s-channel resonances can be performed. In this sense, neutral kaon photoproduction (for example, $\gamma(p, K^0)\Sigma^+$ and $\gamma(n, K^0)\Lambda$) are easier channels to understand. The t-channel exchange (fig. 5(c)) no longer contributes as the photon cannot couple to the neutral K^0. However, K^* t-channel exchange is still expected to contribute (with the photon coupling magnetically to the spin one vector meson, fig. 5(e)). As well as clarifying the contributions of s-channel resonances, neutral kaon photoproduction therefore also provides an insight into contributions from vector meson t-channel exchange.

Furthermore, charged and neutral strangeness channels are intrinsically related via SU(3) symmetry. Eq. 2 shows how this leads to a relation between hadronic coupling constants. For isobar models, where effective Lagrangians use a series of Feynman diagrams to fit to data sets, this can be extended to predict cross sections and polarisation observables of the unmeasured

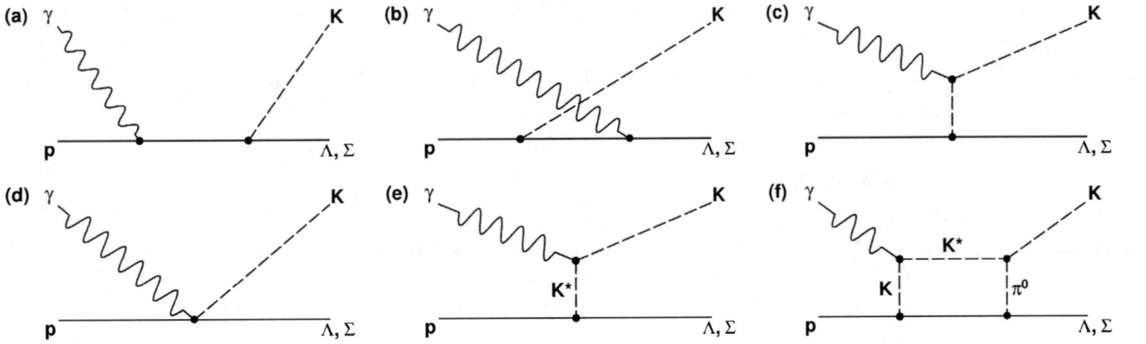

Figure 5. Diagrams contributing to charged and neutral strangeness photoproduction. Resonance contributions can contribute in s- and u-channels ((a) and (b)). t-channel pseudoscaler meson exchange, (c) and the contact term, (d) are proportional to the kaon charge. Vector meson t-channel exchange,(e) can occur in charged and neutral kaon photoproduction. Diagram (f) is sub-threshold K^* production, which couples to K^0 and the pion is reabsorped by the hyperon. See the text for details.

SU(3) partner. Fig. 6 demonstrates how an isobar model fitted to $\gamma(p, K^+)\Lambda$ data was used to determine hadronic coupling strengths and predict the $\gamma(n, K^0)\Lambda$ cross section [14].

$$\begin{aligned} g_{K^+\Lambda_p} &= g_{K^0\Lambda_n} \\ g_{K^+\Sigma^0_p} &= -g_{K^0\Sigma^0_n} \\ g^{V,T}_{K^{*+}\Lambda_p} &= g^{V,T}_{K^{*0}\Lambda_n} \end{aligned} \tag{2}$$

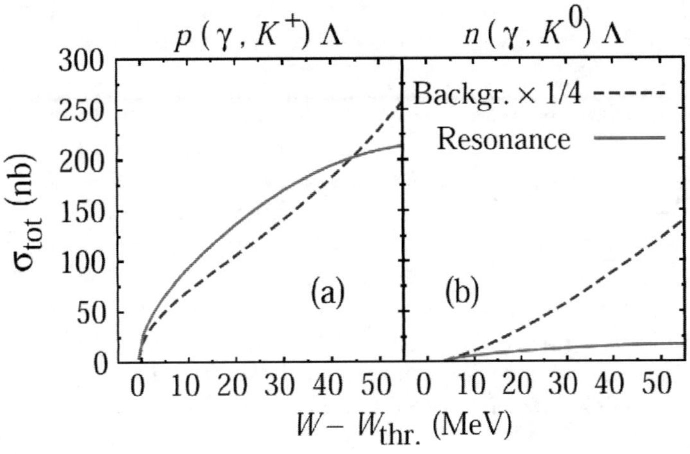

Figure 6. A fitted isobar model to $K^+\Lambda$ photoproduction (left). Relating hadronic coupling constants from eq. 2 allowed the prediction of the $n(\gamma, K^0)\Lambda$ cross section (right). Taken from [14].

Despite the fact that neutral kaon photoproduction is so crucial in understanding the photoproduction process, there is suprisingly little data available. The remainder of this section highlights two recent experiments; $\gamma(p, K^0)\Sigma^+$ cross section [22, 23] and $\gamma(n, K^0)\Lambda$ cross section [28] measurements.

3.1. $\gamma(p, K^0)\Sigma^+$ *cross section*

The experiment was performed with the Crystal Barrel and TAPS experiment at the ELSA facility, Bonn. The Crystal Barrel [19] is a highly segmented NaI calorimeter, with the TAPS detector [20, 21]; a wall of BaF2 crystals, at forward angles. The ELSA accelerator produced an electron beam to an energy of 3.2 GeV. Bremsstrahlung photons were produced to an energy of 2.94 GeV and were incident upon a liquid hydrogen target at the centre of the Crystal Barrel.

The channel was identified from six neutral and one charged particle, from the decays: $K^0 \to \pi^0\pi^0$ and $\Sigma^+ \to p\pi^0$. A kinematic fit tested event by event the hypothesis: $\gamma p \to p\pi^0\pi^0\pi^0$. Events were selected which were consistent with K^0 and Σ^+ reconstructed invariant masses (fig. 7). The nearly uniform, 4π acceptance made the analysis ideal for the extraction of differential cross sections. For further description of the experiment and analysis, see [22, 23].

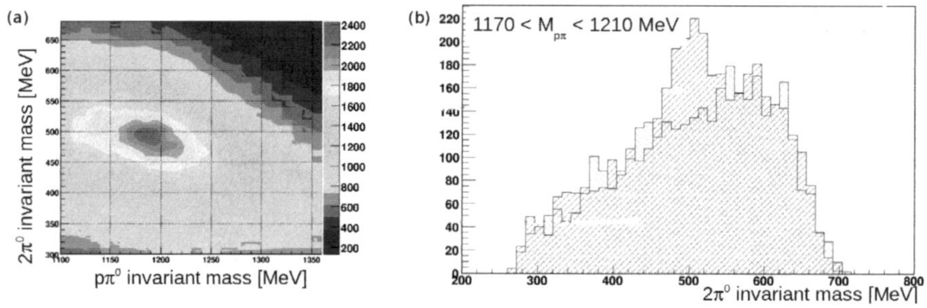

Figure 7. (a) Reconstructed K^0 mass (from the momenta of two decay π^0) versus Σ^+ (from the momenta of a decay proton and π^0). A peak at the expected masses is visible. (b) $2\pi^0$ invariant mass selected over the region of the Σ^+ mass on the x-axis from (b). A peak at the K^0 mass (red hatched region) is evident above a background. The background is mainly from incorrelated $2\pi^0$ and is described well by simulation (green hatched region). Taken from [23].

Fig. 8 shows the differential cross section versus centre of mass polar angle for different $\sqrt{s}$ energy bins. Near threshold, the cross section is flat, suggesting only s-channel contributions. This becomes increasingly forward peaked with energy, suggesting increased contributions from t-channel mechanisms. There is a large "cusp" effect however around 1800 MeV, where the cross section reduces approximately by a factor of four at forward angles. Beyond this energy, the cross section remains flat. It was suggested that K^* t-channel exchange (fig. 5(e)) contributes to the reaction mechanism below this energy. Above the energy where K^* exchange can be produced on-shell, this t-channel mechanism no longer contributes. Fig. 9 shows the total cross section, where the cusp effect at forward angles is still evident. It is clear that the existing Kaon-MAID parameterisation [33] does not describe the data well. The fit to data below K^* threshold is improved by changing the coupling to the $S_{31}(1900)$ state to $G_1 = 0.3$ and $G_2 = 0.3$, and reducing the Born couplings from 1.0 to 0.7. Beyond K^* threshold, the K^* t-channel exchange has been "switched off", matching the data well.

It was speculated that below threshold, the diagram fig. 5(f) contributes to the $K^0\Sigma^+$ channel. K^* couples to a K^0 and a π, and the π is subsequently reabsorbed by the hyperon. Above K^* production threshold however, the K^* is produced as a free particle. In this sense, the $K^0\Sigma^+$

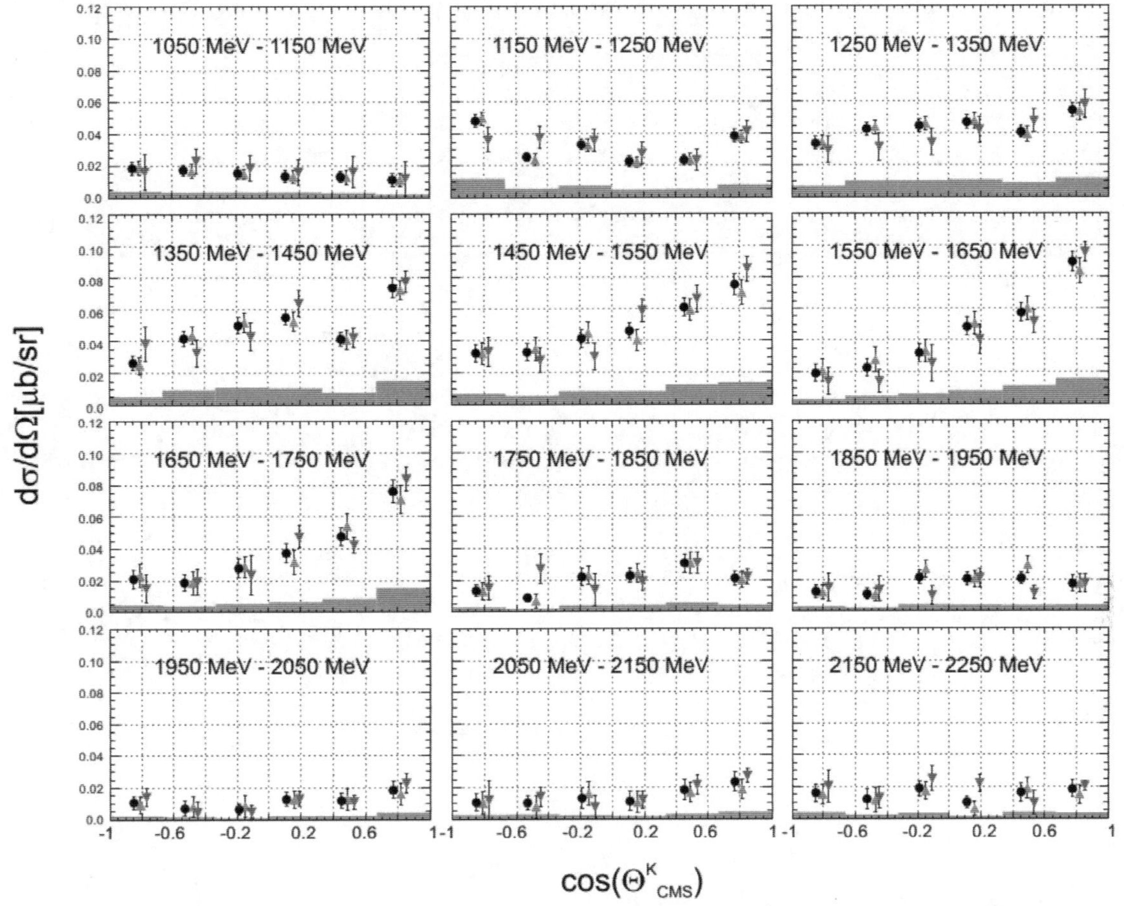

Figure 8. $\gamma(p, K^0)\Sigma^+$ differential cross section as a function of K^0 centre of mass polar angle for 100 MeV bins of photon energy (values inset). Red, down pointing triangles are for data with six neutral particles and one charged particle identified. Green, up pointing triangles are for only six neutral particles identified. Black circles are for all data. Taken from [22].

feeds off the $K^*\Sigma^0$ channel below threshold. Above K^* threshold, the dip in the cross section for $K^0\Sigma^+$ contributes to the $K^*\Sigma^0$ cross section. There is evidence of this in fig. 9, where summing the cross sections of the two channels agrees well with the standard Kaon-MAID parameterisation [33].

Fig. 5(f) could be interpreted as a strong meson-baryon interaction in a relative s-wave. If the reduction in cross section occurs between the $K^*\Lambda$ and $K^*\Sigma^0$ thresholds, there maybe an intermediate strongly bound $K^*\Sigma^0$ state, with J^π $(1/2)^-$ or $(3/2)^-$. Dynamically generated resonances have recently been investigated via the interaction of the nonet of vector mesons and baryon octet [30, 31]. Using a coupled channeled, chiral unitary approach, an isospin 1/2 doublet was predicted close to the K^* threshold [31].

To further understand $K^0\Sigma^+$ photoproduction mechanism, further polarisation observables are required. For example, the beam-target double polarisation observable, E, is sensitive to changes in s-channel contributions. The introduction of a meson-baryon dynamically generated resonance at a given $\sqrt{s}$ could potentially change the sign of E. Furthermore, the loss or addition of a large t-channel contribution would change the magnitude of E, as a pure t-channel

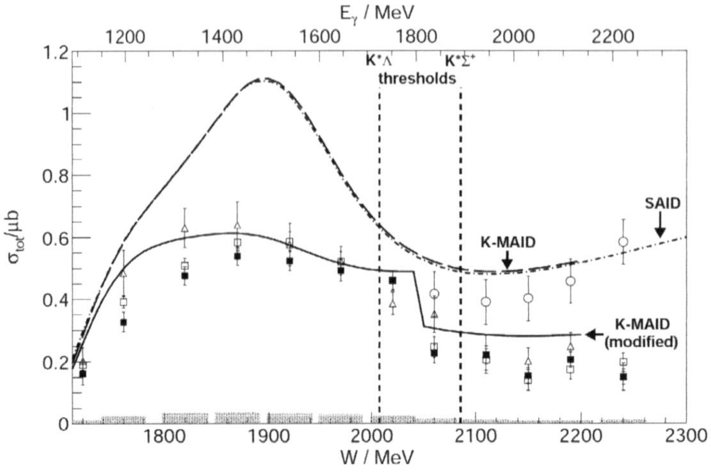

Figure 9. Total $K^0\Sigma^+$ cross section. Recent Crystal Barrel data [22, 23] are full squares (systematic error on the x-axis). Open squares are previous data from the Crystal Barrel [24], and open triangles are data from the SAPHIR experiment [27]. The open circles are the summed cross sections of $K^0\Sigma^+$ and $K^*\Sigma^0$ data from Nanova *et al.* [26]. The Kaon-MAID and modified Kaon-MAID [33] fits are described in the text (solid and dashed lines), and the dot-dashed line is the SAID parameterisation [25].

mechanism would exhibit zero asymmetry.

3.2. $\gamma(n, K^0)\Lambda$ *cross section*

The first cross section data for $\gamma(n, K^0)\Lambda$ using a deuterium target was measured with the Neutral Kaon Spectrometer at the Laboratory of Nuclear Science (LNS) at Tohoku University [28] (a previous measurement used a carbon target [32], but was limited due to the many body nature of the target). Bremsstrahlung photons with an energy range of 0.8-1.1 GeV and a measured energy resolution of 10 MeV were incident upon the target. The NKS identified and measured the momenta of particles using a 0.5 T magnetic dipole and a series of drift chambers, hodoscopes and electron veto scintillation counters (see [28, 29] for more details). The experiment had an acceptance of approximately π steradians, covering forward angles. K^0 were identified via the decay $K^0 \to \pi^+\pi^-$, where the K^0 invariant mass was reconstructed from the pion momenta.

Fig. 10 shows the measured inclusive spectra for K^0Y (where Y is predominantly Λ but also Σ^0 and Σ^+). Three different models were fitted to the data; Kaon-MAID [33], Saclay-Lyon A isobar model [34], and a phenomenological model which fitted Legendre polynomials. The isobar models assumed isospin symmetry to extract the hadronic coupling constants. Kaon-MAID used a simultaneous fit to $K^0\Sigma^+$ data as a constraint. Couplings to t-channel K_1 and K_2 mesons were left as free parameters. It is clear that the Kaon-MAID solution over estimates the cross section and near threshold is more forward peaked. The SLA model, which only fixed the K_1 and K_2 couplings from the lower energy plot has a better agreement. The differences between these two models is predominantly due to different angular distributions, with the SLA model more backward peaked. The PH1 model is a fit to data of a series of Legendre polynomials. PH2 is the same model but with the lowest order term given the opposite sign, producing the opposite angular distribution. The PH2 model is seen to match the Kaon-MAID solution more closely.

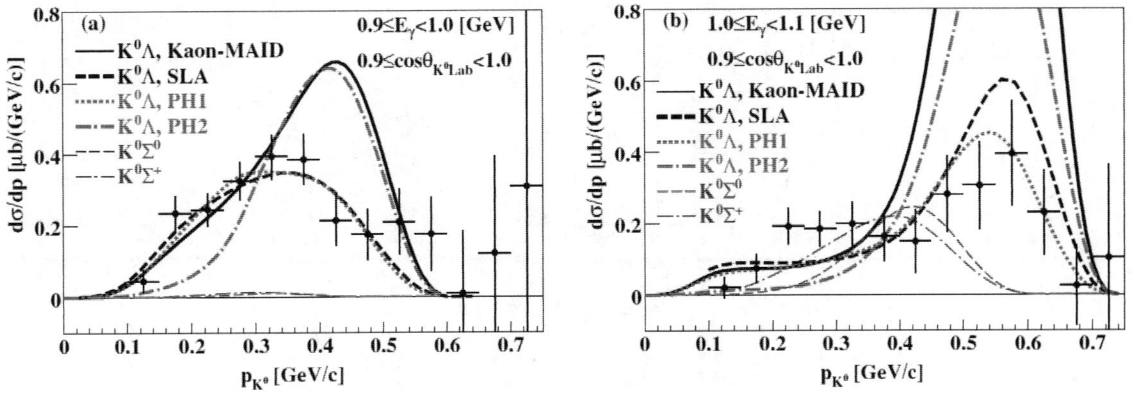

Figure 10. $\gamma(n, K^0)\Lambda$ differential cross section as a function of K^0 momentum for photon beam energy (a) 0.9-1.0 GeV and (b) 1.0-1.1 GeV, for laboratory frame polar angles smaller than 25^0. Solid black line is a Kaon-MAID fit [33], dotted black line the Saclay-Lyon A isobar model [34], dotted red and dot-dashed red are phenomenological models described in the text, and the blue dashed and dot dashed lines are Kaon-MAID fits to $K^0\Sigma^0$ and $K^0\Sigma^+$. Taken from [28].

In conclusion, the data suggests a more backward peaking angular distribution than what was predicted with the Kaon-MAID parameterisation, however the region of acceptance and statistics (and separating the reaction channels) can still be greatly improved in an attempt to understand a crucial channel in strangeness photoproduction.

4. Properties of the $\Lambda(1405)$

The $\Lambda(1405)$ was established in the 1960s between the $\pi\Sigma$ and $N\overline{K}$ thresholds, however the nature of the state is still not well understood. It is difficult to describe in a constituent quark model as it has a lighter mass than the non-strange N(1535), and the mass difference to the spin-orbit partner, $\Lambda(1520)$, is too large.

Recent advances in chiral unitary formulism for meson baryon interactions [36, 35, 37] described the $\Lambda(1405)$ as a two-pole structure, with the contributions interfering on the real energy axis (fig. 11). The $\Lambda(1405)$ can only be observed via the decays: $\Lambda(1405) \to \pi\Sigma$ with $I = 0$, however it was found that the coupling of the two poles to different meson-baryon channels are of different strength, leading to a difference in the $\Lambda(1405)$ line shape depending upon the decay it is observed via (fig 12(a)).

Recent data from CLAS at Jefferson Lab was able to reconstruct the $\Lambda(1405)$ from all three decay modes [38] (fig. 12(b)). It is clear that the line shape differs depending upon the decay mode, however they do not agree with the theoretical prediction in fig. 11, where the $\Sigma^-\pi^+$ line shape is at a higher mass. Resolving this, and understanding the cross section dependence upon the momentum transfer are still open questions.

5. Hypernuclei

Hypernuclei provide a "natural laboratory", where nucleon-nucleon and nucleon-hyperon interactions can be directly observed. Analysis of the bound hyperon also gives a unique insight into the nucleon interior as it is not bound by the Pauli exclusion principle to specific quantum states as with other nucleons. Most hypernuclei have been produced by stopping K^+ or K^- in flight. Recently however, hyperon spectroscopy has been performed with electron beams at Jefferson Lab [39]. Electromagnetic probes allows the creation of neutron rich hypernuclei

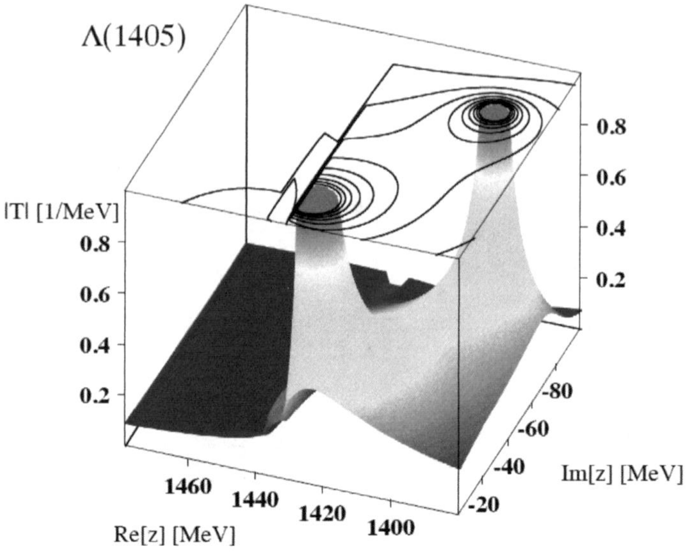

Figure 11. The two pole structure of the $\Lambda(1405)$. Predicted by a chiral unitary formulism [36, 35, 37], the absolute value of the scattering matrix, $|T|$, is plotted in the second Riemann sheet, with poles close together on the real energy axis. Taken from [36].

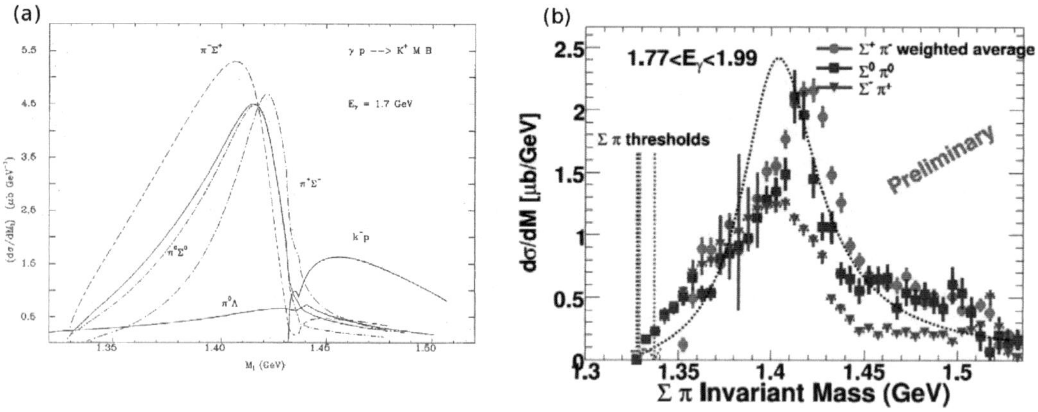

Figure 12. (a) The predicted $\Lambda(1405)$ lineshape for different decay channels (inset) as predicted by a chiral unitary formulism [36, 35, 37]. Taken from [37]. (b) The measured $\Lambda(1405)$ line shape for all decay channels (inset) with the CLAS detector [38]. The dashed line is a relativistic Breit-Wigner shape using mass and width values from the PDG. Taken from [38].

(potentially with halo structure), and due to contributions of spin flip amplitudes, excites states with significant cross sections which are hard to observe using hadronic probes. It was proposed by Shyam, Lenske and Mosel [40] that for these reasons, and also that a photon can penetrate deep into the nuclear medium to create deeply bound hypernuclei, that the photoproduction of hypernuclei would be an important addition to existing analysis. This would allow further investigation of the spin flip amplitude and impose constraints upon the nucleon-hyperon interaction.

It is reasonable to assume that the BGO-OD experiment would be able to provide data with significant statistics and accuracy to improve or consolidate all previous results discussed in this paper. It is therefore worth considering what can be achieved with the BGO-OD experiment in hypernuclei research.

After the production of a hypernuclei, the hyperon usually occupies an excited state within the nucleus. For the hyperon to reach the ground state, photons and nucleons may be emitted. At the ground state, the hyperon will decay weakly. Within the nuclear medium, the decay can be observed as mesonic ($\Lambda \rightarrow \pi N$), which is equivalent to what is observed for free unbound Λ, or non-mesonic, $\Lambda N \rightarrow NN$, where the pion is reabsorbed by a nucleon. The ratio of the two decay modes is shown as a function of atomic number in fig. 13(a).

To identify a hypernuclei state using the BGO-OD, the first step should be to identify the K^+. For the Λ to be bound in the nucleus, the K^+ must have nearly all available momentum and therefore should be detected in the forward spectrometer. The weak decay of the Λ within the nucleus could be observed via the mesonic mode, where the baron and meson can be detected and the Λ invariant mass reconstructed. A further selection cut could reject events where momentum is conserved between the incident momentum and the sum of the K^+ and Λ momenta, in order to reject events where the Λ was not bound in the nucleus. To verify the hypernucleus, and to identify the state occupied by the hyperon, the BGO-Ball would identify a low energy photon in timing coincidence with the Λ weak decay. This photon is from the de-excitation of the hyperon and would be of the order of 1-10 MeV. The resolution would not be sufficient to search for new states, but if a state was already known it could be identified. A similar approach was used recently by Tarbert and Watts *et al.* with the Crystal Ball detector [42], where π^0 photoproduction from a ^{12}C target, leaving the ^{12}C nucleus in 4.4 MeV state was observed (incoherent neutral pion photoproduction). Fig. 13(b) shows the angle of the reconstructed pion and the energy deposition of a third photon (not the two π^0 decay photons). There is a low energy tail arising from misidentified electromagneitc showers from the pion decay photons, however there is also a clear signal from the 4.4 MeV photon from the de-excitation of the ^{12}C nucleus.

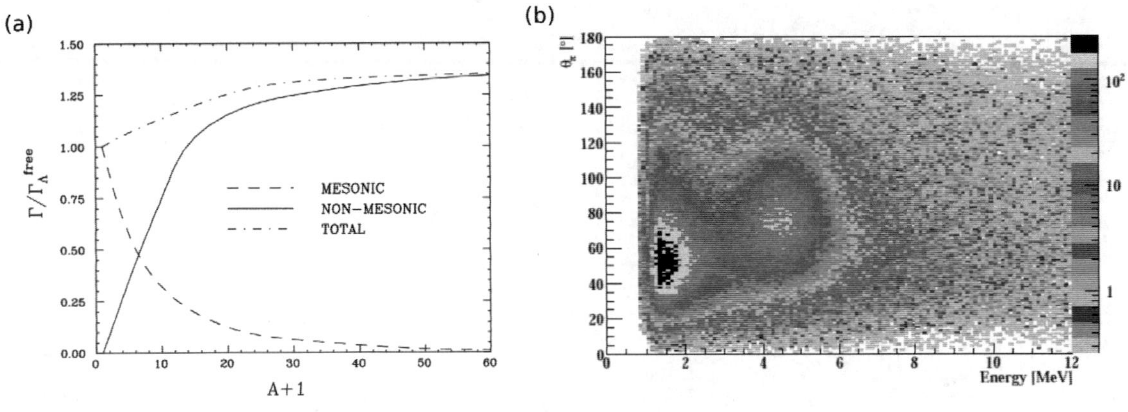

Figure 13. (a) Mesonic and non-mesonic decay widths of bound Λ within a hypernuclei, compared to a free Λ decay as a function of nucleon number. Taken from [41]. (b) Identification of a 4.4 MeV photon from the de-excitation of a ^{12}C following π^0 photoproduction. Further description in the text. Taken from [42].

From fig. 13(b), it is clear that a low atomic mass nucleus is favourable if the mesonic decay mode is used to identify hypernuclei. Rather than calculating absolute cross sections, ratios of yields of different hyperon states could be observed. This could be compared to ratios and cross sections of the same states created with hadronic and electron probes and provide further constraints upon spin flip transitions and hyperon-nucleon interactions.

6. Summary and conclusions

The abundance of recently accrued data imposes large constraints upon strangeness photoproduction mechanisms. It is expected that further analysis of recently published data will narrow down resonances which contribute to the excitation spectrum in $K^+\Lambda$ photoproduction. Neutral kaon photoproduction is still in its infancy in comparison, but is crucial in our understanding of resonance structure, given the SU(3) symmetry with charged photoproduction, and the contribution from vector meson t-channel exchange. The nature of the $\Lambda(1405)$ is beginning to reveal itself to experiment, however at present there is very little understood with how the cross section varies with respect to momentum transfer.

Many results presented were measured from only a few experiments. Consequentially, certain kinematic regions have poor or no statistics due to the acceptance regions of experimental setups. It is therefore paramount to consolidate and add to existing data sets where possible with new experiments and to extend the data sets of polarisation observables where possible.

Hypernuclei research with the BGO-Ball could potentially provide unique measurements which are unattainable at any other accelerator facility. Practical issues, such as whether the cross section is sufficient for reasonable statistics, or low energy photons from the beam dump contaminating the event selection remain open questions.

References

[1] Schmieden H 2009 *Int. J. Mod. Phys.* **15** 1043
[2] Hillert W 2006 *Eur. Phys. J.* A **28** 139
[3] 2012 *http://b1.physik.uni-bonn.de/*
[4] Santopinto E 2005 *Phys. Rev.* C **72** 022201
[5] Capstick S and Roberts W 1998 *Phys. Rev.* D **58** 074011
[6] Jude T C 2010 PhD Thesis *The University of Edinburgh*
[7] Barker I S and Donnachie A 1975 *Nuc. Phys.* B **95** 347
[8] Glander K H *et al.* 2004 *Eur. Phys. J.* A **19** 251
[9] Tran M Q *et al.* 1998 *Phys. Rev. Lett.* B **445** 20
[10] Bradford R and Schumacher R A *et al.* 2006 *Phys. Rev.* C **73** 035202
[11] R. Schumacher 2007 *Eur. Phys. J.* A **35** 299
[12] McCracken M E, Bellis M, Meyer C A and Williams M *et al.* 2010 *Phys. Rev.* C **81** 025201
[13] Mart T and Bennhold C *Phys. Rev.* C **61** 012201
[14] Mart T 2011 *Phys. Rev.* C **83** 048203
[15] Lleres A *et al.* 2007 *Eur. Phys. J.* A **31** 79
[16] Paterson C 2008 PhD Thesis *The University of Glasgow*
[17] Bradford R *et al.* 2007 *Phys. Rev.* C **75** 035205
[18] Borasoy B, Bruns P C, Meissner U G and Nissler R 2007 *Eur. Phys. J.* A 16
[19] Aker E *et al.* 1992 *Nuc. Inst. Meth.* A **321** 69
[20] Novotny R *et al.* 1991 *IEEE Trans. Nuc. Sci.* **38** 392
[21] Gabler A R *et al.* 1994 *Nuc. Inst. Meth.* **346** 168
[22] Ewald R 2008 PhD thesis *Universität Bonn*
[23] Ewald R *et al.* 2011 Anomaly in the $K^0_S\Sigma^+$ photoproduction cross section off the proton at the K^* threshold *Preprint* arXiv:1112.0811v1 [nucl-ex]
[24] *et al.* 2008 *Eur. Phys. J.* A **35** 39
[25] Arndt R A, *et al.* 2011 http://gwdac.phys.gwu.edu
[26] Nanova M *et al.* for the Crystal-Barrel/TAPS Collab. 2008 *Eur. Phys. J.* A **35** 333
[27] Lawall R *et al.* 2005 *Eur. Phys. J.* A **24** 275
[28] Tsukada K *et al.* 2008 *Phys. Rev.* C **78** 014001

[29] Maruyama *et al.* for the TAGX Collaboration 1996 *Nuc. Inst. Meth.* A **335** 1996
[30] Oset E and Ramos A 2010 *Eur. Phys. J.* A **44** 2010
[31] Oset E, Garzon E J, Ju Jun Xie, Gonzalez P, Ramos A and Martínez Torres A 2011 Dynamically generated hadron resonances *Preprint* arXiv:1103.0807v1 [nucl-th]
[32] Watanabe T *et al.* 2007 *Phys. Lett.* B **651** 269
[33] Mart T, Bennhold C, Haberzettl H and Tiator L 2010 An effective Lagrangian Model for Kaon Photo- and Electroproduction on the Nucleon *wwwkph.kph.uni-mainz.de/MAID//kaon/*
[34] Mizutani T, Fayard C, Lamot G H and Saghai B 1998 *Phys. Rev.* C **58** 75
[35] Jido D, Oller J A, Oset E, Ramos A and Meissner U G 2003 *Nuc. Phys.* A **725** 181
[36] Hyodo T and Jido D 2012 *Prog. Part. Nuc. Phys.* **67** 55
[37] Nacher J C, Oset E, Toki H, Ramos A and Meissner U G 1999 *Phys. Lett.* B **455** 55
[38] Moriya K and Schumacher R, for the CLAS collaboration 2010 *Nuc. Phys.* A **835** 325
[39] Hashimoto O, Nakamura S N, *et al.* 2008 *Nuc. Phys.* A **804** 125
[40] Shyam R, Lenske H and Mosel U 2008 *Phys. Rev.* C **77** 052201
[41] Alberico W M and Garbarino G 2002 *Phys. Reports* **369** 1
[42] Tarbert C M, Watts D P, *et al.* 2008 *Phys. Rev. Lett.* **100** 132301

Journal of Physics: Conference Series **349** (2012) 012013 doi:10.1088/1742-6596/349/1/012013

KLOE results on hadronic cross section

Giuseppe Mandaglio on behalf of the KLOE/KLOE-2 Collaboration‡

E-mail: gmandaglio@unime.it
Dipartimento di Fisica, Università di Messina, I-98166 Messina, Italy
INFN, Sezione di Catania, I-95123 Catania, Italy

Abstract. The KLOE experiment at the $\phi - factory$ DAΦNE is the first to have exploited Initial State Radiation (ISR) to precisely determine the $e^+e^- \rightarrow \pi^+\pi^-(\gamma)$ cross section below 1 GeV, representing the 70% of the leading order contribution to the muon anomaly. The leading order contribution $a_\mu^{\rm hlo}$ is presently the main source of uncertainty in the theoretical evaluation of the muon anomaly, and it can be evaluated by dispersion integral using the experimental measurement of hadronic cross section. A persistent discrepancy of about 3 σ between standard model (SM) prediction and experimental measurements of the muon anomalous magnetic moment has been up to now observed. The KLOE collaboration published two measurements of the $\pi^+\pi^-$ cross section with the photon in the initial state emitted at small polar angle in Phys. Lett. B vol. 606 pg. 12 and vol. 670 pg. 285, and an independent measurement with the photon emitted at large polar angle in Phys. Lett. B vol. 700 pg. 102. These measurements were normalized to the DAΦNE luminosity. Recently, a new analysis deriving the pion form factor directly from measuring the bin-by-bin $\pi^+\pi\gamma$ and $\mu^+\mu\gamma$ final states ratio has been performed. In this paper, the preliminary results of this new measurement and the comparison to the previous published ones, the impact on the evaluation of the hadronic contribution to the muon anomaly, the preliminary $\mu^+\mu\gamma$ cross section measurement and the comparison with the PHOKHARA-MC prediction are presented.

‡ The KLOE Collaboration: F. Ambrosino, A. Antonelli, M. Antonelli, F. Archilli, I. Balwierz, G. Bencivenni, C. Bini, C. Bloise, S. Bocchetta, F. Bossi, P. Branchini, G. Capon, T. Capussela, F. Ceradini, P. Ciambrone, E. Czerwiński, E. De Lucia, A. De Santis, P. De Simone, G. De Zorzi, A. Denig, A. Di Domenico, C. Di Donato, B. Di Micco, M. Dreucci, G. Felici, S. Fiore, P. Franzini, C. Gatti, P. Gauzzi, S. Giovannella, E. Graziani, M. Jacewicz, W. Kluge, J. Lee-Franzini, P. Lukin, M. Martemianov, M. Martini, P. Massarotti, S. Meola, S. Miscetti, G. Morello, M. Moulson, S. Müller, M. Napolitano, F. Nguyen, M. Palutan, A. Passeri, V. Patera, I. Prado Longhi, P. Santangelo, B. Sciascia, M. Silarski, T. Spadaro, C. Taccini, L. Tortora, G. Venanzoni, R.Versaci, G. Xu, J. Zdebik, and, as members of the KLOE-2 collaboration: D. Babusci, D. Badoni, V. Bocci, A. Budano, S. A. Bulychjev, L. Caldeira Balkeståhl, P. Campana, E. Dané, G. De Robertis, D. Domenici, O. Erriquez, G. Fanizzi, G. Giardina, F. Gonnella, F. Happacher, B. Höistad, L. Iafolla, E. Iarocci, T. Johansson, A. Kowalewska, V. Kulikov, A. Kupsc, F. Loddo, G. Mandaglio, M. Mascolo, M. Matsyuk, R. Messi, D. Moricciani, A. Ranieri, C. F. Redmer, I. Sarra, M. Schioppa, A. Sciubba, W. Wiślicki, M. Wolke

1. Introduction

The anomalous magnetic moment of the muon, a_μ , is one of the best known quantities in particle physics. Recent theoretical evaluations [1–3] find a discrepancy of about 3 standard deviations from the value obtained from the g-2 experiment at Brookhaven [4]. The main contribution to the uncertainty on the theoretical estimates come from the leading order hadronic contribution $a_\mu^{\rm had,lo}$ and the hadronic light-by-light contribution $a_\mu^{\rm LbyL}$ [5]. The $a_\mu^{\rm had,lo}$ contribution is not calculable by perturbative QCD at low energies but it can be evaluated using measured hadronic cross sections by the dispersion integral

$$a_\mu^{\rm had} = \frac{1}{4\pi^3}\sum_f \int_{s_{th}(f)}^{\infty} K(s)\sigma(e^+e^- \to f)ds, \tag{1}$$

where $K(s)$ is the QCD kernel function that approximately varies as $1/s$, with s the squared center-of-mass collision energy, and $s_{th}(f)$ is the production threshold.

The use of Initial State Radiation (ISR) has opened a new way to measure these cross sections at particle factories operating at fixed energies [6]. The region below 1 GeV, which is accessible with the KLOE experiment, is dominated by the $\pi^+\pi^-$ final state and contributes with $\sim$ 70% to $a_\mu^{\rm had,lo}$, and $\sim$ 60% to its uncertainty. Therefore, an improved precision on the $\pi^+\pi^-$ cross section would result in a reduction of the uncertainty on the leading order hadronic contribution to a_μ, and in turn to the SM prediction for a_μ.

2. Measurement of $\sigma_{\pi\pi}$ with ISR

The KLOE detector operates at the DAΦNE e^+e^- collider of the INFN Laboratori Nazionali di Frascati. It consists (Fig. 1) of a huge drift chamber providing high momentum resolution($\sigma_p/p \leq 0.4\%$) [7] on reconstructed tracks and Pb-scintillating fibers calorimeter with excellent time ($\sigma_t \sim 54$ ps/$\sqrt{E\ [\rm GeV]}$ $\oplus$100 ps) and good energy ($\sigma_E/E \sim 5.7\%/\sqrt{E\ [\rm GeV]}$) resolution [8].

DAΦNE is a $\phi-factory$ (see Fig. 2)running at $\sqrt{s} \simeq M_\phi$, and has delivered ca. 2.5 fb^{-1} of integrated luminosity to the KLOE experiment. So far KLOE has reported two measurements of the $\pi^+\pi^-$ cross section between 0.35 and 0.95 GeV^2 called KLOE05 [9] and KLOE08 [10] in the following. In addition, about 250 pb^{-1} of data have been collected at $\sqrt{s} \simeq 1$ GeV, 20 MeV below the ϕ-meson resonance, from which a new measurement of $\pi^+\pi^-$ cross section was obtained (KLOE10 [11]). Running at energies below the ϕ-meson resonance reduces considerably the background from the copious ϕ-meson decay products, including scalar mesons. As DAΦNE was designed to operate at a fixed energy around M_ϕ, the differential cross section $d\sigma(e^+e^- \to \pi^+\pi^-\gamma)/dM^2_{\pi\pi}$ is measured, and the total cross section $\sigma_{\pi\pi} \equiv \sigma_{e^+e^-\to\pi^+\pi^-}$ is evaluated using the formula [6]:

$$s \cdot \frac{d\sigma_{\pi\pi\gamma}}{dM^2_{\pi\pi}} = \sigma_{\pi\pi}(M^2_{\pi\pi})\ H(M^2_{\pi\pi}, s)\ , \tag{2}$$

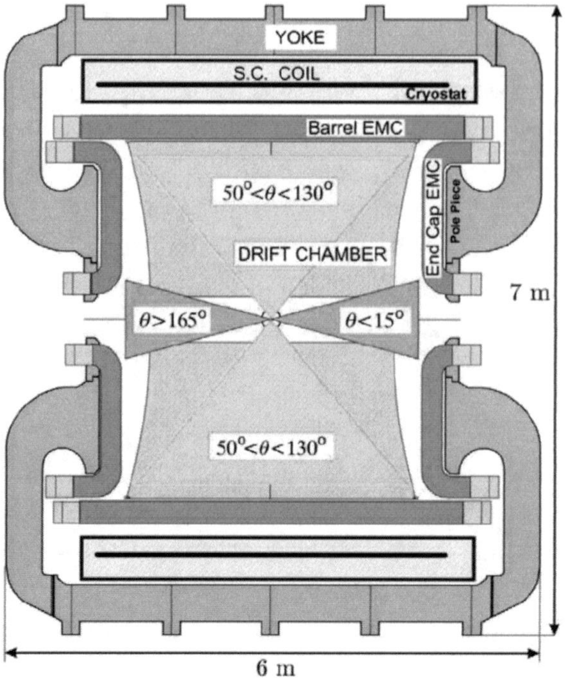

Figure 1. Left: Schematic view of the KLOE detector with selection regions; Right: Pion form factor $|F_\pi|^2$ obtained in KLOE10 and the previous (KLOE08) analysis.

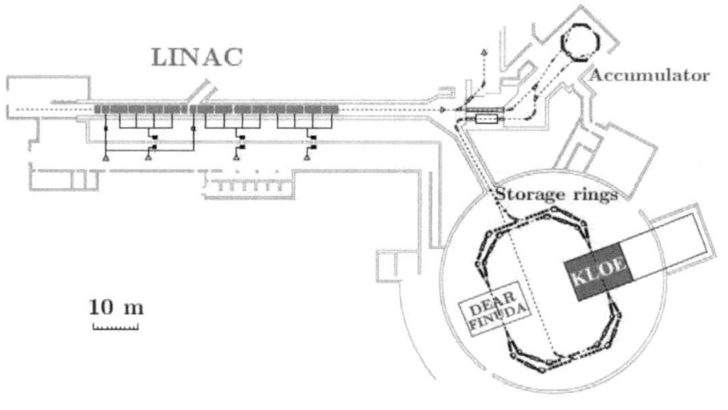

Figure 2. The DAΦNE complex.

with s the squared e^+e^- center of mass energy, and H the radiator function obtained from theory describing the photon emission in the initial state. An alternative method to extract the $\pi^+\pi^-$ cross section uses the $\pi^+\pi^-\gamma/\mu^+\mu^-\gamma$ ratio [12]:

$$\sigma_{\pi\pi(\gamma)} = \sigma_{\mu\mu(\gamma)} \frac{\mathrm{d}\sigma_{\pi\pi\gamma}/\mathrm{d}s'}{\mathrm{d}\sigma_{\mu\mu\gamma}/\mathrm{d}s'} = \frac{4\pi\alpha^2}{3s'}(1 + 2m_\mu^2/s')\beta_\mu \frac{\mathrm{d}\sigma_{\pi\pi\gamma}/\mathrm{d}s'}{\mathrm{d}\sigma_{\mu\mu\gamma}/\mathrm{d}s'}, \qquad (3)$$

with s' the four-momentum square of the virtual photon, *i.e.* the e^+e^-center-of-mass energy squared after ISR emission, m_μ the muon mass, β_μ, β_π the muon and pion velocities in the center-of-mass frame, $d\sigma_{\pi\pi\gamma}/ds'$, $d\sigma_{\mu\mu\gamma}/ds'$ the $e^+e^- \to \pi^+\pi^-\gamma$, $e^+e^- \to \mu^+\mu^-\gamma$ differential cross sections, respectively. In both Eqs.(2) and 3) the Final State Radiation (FSR) terms are neglected, but are taken into account properly in the analyses. The differential cross sections ratio in Eq. (3) allows us to completely cancel important contributions to the total uncertainty which are instead present in the measurement performed with Eq. (2), namely the integrated luminosity, the radiation function H and the vacuum polarization.

3. Measurements of Pion form factor normalized to Luminosity from Bhabha events

The first two KLOE published analyses (KLOE05 and KLOE08) used selection cuts in which the photon is emitted within a cone of $\theta_\gamma < 15°$ around the beam line (narrow cones in Fig. 1) and the two charged pion tracks have $50° < \theta_\pi < 130°$ (wide cones in Fig. 1). In this configuration, the photon is not explicitly detected, and its direction is reconstructed from the two detected charged pion tracks' momenta as: $\vec{p}_\gamma \simeq \vec{p}_{\rm miss} = -(\vec{p}_{\pi^+} + \vec{p}_{\pi^-})$. These cuts guarantee a high statistics for ISR signal events, and a reduced contamination from the resonant process $e^+e^- \to \phi \to \pi^+\pi^-\pi^0$, in which the π^0 mimics the missing momentum of the photon(s), and from the final state radiation process $e^+e^- \to \pi^+\pi^-\gamma_{\rm FSR}$. As a consequence of the event selection, the highly energetic photon emitted at small angle forces the pions also to be also at small angles (and thus outside the selection cuts), resulting in a kinematical suppression of events with $M^2_{\pi\pi} < 0.35$ GeV^2.

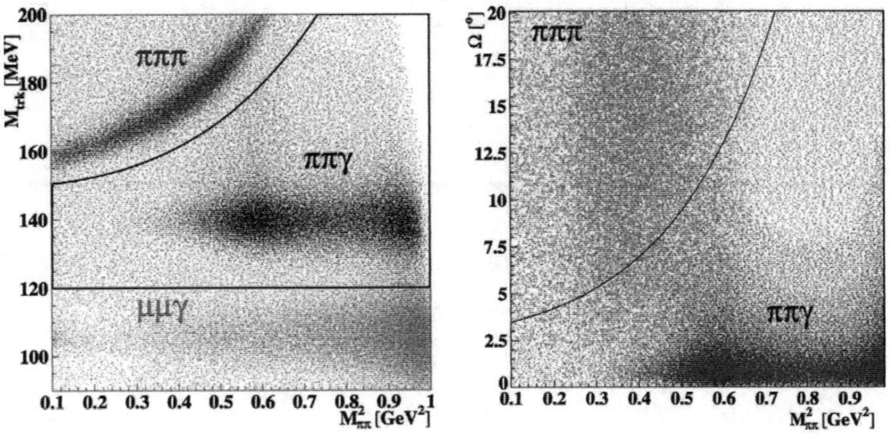

Figure 3. Left: MC simulation of M_{TRK} vs. $M^2_{\pi\pi}$. $\pi\pi\gamma$ and $\mu\mu\gamma$ events are located around m_π and m_μ respectively, while $\pi^+\pi^-\pi^0$ events occupy a region in the upper left of the plot. Right: MC simulation of Ω vs. $M^2_{\pi\pi}$. The black lines represent the cuts used in the analysis.

To access the two pion threshold, a new analysis has been performed (KLOE10) requiring events with photon at large polar angles with $50° < \theta_\gamma < 130°$ (wide cones in

Fig. 1), in the same angular region as the pions. The drawback in using such acceptance cuts is about a factor 5 reduction in statistics, as well as an increase of background events from final state radiation and from ϕ radiative decays compared to the small polar angle photon acceptance criterion. The model dependence of the ϕ radiative decays to the $f_0(980)$ and $f_0(600)$ scalars together with $\phi \to \rho\pi \to (\pi\gamma)\pi$, has a strong impact on the measurement [13]. To significantly reduce the contamination from $f_0\gamma$ and $\varrho\pi$ decays of the ϕ-meson, data were taken at $\sqrt{s} = 1$ GeV below M_ϕ of about 5 times the ϕ-meson decay width ($\Gamma_\phi = 4.26 \pm 0.04$ MeV [14]) outside the narrow peak of the ϕ resonance.

This reduces the contamination effect due to contributions from $f_0\gamma$ and $\rho\pi$ decay ϕ-meson to within 1%. Contamination from the processes $\phi \to \pi^+\pi^-\pi^0$ and $e^+e^- \to \mu^+\mu^-\gamma$ are rejected cutting the kinematical variables trackmass M_{TRK} (see Fig. 3 Left) and Ω^2 (see Fig. 3 Right). The trackmass is defined using conservation of 4-momentum under the hypothesis that the final state consists of two charged particles with equal mass M_{trk} and one photon: $(\sqrt{s} - \sqrt{|\vec{p}_+|^2 + M^2_{mtr}} - \sqrt{|\vec{p}_-|^2 + M^2_{mtr}})^2 - (\vec{p}_+ + \vec{p}_-)^2 = 0$. The Ω variable is the angle between the directions of the detected photon and the reconstructed missing momentum $\vec{p}_{\text{miss}}$.

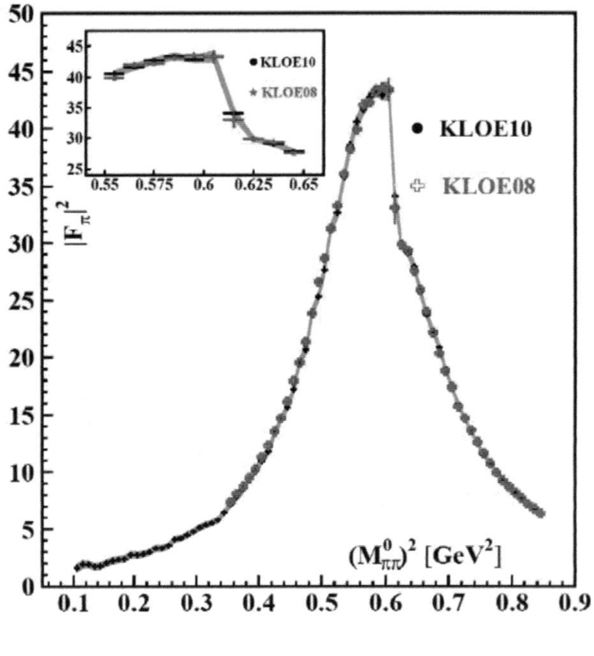

Figure 4. Pion form factor obtained with KLOE08 and KLOE10 analysis data points include statistical errors, the grey band shows the statistical and systematic uncertainties added in quadrature. KLOE08 points include the combined statistical and systematic uncertainties.

To efficiently suppress the high rate of radiative Bhabha scattering events, a identification particle estimator based on calorimeter information and time-of-flight is used. The residual background content is found by fitting the M_{TRK} spectrum (see 5) of the selected data sample with a superposition of Monte Carlo (MC) distributions describing the signal and background sources. The fit parameters are the fractional normalization factors for the MC distributions, obtained in intervals of $M^2_{\pi\pi}$(see Fig. 5).

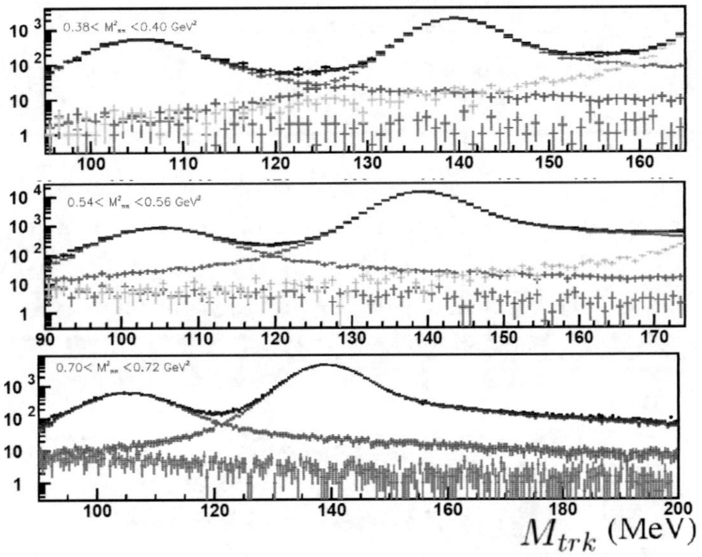

Figure 5. Results of the background fit procedure for different slices in $M^2_{\pi\pi}$: experimental data (black); sum of all Monte Carlo contributions (blue); and Monte Carlo $\pi\pi\gamma$ and $\mu\mu\gamma$ (red), $\pi\pi\pi$ (green), and $ee\gamma$ (magenta) contributions.

Using Eq. (2) the pion form factor $|F_\pi|^2$ is extracted and compared with the previous published one (KLOE08), showing excellent agreement (see Fig. 4). The dipion contribution to the muon anomaly $a_\mu^{\pi\pi}$ is then extracted from the bare cross section, *i.e.* corrected for the running of electromagnetic coupling α_{em} and inclusive of FSR.

The bare cross section $\sigma^{bare}_{\pi\pi(\gamma)}$ is used to determine $a_\mu^{\pi\pi}$ via a dispersion integral:

$$a_\mu^{\pi\pi} = \frac{1}{4\pi^3}\int_{s_{min}}^{s_{max}} \mathrm{d}\, s' \; \sigma^{bare}_{\pi\pi(\gamma)}(s')\, K(s') \;, \tag{4}$$

with $s_{min} = 0.10$ GeV^2 and $s_{max} = 0.85$ GeV^2 the lower and upper bounds in the present analysis, and $K(s)$ the kernel function described in [15].

The final results [11]:

$$\Delta a_\mu^{\pi\pi}(0.1 - 0.85 \text{ GeV}^2) =$$
$$(478.5 \pm 2.0_{\text{stat}} \pm 4.8_{\text{exp}} \pm 2.9_{\text{theo}}) \cdot 10^{-10}. \tag{5}$$

The evaluation of $\Delta a_\mu^{\pi\pi}$ in the range between 0.35 and 0.85 GeV^2 and its comparison with KLOE08 [10] result is reported in Table 1.

Combining the results from the KLOE measurements one obtains:

$$a_\mu^{\pi\pi}(0.1 - 0.95 \text{ GeV}^2) = (488.6 \pm 5.0) \cdot 10^{-10}. \tag{6}$$

The KLOE experiment covers 70% of the leading order contribution to the muon anomaly with 1% total error.

Analysis	$\Delta a_\mu^{\pi\pi}(0.35 - 0.85\ \text{GeV}^2) \times 10^{-10}$
KLOE10	$376.6 \pm 0.9_{\text{stat}} \pm 2.4_{\text{exp}} \pm 2.1_{\text{theo}}$
KLOE08	$379.6 \pm 0.4_{\text{stat}} \pm 2.4_{\text{exp}} \pm 2.2_{\text{theo}}$

Table 1. Comparison between KLOE10 and KLOE08 measurements.

The KLOE measurement has been compared with the results from e^+e^- experiments available in literature [11]. A reasonable agreement has been found with CMD-2 and SND experiments at Novosibirsk [16–18] (especially below the ρ), while some deviation is found with respect to new BaBar measurement [19], especially above 0.65 GeV where the BaBar result is higher by 2-3% (see Fig. 6).

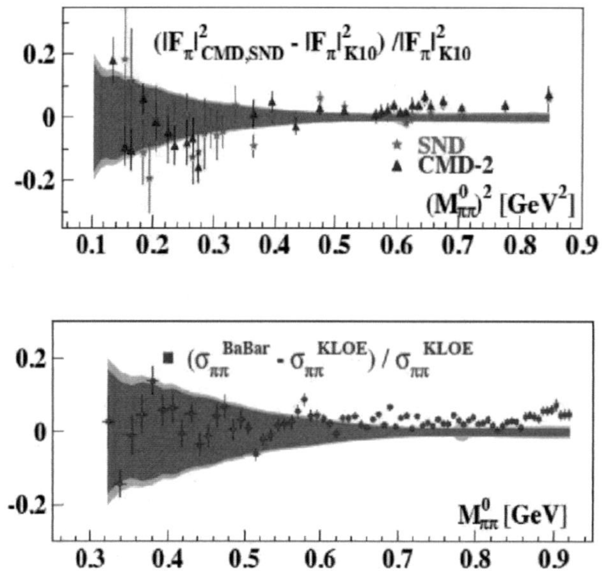

Figure 6. Top: Fractional difference between CMD-2 [16, 17] or SND[18] and KLOE10. Bottom: Fractional difference between BaBar[19] and KLOE10. The dark (light) grey bands show statistical (total) error of KLOE10 result.

4. Measurement of the pion form factor from the $\pi\pi\gamma/\mu\mu\gamma$ ratio

Equation (3) has been used to extract the pion form factor via a bin-by-bin ratio between the observed pion and muon ISR differential cross sections. This new approach has several benefits especially radiative for corrections with respect to the previous published results. The integrated luminosity as well as the radiation function H and the vacuum polarisation cancel completely in the $\pi\pi\gamma/\mu\mu\gamma$ ratio. In addition the ratio of the acceptance enters in eq. (3) giving corrections of the order of few percents. The

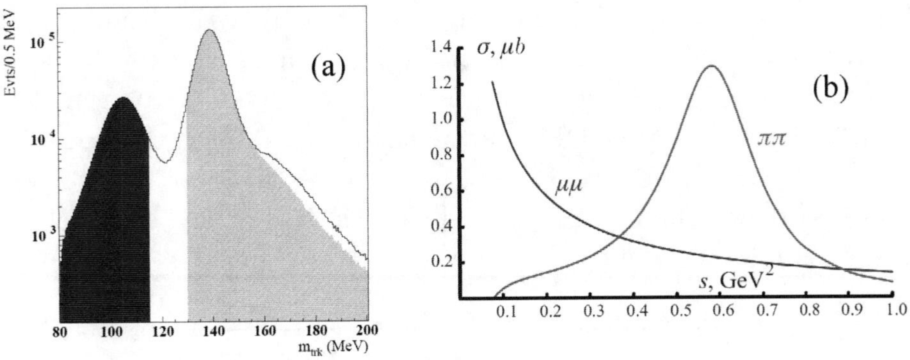

Figure 7. (a) Discrimination of the $\mu\mu\gamma$ (filled left histogram) from the $\pi\pi\gamma$ (filled right histogram) events on the M_{TRK} variable. (b)$\pi^+\pi^-$ and $\mu^+\mu^-$ cross-section as a function of s.

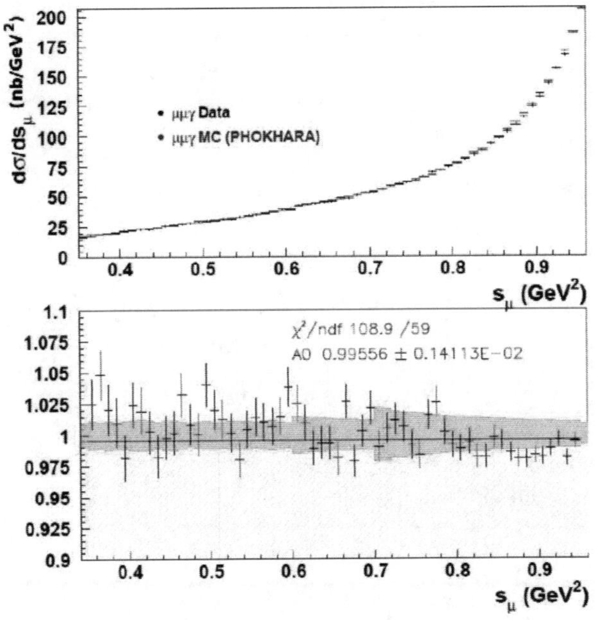

Figure 8. Top: comparison of data (black points) and MC (blue points)for the $\mu\mu\gamma$ absolute cross section as a function of $M^2_{\mu\mu}$ (GeV). Bottom: data to MC ratio. The green band shows the total experimental and theoretical systematic error.

additional analysis on muon events has to be performed at subpercent level, not a trivial task especially for the pion/muon separation(see Fig. 7 (a)). In fact, due to the ρ-resonance enhancement, the $\pi^+\pi^-$ cross section is up to one order of magnitude larger than the $\mu\mu$ cross section around 0.6 GeV^2 (see Fig. 7 (b)).

239.2 pb^{-1} KLOE data, the same sample used for KLOE08 measurement, is analyzed with the small angle photon selection. While the analysis for $\pi\pi\gamma$ is essentially the same as for KLOE08, the analysis for $\mu\mu\gamma$ is completely new and is based on following main features: 1) separation between $\mu\mu\gamma$ and $\pi\pi\gamma$ events performed assuming the final state of two charged particles with equal mass M_{TRK} and one photon: the $M_{TRK} <$ 115 MeV ($M_{TRK} >$ 130 MeV) selection leads to 9×10^5 (3.1×10^6) candidate $\mu\mu\gamma$ ($\pi\pi\gamma$) events. This selection, KLOE11 in the following, has been checked using other techniques, such as a kinematic fit or tighter cuts on the quality of the charged tracks, all giving to consistent results; 2) trigger, particle identification and tracking efficiencies measured with data control samples.

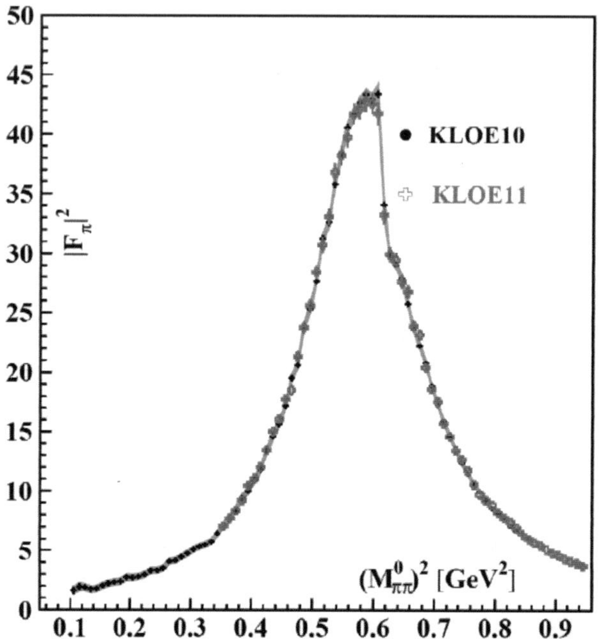

Figure 9. Pion form factor obtained with $\pi\pi\gamma$ /$\mu\mu\gamma$ ratio (KLOE11) compared to the previous (KLOE10) analysis.

The $\mu\mu\gamma$ cross section measurement has been performed by subtracting the residual background to the observed spectra and dividing it for the selection efficiency and integrated luminosity. The luminosity was evaluated by using large Bhabha events to be 239.2 pb^{-1}, with a systematic error of 0.3 %[20] . This preliminary result was compared with the value estimated by using PHOKHARA MC [12], and a good agreement within the total (theoretical and experimental)systematic error has been found (Fig.8). Then a preliminary result on the pion form factor has been extracted and compared with the one from KLOE10, showing good agreement (see Fig. 9). As a consequence of the comparison reported in Figure 4 we can deduce it is also in good agreement with KLOE08 measurement. The new measurement represents an important check of previous measurements, because this last one does not rely on the value of the integrated luminosity and on the radiator knowledge.

The preliminary value for $a_\mu^{\pi\pi}$ has been computed and compared with previous KLOE results (see Table 2). These results are consistent with the previous estimation

Analysis	$\Delta a_\mu^{\pi\pi}(0.35-0.85\ \mathrm{GeV}^2)\times 10^{10}$
KLOE11	$376.4 \pm 1.2_{\mathrm{stat}} \pm 4.1_{\mathrm{sys+theo}}$
KLOE10	$376.6 \pm 0.9_{\mathrm{stat}} \pm 3.3_{\mathrm{sys+theo}}$
	$\Delta a_\mu^{\pi\pi}(0.35-0.95\ \mathrm{GeV}^2)\times 10^{10}$
KLOE11	$384.1 \pm 1.2_{\mathrm{stat}} \pm 4.0_{\mathrm{sys}} \pm 1.2_{\mathrm{theo}}$
KLOE08	$387.2 \pm 0.5_{\mathrm{stat}} \pm 2.4_{\mathrm{sys}} \pm 2.3_{\mathrm{theo}}$

Table 2. Comparison between the new preliminary result KLOE11 and the published KLOE10 and KLOE08.

thus confirming the current disagreements of about 3σ between the a_μ experimental determination and its Standard Model prediction.

5. Conclusions

A preliminary evaluation of dipion cross section, pion form factor, and leading order contribution to muon anomaly by using the $\pi\pi\gamma\ /\mu\mu\gamma$ ratio measurement (KLOE11) has been performed, analyzing the 239.2 pb^{-1} KLOE data. The comparison of this preliminary measurement with the previous KLOE08[10] and KLOE10[11] showed a good agreement and represents a powerful check, the $\pi\pi\gamma\ /\mu\mu\gamma$ result not depending on the integrated luminosity and on the radiator function knowledge. We also presented the evaluation of $\sigma(e^+e^- \to \mu^+\mu(\gamma))$ normalized to integrated luminosity showing a good agreement with the PHOKHARA MC [12] prediction. The present preliminary evaluation of hadronic leading order contribution a_μ^{hlo} confirms the 3σ discrepancy between SM prediction and experimental measurement [4]. This result remarks the relevance not only of the expected future improvements on the experimental measurements of the muon anomaly, in program at JLAB and JPARC laboratories, but also of its theoretical prediction through the $\pi^+\pi^-$ cross section measurement, with the contribution of the KLOE-2 experiment[5, 21, 22] for the a_μ^{hlo} contribution. KLOE-2 will also contribute to the reduction of the a_μ uncertainty from the hadronic light-by-light contribution[5].

References

[1] Jegerlehner F and Nyffeler A 2009 *Phys. Rept* **477** 1
[2] Davier M, Hoecker A, Malaescu B and Zhang Z 2011 *Eur. Phys. J.* C **71** 1515
[3] Hagiwara K, Liao R, Martin A D, Nomura D and Teubner T 2011 *J. Phys. G* **38** 085003
[4] Bennett G W *et al.* 2006 *Phys. Rev.* D **73** 072003
[5] Mascolo M 2012 *J. Phys:Conf. Ser.* [M. Mascolo on behalf of KLOE-2 Collaboration - presented in this Workshop]
[6] Binner S, Kühn J H and Melnikov K 1999 *Phys. Lett.* B **459** 279
[7] Adinolfi M *et al.* 2002 *Nucl. Inst. Meth.* A **488** 51
[8] Adinolfi M *et al.* 2002 *Nucl. Inst. Meth.* A **482** 364

[9] Aloisio A *et al.* 2005 *Phys. Lett.* B **606** 12 [KLOE Collaboration]
[10] Ambrosino F *et al.* 2009 *Phys. Lett.* B **670** 285 [KLOE Collaboration]
[11] Ambrosino F *et al.* 2011 *Phys. Lett.* B **700** 102 [KLOE Collaboration]
[12] Czyż H, Grzelińska A, Kühn J and Rodrigo G 2005 *Eur. Phys. J.* C **39** 411
[13] Ambrosino F *et al.* 2006 *Phys. Lett.* B **634** 148 [KLOE Collaboration]
[14] Amsler C *et al.* 2008 *Phys. Lett.* B **667** 1 [Particle Data Group]
[15] Brodsky S and de Rafael E 1968 *Phys. Rev.* **174** 1835
[16] Akhmetshin R R *et al.* 2007 *Phys. Lett.* B **648** 28
[17] Akhmetshin R R *et al.* 2006 *JETP Lett.* **84** 413
[18] Achasov M *et al.* 2006 *J. Exp. Theor. Phys.* **103** 380
[19] Aubert B *et al.* 2009 *Phys. Rev. Lett.* **103** 231801 [BaBar Collaboration]
[20] Ambrosino F *et al.* 2006 *Eur. Phys. J.* C **47** 589 [KLOE Collaboration]
[21] Bossi F *et al.* 2008 *Rivista Del Nuovo Cimento* **31** 531 [KLOE Collaboration]
[22] Amelino-Camelia G *et al.* 2010 *Eur. Phys. J.* C **68** 619 [KLOE KLOE-2 Collaboration]

Journal of Physics: Conference Series **349** (2012) 012014 doi:10.1088/1742-6596/349/1/012014

$\gamma\gamma$ physics at KLOE/KLOE-2

Matteo Mascolo (on behalf of the KLOE/KLOE-2 collaboration)
Dipartimento di Fisica, Università di Roma Tor Vergata, Roma, Italy
INFN, Sezione di Roma Tor Vergata

E-mail: matteo.mascolo@roma2.infn.it

Abstract. Two-photon physics from e^+e^- collision is studied at DAΦNE. The processes $e^+e^- \to e^+e^-X$, with X being either the η meson or $\pi^0\pi^0$ have been studied with beams colliding at $\sqrt{s} \simeq 1$ GeV, below the ϕ resonance peak and without tagging of the outgoing e^+e^-. Preliminary results are presented on the observation of the $\gamma\gamma \to \eta$ process, with both $\eta \to \pi^+\pi^-\pi^0$ and $\eta \to \pi^0\pi^0\pi^0$ channels, and the evidence for $\gamma\gamma \to \pi^0\pi^0$ production at low $\pi^0\pi^0$ invariant mass. The process $e^+e^- \to e^+e^-\pi^0$ will be studied at KLOE-2 running the machine at ϕ peak ($\sqrt{s} = 1.02$ GeV) thanks to new lepton tagger detectors. In particular, the possibility to measure the width $\Gamma_{\gamma\gamma\to\pi^0}$ and the $\pi^0\gamma\gamma^*$ form factor, $F(Q^2)$, at low invariant masses of the virtual photon (in the space-like region) is considered. The feasibility of these measurements is estimated on the basis of a Monte Carlo simulation, whose results are presented here.

1. Introduction

1.1. *Two-photon physics at DAΦNE*

The term "$\gamma\gamma$ physics" stands for the study of the reaction (of order α^4) $e^+e^- \to e^+e^-\gamma^*\gamma^* \to e^+e^-X$, where X is some arbitrary final state allowed by conservations laws . Since the two photons are in a $C = +1$ state and the value $J = +1$ is excluded by the Landau-Yang theorem [1] if they can be considered *quasi-real*, $\gamma\gamma$ scattering at e^+e^- colliders give access to states with $J^{PC} = 0^{\pm+}, 2^{\pm+}$, not directly coupled to one photon $J^{PC} = 1^{--}$. The number of produced events, N_{eeX}, can be estimated from the expression:

$$N_{eeX} = L_{ee} \int dW_{\gamma\gamma} \frac{dL}{dW_{\gamma\gamma}} \sigma(\gamma\gamma \to X)$$

where L_{ee} is the e^+e^- integrated luminosity, $W_{\gamma\gamma}$ the center of mass energy ($W_{\gamma\gamma} = M_X$), $\frac{dL}{dW_{\gamma\gamma}}$ the $\gamma\gamma$ flux and $\sigma(\gamma\gamma \to X)$ the cross section to a given final state, X.

The photon flux at DAΦNE ($\sqrt{s} = 1.02$ GeV) is shown in Fig. 1, where accessible final states are also indicated.

1.2. *$\gamma\gamma$ collisions with KLOE at DAΦNE*

The cross section $\sigma(\gamma\gamma \to X)$ for light mesons production (namely $X = \pi^0, \eta, \eta'$) has been studied over the years by many e^+e^- colliders, from PETRA to CESR to LEP, mostly at a center of mass energy $7 < \sqrt{s} < 90$ GeV. However, in the low energy region ($m_\pi < W_{\gamma\gamma} < 700$

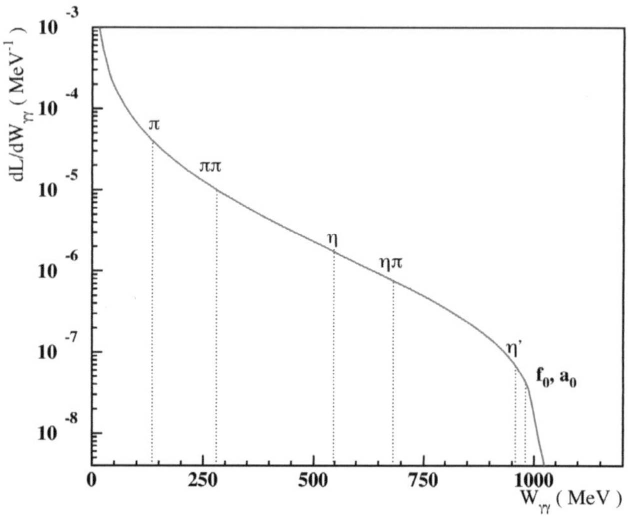

Figure 1. Photon-photon flux at DAΦNE as a function of $W_{\gamma\gamma}$ for a machine integrated luminosity, L_{ee}, of $1\,\text{fb}^{-1}$ and a center of mass energy of $\sqrt{s} =$ 1.02 GeV.

MeV) the experimental situation is unsatisfactory because of large statistical and systematic uncertainties due to small data samples, large background contributions, very small detection efficiency and particle identification ambiguities for low-mass hadronic systems.

DAΦNE is a e^+e^- collider designed to operate at the center of mass energy of $\sqrt{s} \simeq 1.02$ GeV, namely the ϕ meson mass. It has provided to the KLOE experiment an integrated luminosity of about $2.5\,\text{fb}^{-1}$ on peak of the ϕ meson from year 1999 up to 2005 and also $250\,\text{pb}^{-1}$ at the off-peak center of mass energy of $\sqrt{s} = 1$ GeV in 2006.

The KLOE detector (Fig. 3) consists of a large volume drift chamber [2] (~ 3.3 m length and ~ 2 m radius) surrounded by an electromagnetic lead-scintillating fibers calorimeter (EmC) [3] with excellent time ($\sigma_t = 57$ ps$/\sqrt{E(\text{GeV})}\oplus 100$ ps) and good energy ($\sigma_E/E = 5.7\%/\sqrt{E(\text{GeV})}$) resolution. The whole detector is surrounded by an iron yoke coupled to a superconducting coil providing an axial field of nearly ~ 0.5 T. This allows the reconstruction of charged particles momenta with resolution $\sigma_p/p \simeq 0.4\%$ ($\sigma_p/p \simeq 1\%$) for large (small) angle tracks coming from the collision point. The trigger system [4] requires at least two energy deposits above threshold in the calorimeter, not in the same end cap: the threshold is about 50 MeV for the barrel and 150 MeV for each end cap.

2. KLOE measurements

The $\gamma\gamma$ processes studied with KLOE are $e^+e^- \to e^+e^-\eta$ and $e^+e^- \to e^+e^-\pi^0\pi^0$ with e^+e^- in the final state going undetected along the beam pipe. The sample used for the present analyses consists of $240\,\text{pb}^{-1}$ (integrated luminosity) data collected at $\sqrt{s} = 1$ GeV, which allows the reduction of the background coming from ϕ decays. Data are processed with a dedicated filter asking for at least two prompt photons, clusters not associated to tracks and propagating with $|\vec{r} - ct| \sim 0$, with energy $E > 15$ MeV and polar angle $20° < \theta < 160°$, the most energetic one with $E > 50$ MeV, the fraction of energy carried by photons $R = (\sum_\gamma E_\gamma)/E_{cal} > 0.3$ and the total energy in the calorimeter $100 < E_{cal} < 900$ MeV. The latter requirement rejects low energy background and high rate processes like $e^+e^- \to e^+e^-$, $e^+e^- \to \gamma\gamma$.

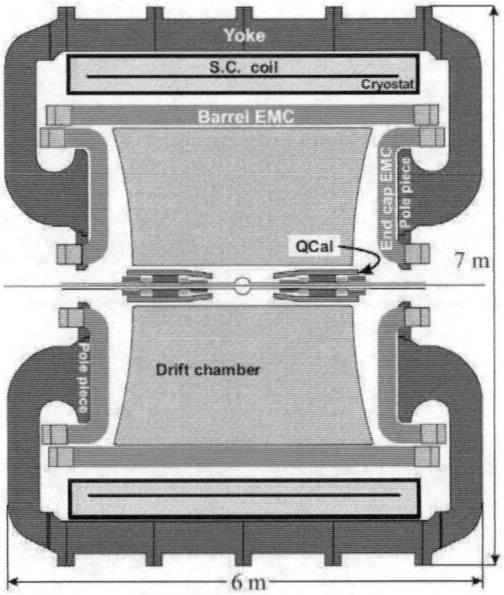

Figure 2. A schematic view of the KLOE detector.

Figure 3. The KLOE detector.

2.1. Observation of $\gamma\gamma \to \eta$, with $\eta \to \pi^+\pi^-\pi^0$

The selection of these events asks for two photons constrained to originate from a π^0 decay and two tracks with opposite curvature coming from the collision point. The sum of the track momenta is constrained to be $|\vec{p}_1| + |\vec{p}_2| < 700$ MeV in order to suppress higher momentum values, typical of $e^+e^- \to \pi^+\pi^-\pi^0$ or $e^+e^- \to \omega(\to \pi^+\pi^-\pi^0)\pi^0$. The charged pion mass is assigned to the two tracks and a least squares function based on Lagrange multipliers imposes that $\pi^+\pi^-\pi^0$ come from an η decay. Therefore most background events are suppressed, except for the irreducible process $e^+e^- \to \eta\gamma \to \pi^+\pi^-\pi^0\gamma$, for which the requirement to have two and only two photons constrains the 3π system to be emitted with longitudinal momentum $p_L \simeq 350$ MeV and recoil missing mass $M_{miss} \simeq 0$; the monochromatic photon, $E_\gamma = 350$ MeV, is lost in the beam pipe. On the other hand, signal events are characterized by the following parabolic relation between M^2_{miss} and p_L:

$$M^2_{miss} \approx s + m^2_\eta - 2E_T\sqrt{s} - \frac{p^2_L}{E_T}\sqrt{s} \qquad E_T = \sqrt{p^2_T + m^2_\eta} \approx m_\eta \qquad (1)$$

with negligible transverse momentum, a continuous spectrum in p_L, $-350 < p_L < 350$ MeV, and high M^2_{miss} values. Fig. 4 (right) shows this difference in the correlation between the squared missing mass, M^2_{miss}, and the longitudinal momentum, p_L, for the signal (top) and for the irreducible $e^+e^- \to \eta\gamma \to \pi^+\pi^-\pi^0\gamma$ background (bottom).

Further criteria are applied for suppressing processes with photons and e^+e^- as tracks in the final state, such as $\eta(\to 3\pi^0)\gamma$ with photon conversion or radiative Bhabha events, $e^+e^-\gamma(\gamma)$. Event distributions in p_L and M^2_{miss} from data are independently fitted with the superposition of MC shapes for signal and background (Fig. 4 (left)). A preliminary value for the cross section is $\sigma_{\gamma\gamma\to\eta\to\pi^+\pi^-\pi^0}(1 \text{ GeV}) = 41.7 \pm 4_{stat}$ pb. Systematics are under evaluation.

The cross section of the irreducible process $e^+e^- \to \eta\gamma \to \pi^+\pi^-\pi^0\gamma$ is evaluated using the same sample of data and asking for three photons in the final state. A kinematic fit is performed, requiring energy and momentum conservation, and the improved variables are used to fit the distribution of the recoil photon energy and the distribution of the invariant mass of the $\pi^+\pi^-\pi^0$ system. The preliminary result is $\sigma_{e^+e^-\to\eta\gamma}(1 \text{ GeV}) = 0.866\,(9)_{stat}\,(93)_{syst}$ nb, where the systematic uncertainty is given by residual $e^+e^- \to \omega\pi^0 \to \pi^+\pi^-\pi^0\pi^0$ background.

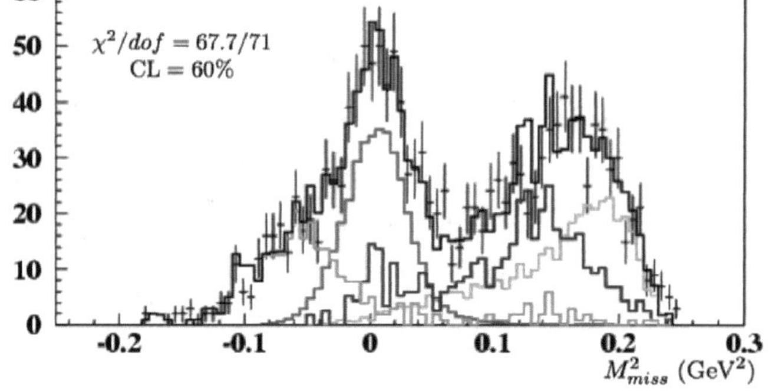

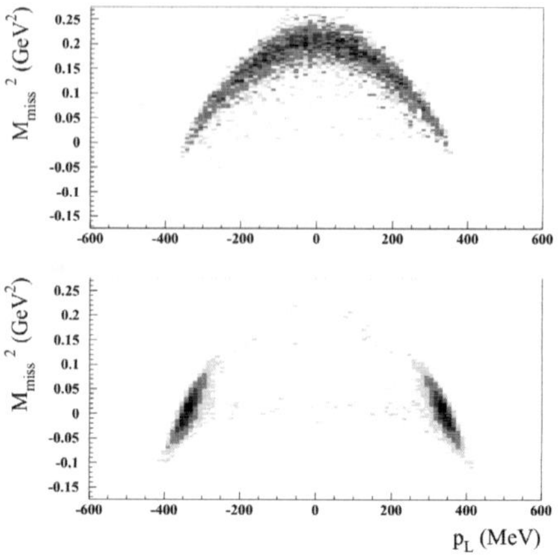

Figure 4. Left: fit in dN/dM^2_{miss}: signal (cyan), $\eta\gamma$ (red) and $e^+e^-\gamma(\gamma)$ (green) are well visible; Right: correlation between the squared missing mass, M^2_{miss}, and the longitudinal momentum, p_L, for the signal (top) and for the irreducible background $e^+e^- \to \eta\gamma \to \pi^+\pi^-\pi^0\gamma$ (bottom).

2.2. Observation of $\gamma\gamma \to \eta$, with $\eta \to \pi^0\pi^0\pi^0$

The main backgrounds for this analysis are annihilation processes with at least four prompt photons in the final state (i.e. $e^+e^- \to \eta\gamma$, $e^+e^- \to K_SK_L$, $e^+e^- \to \omega\pi^0$), for which accidental or split calorimeter clusters increase the photon multiplicity. Events with six and only six prompt photons in the final state and with no tracks in the drift chamber are selected. The photons are paired choosing the combination which minimizes the χ^2-like variable

$$\chi^2_{6\gamma} = \frac{(m_{\pi^0} - m_{ij})^2}{\sigma^2_{ij}} + \frac{(m_{\pi^0} - m_{mn})^2}{\sigma^2_{mn}} + \frac{(m_{\pi^0} - m_{kl})^2}{\sigma^2_{kl}}, \tag{2}$$

where m_{ij} is the invariant mass of each pair of photons and σ_{ij} the resolution. Then, a kinematic fit is performed asking for the six photons invariant mass to be equal to the mass of the η meson. A cut is then applied on the energy of the most energetic photon, to reject $e^+e^- \to \eta\gamma$ events in which the monochromatic photon is detected. The squared missing mass, M^2_{miss}, distribution is well described by the $\gamma\gamma \to \eta \to \pi^0\pi^0\pi^0$ and $\eta(\to \pi^0\pi^0\pi^0)\gamma$ MC shapes. Fig. 5, left, shows the fit in M^2_{miss}.

The preliminary result obtained from the fit is $\sigma_{\gamma\gamma\to\eta\to3\pi^0}(1\text{ GeV}) = 37.0 \pm 1.4_{stat} \pm 2.2_{syst}$ pb. We also quote a preliminary value for the cross section of the irreducible background $e^+e^- \to \eta\gamma$, $\sigma_{e^+e^-\to\eta\gamma}(1\text{ GeV}) = 0.875\,(18)_{stat}\,(35)_{syst}$ nb, in agreement with that measured in the $e^+e^- \to \eta\gamma \to \pi^+\pi^-\pi^0\gamma$ analysis. This is shown in Fig. 5, right, among other experimental results [5] as a function of $\sqrt{s}$.

3. KLOE measurement: $\gamma\gamma \to \pi^0\pi^0$

The major sources of background for this analysis is represented by annihilation processes with four or more prompt photons in the final state: $e^+e^- \to K_SK_L$, $e^+e^- \to \eta\gamma$, $e^+e^- \to \omega\pi^0$, $e^+e^- \to f_0(980)(\to \pi^0\pi^0)\gamma$. In addition, due to the possibility of cluster splitting, the $e^+e^- \to \gamma\gamma$ process is also considered as a source of background. Selected events have no tracks in the drift chamber and four prompt photons in the final state with polar angle $23° < \vartheta < 157°$ and energy $E_\gamma > 15$ MeV. The photons are paired choosing the combination which minimizes the χ^2-like variable of Eq. (2) in the case of four photons coming from two neutral pions, $\chi^2_{4\gamma}$. Events with $\chi^2_{4\gamma} > 4$ are rejected: the effect of this selection is shown in Fig. 6, left. To reject

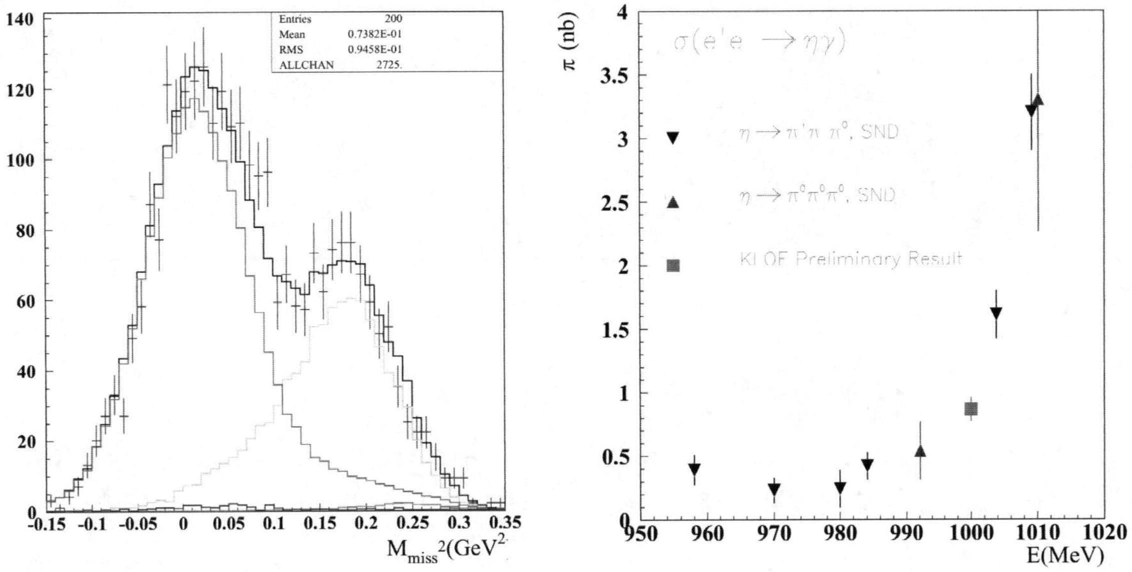

Figure 5. $\gamma\gamma \to \eta \to \pi^0\pi^0\pi^0$ analysis. Left: squared missing mass distributions for data (points whith error bars) and backgrounds (histograms) normalized according to the fit results. Colour code: red = $e^+e^- \to \eta\gamma$, light blue = signal, blue = $e^+e^- \to \omega\pi^0$. Right: preliminary KLOE value for $\sigma(e^+e^- \to \eta\gamma)$ at $\sqrt{s} = 1$GeV (red point), and SND results [5] for several values of $\sqrt{s}$.

$e^+e^- \to K_S K_L$ events, where a large amount of non-prompt energy is released in the detector a tighter cut on the energy fraction carried by photons, $R = (\sum_\gamma E_\gamma)/E_{cal} > 0.75$, is applied. The four photon invariant mass spectrum for the selected data sample is shown in Fig. 6, right, together with the normalized background Monte Carlo simulations. The excess of events with respect to the expected annihilation processes, in the low invariant mass region, is an indication of the $\gamma\gamma \to \pi^0\pi^0$ production.

The determination of the differential cross section $d\sigma(\gamma\gamma \to \pi^0\pi^0)/dM_{4\gamma}$ and of other possible residual backgrounds is in progress.

4. KLOE analyses: conclusions

From an integrated luminosity of 240 pb^{-1} of data collected at DAΦNE operating at $\sqrt{s} \simeq 1$ GeV, the following preliminary results in $\gamma\gamma$ analyses are achieved:

- unambiguous signature of both $\gamma\gamma \to \eta$ and $\gamma\gamma \to \pi^0\pi^0$ events, without any $e^\pm$ tagger;
- $\gamma\gamma \to \eta$ events are observed through both $\eta \to \pi^+\pi^-\pi^0$ and $\eta \to \pi^0\pi^0\pi^0$ channels, with independent systematics;
- from the same data sample, an exploratory research shows a structure at small values of the $M_{4\gamma}$ spectrum, where the process $e^+e^- \to e^+e^-\pi^0\pi^0$ is expected.

The cross section $\sigma(e^+e^- \to \eta\gamma)$ at $\sqrt{s} = 1$ GeV has been determined as well, with accuracy better than the closer data points.

These results are encouraging also in view of the forthcoming data taking campaign of the KLOE-2 project [6]

Journal of Physics: Conference Series **349** (2012) 012014 doi:10.1088/1742-6596/349/1/012014

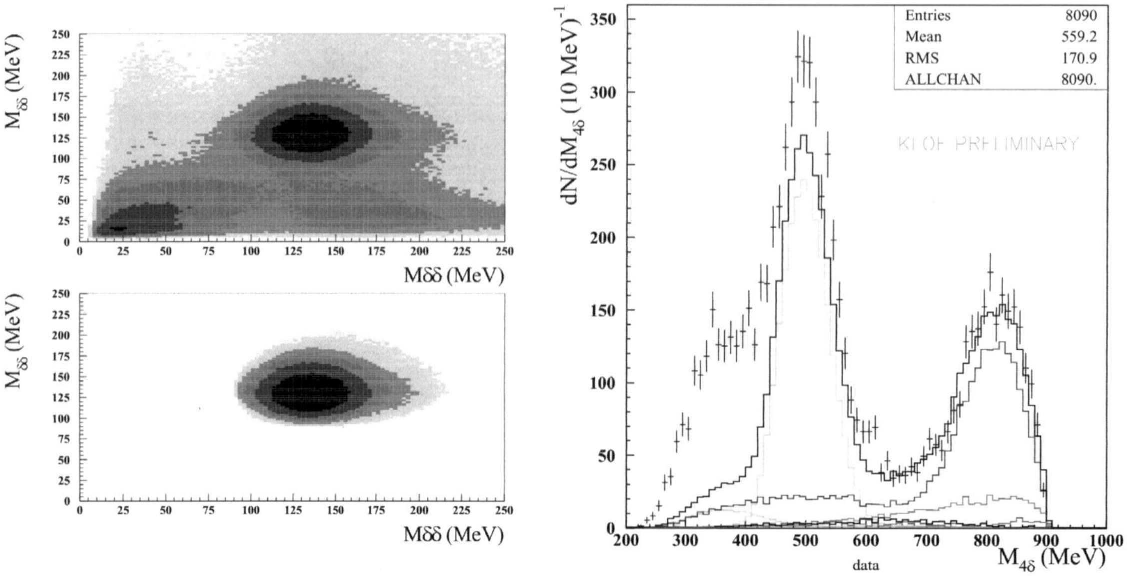

Figure 6. $\gamma\gamma \to \pi^0\pi^0$ analysis. Left: scatter plots in two photons pairs invariant masses before (up) and after (down) rejecting bad $\chi^2_{4\gamma}$ events; selected events have both photons pairs invariant masses centered around π^0 mass value. Right: 4γ invariant mass spectrum for data (points with error bars) and backgrounds. Colour code: light blue = $e^+e^- \to K_S K_L$, blue = $e^+e^- \to \omega\pi^0$, violet = $e^+e^- \to f_0(980)\gamma$, green = $e^+e^- \to \eta\gamma$, red = $e^+e^- \to \gamma\gamma$.

5. Two-photon physics at KLOE-2

After recent luminosity upgrade [7] performed on DAΦNE collider, a new data taking campaign has been approved for the KLOE experiment, together with several detector upgrades. One of the most significant features, characterising this new project called KLOE-2, is the presence of a small angle tagging system composed of two couples of new detectors (one couple for each accelerator's arm) referred to as the Low Energy Tagger (LET) and High Energy Tagger (HET). These detectors will allow the study of $\gamma\gamma$ physics running the collider at ϕ-peak in the center of mass energy. A detailed description of all new detectors and of the KLOE-2 scientific program can be found in Ref. [8].

5.1. The taggers

The $\gamma\gamma$ processes differ from the ϕ-decay because of the presence of e^+e^- in the final state. The detection of these final leptons is absolutely mandatory to close the kinematics of the $\gamma\gamma$ interactions, making them independent from backgrounds. Table 1 lists all interesting $\gamma\gamma$ physics channels together with at least one ϕ-decay which could produce an identical final state in the detector once outgoing e^+e^- from $\gamma\gamma$ events go undetected along the beam pipe.

In order to disentangle signal from background a dedicated lepton tagging system has been designed and built to catch those e^+e^- which are lost along the beam pipe. According to DAΦNE layout, only two possible positions were found to install new detectors: one next to the interaction point (where LET detector is placed) and the other one after the first bending dipole magnet of the machine (for the HET detector). The LET is placed inside the KLOE detector ($\sim$ 1 m away from IP) to catch lower energy leptons, i.e. leptons which emitted a high energy photon. The HET is placed 11 m away from IP and detects higher energy leptons, that is those leptons which have lost such a tiny amount of energy that they manage to reach the exit of the

$\gamma\gamma$ channel	Missing	Signal Events	ϕ-decay	Missing	BG events
$e^+e^- \to e^+e^-\pi^0\pi^0$	e^+e^-	2×10^4	$K_S(\pi^0\pi^0)K_L$	K_L	$\sim 10^9$
$e^+e^- \to e^+e^-\pi^+\pi-$	e^+e^-	2×10^6	$K_S(\pi^+\pi^-)K_L$	K_L	$\sim 2 \times 10^9$
$e^+e^- \to e^+e^-\pi^+\pi-$	e^+e^-	2×10^6	$\pi^+\pi^-\pi^0$	π^0	$\sim 1 \times 10^9$
$e^+e^- \to e^+e^-\eta$	e^+e^-	1×10^6	$\eta(\gamma\gamma)\gamma$	γ	$\sim 10^8$
$e^+e^- \to e^+e^-\pi^0$	e^+e^-	4×10^6	$\pi^0(\gamma\gamma)\gamma$	γ	$\sim 5 \times 10^8$

Table 1. Possible background processes for $\gamma\gamma$ physics coming from ϕ meson decay. Number of events are estimated for an integrated luminosity of $1\,\text{fb}^{-1}$.

dipole.

In order to understand what kind of detector is needed in the region of interest the tracking of the off-energy particles along DAΦNE optics with BDSIM[9] (a GEANT4 extension toolkit capable of simulating nominal and off-energy particle transport in the accelerator beam line) is performed. Trajectories of particles coming from the IP with energies from 5 MeV to 510 MeV (which is the beam nominal energy of DAΦNE) are calculated in step of 0.5 MeV, in order to infer the energy distribution of leptons as a function of their position.

Energy distribution of leptons hitting the beam pipe wall in the LET region shows that there is essentially no correlation between their energy and z-coordinate (the beam line coordinate), hence a calorimetric choice for this detector is the most suitable one (for details see Ref. [10]).

Concerning HET detector instead, simulation shows that off-energy leptons do not hit the vacuum chamber wall in the corresponding region, being simply deviated from the nominal orbit in the x-coordinate direction (the one pointing to the collider center). In Fig. 7 (left) the distance between the tracked lepton and the nominal orbit is plotted as a function of the energy of the particle. The linear correlation between these quantities is striking. Moreover from the fit parameters, it possible to see that in order to obtain an energy resolution of about 0.6 MeV, a spatial resolution of 1 mm is sufficient.

Another important feature of particles surviving the travel up to the exit of the dipole magnet is that they are the ones having a narrower angular distribution at IP (see Fig. 7 (right)). This will justify the fact that photons associated with tagged leptons have very low q^2 (see section 6.1).

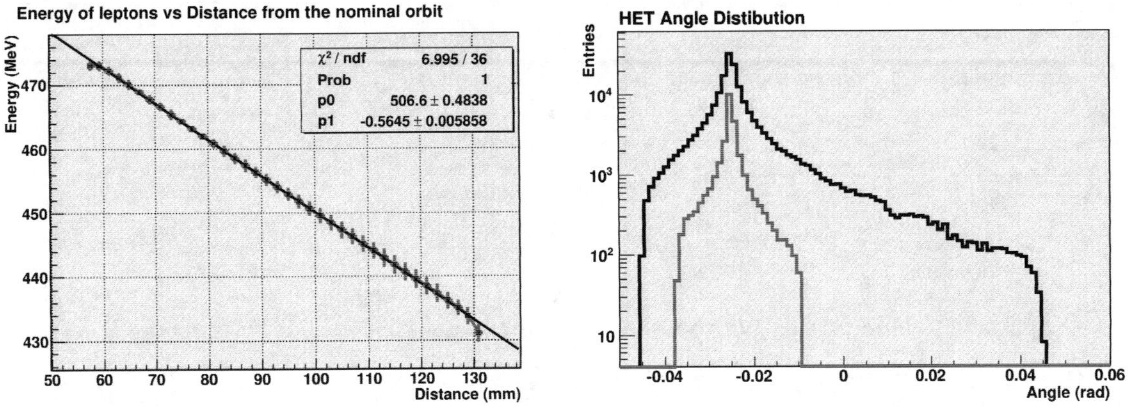

Figure 7. Tracking to HET region. Left: lepton energy versus its trajectory distance (in x direction) from nominal orbit in HET region. Right: Comparison between lepton angular distribution at IP for all leptons (black) and for the ones reaching the HET region (red).

The presence of a flange just after the dipole, together with the discussed favorable lepton displacement in the HET region, suggest that the optimal choice for this second tagging station is to install a position detector here.

Physical requirements for the HET detector are the following: good time resolution to disentangle each bunch coming with a period of $\sim$ 2.7 ns; capability to acquire data at a frequency of 368 MHz in order to allow event reconstruction with KLOE apparatus; radiation hardness to let the detector stand $\sim$ 30 mm from the beam (a closer position would interfere with the proper functioning of the machine) for long term acquisition and tiny size in order to install it through the flange inside the DAΦNE vacuum chamber. The proposed tagger detector consists of a set of 28 plastic scintillators (3 mm $\times$ 5 mm $\times$ 6 mm) plus one additional long scintillator (3 mm $\times$ 5 mm $\times$ 6 mm) for coincidence purposes, coupled to properly shaped light guides which convey light signals out of the beam pipe to high quantum efficiency photomultipliers (PMT). A drawing of the detector inside the beam pipe is represented in Fig. 8 (left), together with a real picture of the detector itself in place (right).

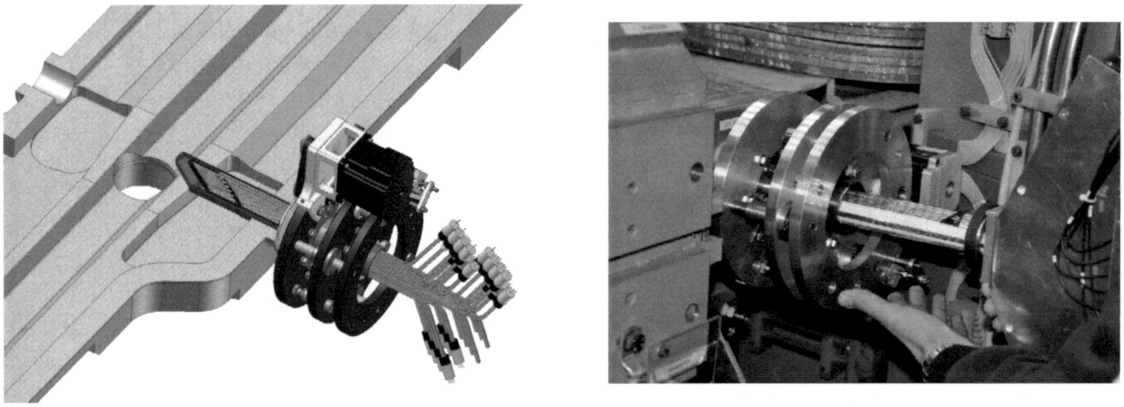

Figure 8. HET detector. Right: a drawing of the High Energy Tagger placed in the accelerator's flange. Left: a picture taken during HET detector insertion. The green object oh the right is the bending dipole magnet.

6. The process $\gamma\gamma \to \pi^0$

Concerning the $\gamma\gamma$ physics program for the KLOE-2 experiment, the collaboration is planning to perform two interesting measurements in single π^0 production process (through $e^+e^- \to e^+e^-\pi^0$ process), that is:

- a precision determination of the $\gamma\gamma \to \pi^0$ decay width: $\Gamma_{\gamma\gamma\to\pi^0}$;
- a measurement of the $\gamma^*\gamma \to \pi^0$ transition form factor, $F(Q^2)$, in the unexplored region of low-momentum transfer.

A complete feasibility study of these two measurements has been performed and is available in Ref. [11].

6.1. The basics of width and form factor measurements

Since the original proposal of Low [12] to measure the width of the neutral pion decay into two photons through the $e^+e^- \to e^+e^-\pi^0$ process, only with the Crystal Ball detector at DESY this measurement has been done using this method, with the result $\Gamma_{\gamma\gamma\to\pi^0} = (7.7 \pm 0.5 \pm 0.5)$ eV [13]. It was stressed ([14] and [15]) that for a precision measurement of π^0 width via "$\gamma\gamma$

fusion" ($\gamma\gamma \to \pi^0$) process, one needs to improve the original Low's proposal. Namely, instead of a no-tag experiment (like [13]) one should perform a lepton double-tagging at small angles. Thanks to the lepton tagging system, and the HET in particular, it is now possible to perform that double tagging at KLOE-2.

One can extract the value of the partial decay width from data, using the formula

$$\Gamma_{\pi^0\to\gamma\gamma} = \frac{N_{\pi^0}}{\epsilon\mathcal{L}} \frac{\tilde{\Gamma}_{\pi^0\to\gamma\gamma}}{\tilde{\sigma}_{e^+e^-\to e^+e^-\pi^0}} \tag{3}$$

where N_{π^0} is the number of detected pions, ϵ accounts for the detection acceptance and efficiency, $\mathcal{L}$ is the integrated luminosity, $\tilde{\Gamma}_{\pi^0\to\gamma\gamma}$ is the π^0 width calculated from the theoretical model and $\tilde{\sigma}_{e^+e^-\to e^+e^-\pi^0}$ the cross section obtained with a high statistics Monte Carlo simulation using the same model as for the $\tilde{\Gamma}_{\gamma\gamma\to\pi^0}$ calculation.

The form factor $F(Q^2)$ can be evaluated through the relation:

$$\frac{F^2(Q^2)}{F^2(Q^2)_{MC}} = \frac{\left(\frac{d\sigma}{dQ^2}\right)_{data}}{\left(\frac{d\sigma}{dQ^2}\right)_{MC}} \tag{4}$$

where $\left(\frac{d\sigma}{dQ^2}\right)_{data}$ is the experimental differential cross section, and $\left(\frac{d\sigma}{dQ^2}\right)_{MC}$ is the Monte Carlo one obtained with the from factor $F(Q^2)_{MC}$.

The π^0 production in the process $e^+e^- \to e^+e^-\pi^0$ is simulated with EKHARA [16] Monte Carlo generator. The simulated signal is given only by the t-channel amplitude ($\gamma^*\gamma^* \to \pi^0$). Since a stand-alone EKHARA version works in the center of mass frame of incident leptons and does not simulate the pion decays, it has been modified to take into account the DAΦNE crossing angle between the incoming beams ($\theta_{e^+e^-} \approx 51.3$ mrad) and the decay of the π^0 into two photons. Tracking simulation shows that only few of the emitted leptons ($\sim 2\%$) reach the HET detectors, providing a realistic estimate of the coincidence acceptance. In the following the coincidence of the HET detectors (briefly HET-HET coincidence) will be required, which selects the energy of the final leptons to be between 420 and 460 MeV (single HET acceptance is $420 < E < 490$ MeV).

6.2. $\gamma\gamma \to \pi^0$: *feasibility of the* π^0 *width measurement*

Fig. 9, left, shows the energy of the emitted π^0 in the $\gamma\gamma$ process: as can be seen, the request of HET-HET coincidence allows the selection of π^0 which are almost at rest (dark region), compared with the no-tag case (light-gray). Since the π^0 decays almost at rest, most of the photons from its decay are emitted with large polar angle (defined as the angle between the direction of the photon and the beam axis), as show in Fig. 9 (right).

In particular, about 95% of the decay photons are emitted above 25° and below 155°, resulting in a large acceptance for photons reaching the KLOE electromagnetic calorimeter.

By requiring both photons in the barrel of the EmC (i.e. between 50° and 130°) an the HET-HET coincidence, a value for the acceptance ϵ_{acc} of 1.2% is obtained with the detector 30 mm far away from nominal orbit (from this moment on x_0 will be the distance of closest plastic scintillator to nominal orbit from the nominal orbit itself). Since the total cross-section is $\sigma_{tot} \approx 0.28$ nb, a visible cross section of about 3.4 pb is obtained within the acceptance cuts. So the integrated luminosity $\mathcal{L}$ at DAΦNE required to reach a 1% statistical error (for $x_0 = 30$ mm) is:

$$\mathcal{L} = \frac{10000}{\sigma_{tot}\epsilon_{acc}\epsilon_{det}} \approx \frac{3}{\epsilon_{det}}\text{fb}^{-1} \tag{5}$$

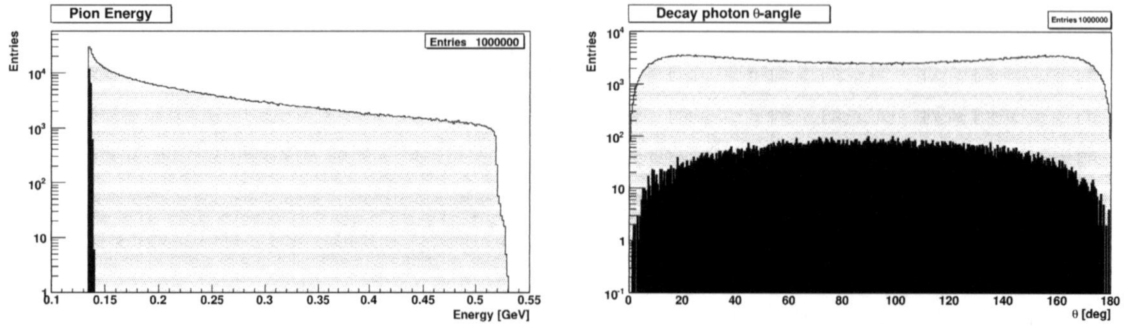

Figure 9. The π^0 energy (left) and the polar angle (right) distribution in the laboratory frame with (dark) and without (light-gray) HET-HET coincidence.

where the efficiency ϵ_{det} due to trigger, reconstruction and analysis criteria is estimated to be about 50%, by means of the KLOE Monte Carlo code (`GEANFI` [17]). Therefore, the required data sample can be obtained during the first phase (about one year) of data taking.

Extraction of the width $\Gamma_{\gamma\gamma\to\pi^0}$ with $\sim 1\%$ accuracy requires a very good control of the systematic errors. From the experimental point of view, the clean signature of the process (the photons from π^0 decay are emitted in back-to-back configuration because of pion momentum, see Fig.9 (left)), the use of the KLOE detector and the HET-HET coincidence should allow to keep systematic effects under control at the required level of precision. A possible background to this measurement, in particular, comes from radiative Bhabha scattering, which could have the same signature of our signal. An extensive simulation of this background (based on $\sim 10^8$ events generated with Babayaga MC generator [18], [19]) shows that no events survive the coincidence of HET detectors for e^+ and e^- and the KLOE acceptance for photon, and therefore the expected contribution is negligible.

Fig. 10(a) and Fig. 10(b) clarify the Bhabha background issue. In particular, the former shows how the reconstructed invariant mass of leptons hitting the two HET stations from background events produces a wide distribution which crosses the corresponding invariant mass distribution peak (around $m_{\pi^0} \simeq 135$ MeV) from signal.

The distributions in Fig. 10(a) are obviously not on scale. Due to the huge difference in the total cross section, the signal-to-background ratio is extremely unfavorable (about 6×10^{-7}) preventing the disentanglement of the two processes to be performed only with the information coming from the HET detector alone (note that since the HET is unable to measure leptons momenta, the invariant mass evaluation is performed assuming a zero angular deviation from the nominal orbit at IP).

If the requirement of at least one photon in the final state (to be detected by KLOE EmC) is added, the disentanglement becomes very simple. In Fig. 10(b), the angular distribution of photons emitted within a Bhabha scattering process is shown in blue. Asking for both positron and electron to be detected by HET the red angular distribution in the same figure is obtained. It's clearly visible that the HET-HET coincidence selects photons which are almost collinear with the beam (i.e. $0° < \theta < 175°$ and $175° < \theta < 180°$) which cannot be detected by the KLOE EmC, whose angular acceptance is $25° < \theta < 155°$. In the case of $\gamma\gamma$ processes on the contrary, the photons coming from the decay of the pion, produced at rest, are emitted back-to-back almost uniformly in the solid angle (cfr Fig.9 (right)). Signal and background are distinguished with very high efficiency (almost 100%) simply requiring at least a photon in the EmC.

The KLOE trigger itself requires two energy releases in the EmC in order to assert the first level trigger. Since the energy distribution of the decay photons is peaked around 70 MeV and

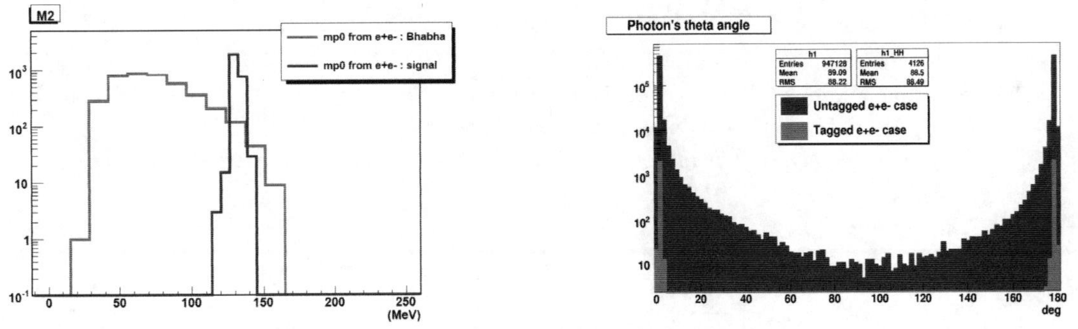

(a) (b)

Figure 10. (a) Reconstructed invariant mass of leptons hitting the HET detectors in coincidence. Blue histogram refers to signal final leptons (generated by EKHARA), red one to Bhabha background (generated by BabaYaga)); (b) Angular distribution of radiative Bhabha photons with (red) and without (blue) the requirement of double tagging (HET-HET coincidence) of the final leptons.

the energy threshold levels of the EmC are comparable with this value, a significant percentage of $\gamma\gamma$ events are lost (especially when the photon is hitting the End Cap region, where threshold is higher). An alternative to increase the trigger efficiency would be to acquire $\gamma\gamma$ events by means of HET-HET coincidence plus a single barrel energy release in KLOE. This kind of trigger logic would not interfere with KLOE data taking and would be, above all, free from radiative Bhabha scattering.

From the theoretical point of view, the systematic errors can arise from the uncertainty in the modeled $\pi^0\gamma^*\gamma^*$ transition form factor. For this kind of effect, numerical simulation with different formulae for the form factor are performed, showing that the imposition of the HET-HET coincidence leads to a significant restriction on the photon virtuality in $\gamma^*\gamma^* \to \pi^0$: for most of the events one has $|q^2| < 10^{-4}\,\mathrm{GeV}^2$ has shown in Fig. 11.

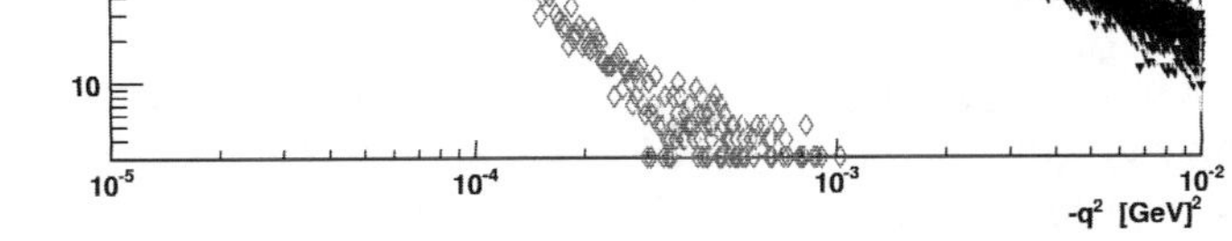

Figure 11. Distribution of the photon virtuality in $\gamma^*\gamma^* \to \pi^0$. The lepton tagging selects the events (red diamonds) with very small virtuality of the corresponding emitted photon.

Thus for the KLOE-2 case the possible effect of the photon virtualities which can influence the accuracy of Eq. (3) is negligible. In particular, the uncertainty in the measurement of $\Gamma_{\gamma\gamma\to\pi^0}$ due to the form factor parametrization in the generator is expected to be less than 0.1%.

6.3. $F(m_{\pi^0}^2, 0, q^2)$: Monte Carlo simulation

By requiring one lepton inside the KLOE drift chamber ($20° < \theta < 160°$, corresponding to $0.01 < |q_1^2| < 0.1\,\mathrm{GeV}^2$) and the other one in the HET detector (corresponding to $|q_2^2| < 10^{-4}\,\mathrm{GeV}^2$ for most of the events, see Fig. 11) one can measure the differential cross section $(d\sigma/dQ^2)_{data}$ where $Q^2 = -q^2$. Using Eq. (4), the form factor $|F(Q^2)|$ can be extracted from this cross section.

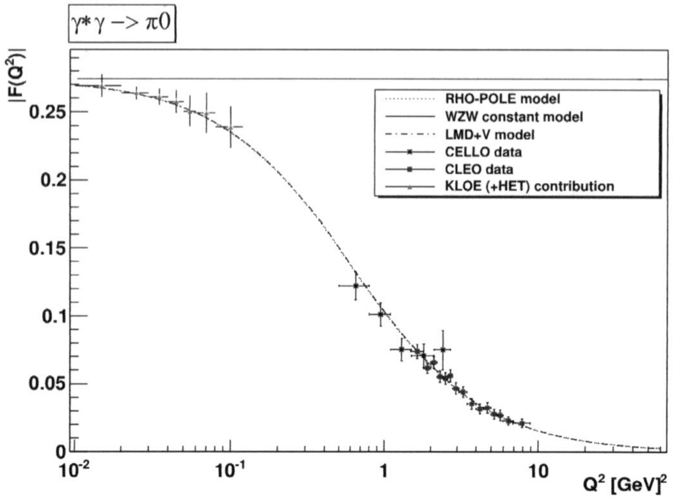

Figure 12. Simulation of KLOE-2 measurement of $F(Q^2)$ with statistical errors for $5\,\mathrm{fb}^{-1}$. Dashed line is the $F(Q^2)$ according to LMD+V model [20], solid line is $F(0)$ given by Wess-Zumino-Witten term. CELLO [21] and CLEO [22] data at higher Q^2 are also shown for illustration.

Fig. 12 shows the expected experimental uncertainty (only statistical) on $F(Q^2)$ achievable at KLOE-2, with an integrated luminosity of $5\,\mathrm{fb}^{-1}$. The detection efficiency in this case is estimated to be about 20%. From the simulation, a statistical uncertainty of less than 6% is expected for each bin.

Once the form factor is measured, one can evaluate also the slope parameter a of the form factor at the origin:

$$a = m_\pi^2 \frac{1}{\mathcal{F}_{\pi^0\gamma^*\gamma^*}(0,0)} \left(\frac{d\mathcal{F}_{\pi^0\gamma^*\gamma^*}(q^2,0)}{dq^2} \right)_{q^2=0} \tag{6}$$

where $\mathcal{F}_{\pi^0\gamma^*\gamma^*}(0,0)$ can be extracted from the $\Gamma_{\gamma\gamma\to\pi^0}$ through:

$$\mathcal{F}^2_{\pi^0\gamma^*\gamma^*}(q_1 = 0, q_2 = 0) = \frac{4}{\pi\alpha^2 m_\pi^3} \Gamma_{\gamma\gamma\to\pi^0}. \tag{7}$$

Slope calculations suffer from model dependence (not accounted for in the error estimation) due to form factor parametrization. The validity of these parametrizations has never been verified, because there are no data at low momenta ($Q^2 < 0.5\,\mathrm{GeV}^2$). Therefore, filling of this experimental gap in Q^2 by the KLOE-2 experiment (at $Q^2 < 0.1\,\mathrm{GeV}^2$) would help to verify the consistency of the form factor parametrizations.

The impact of these measurement on the model parameters (e.g., the normalization of the form factor) and hence on the value and the precision of the contribution of the pion pseudoscalar exchange to the hadronic light-by-light scattering [23] has also been evaluated. According to simulations the proposed measurements can improve the uncertainty on the value $a_\mu^{LbyL;\pi^0}$ of ~ 2 (see Ref. [11] for details).

7. KLOE-2 perspectives: conclusions

A simulation of the KLOE-2 experiment with 1 year of data taking was performed. Numerical results indicate the feasibility of a $\sim 1\%$ statistical error in the measurement of $\Gamma_{\gamma\gamma\to\pi^0}$. Such a precision is better than the current experimental world average and theoretical accuracy. The π^0 electromagnetic transition form factor, $F(Q^2)$, can be measured in the region $0.01 < Q^2 < 0.1\,\mathrm{GeV}^2$ with a statistical error of $< 6\%$ in each bin. This low Q^2 measurement can test the consistency of the models which have been fitted so far to the data from CELLO, CLEO and BaBar [24] at higher Q^2. The proposed measurement with the KLOE-2 experiment can also reduce the error on the hadronic light-by-light contribution to g-2 of the muon by a factor ~ 2.

Bibliography

[1] Yang C N 1950 *Phys. Rev.* **77**(2) 242–245 URL http://link.aps.org/doi/10.1103/PhysRev.77.242
[2] Adinolfi M *et al.* 2002 *Nucl. Instrum. Meth.* A **488** 51
[3] Adinolfi M *et al.* 2002 *Nucl. Instrum. Meth.* A **482** 364
[4] Adinolfi M *et al.* 2002 *Nucl. Instrum. Meth.* A **492** 134
[5] Achasov M N *et al.* (SND Collaboration) 2007 *Phys. Rev.* D **76** 077101
[6] Amelino-Camelia G *et al.* 2010 *The European Physical Journal C - Particles and Fields* **68**(3) 619–681 ISSN 1434-6044 10.1140/epjc/s10052-010-1351-1 URL http://dx.doi.org/10.1140/epjc/s10052-010-1351-1
[7] Zobov M, Alesini D, Biagini *et al.* 2010 *Phys. Rev. Lett.* **104** 174801
[8] Camelia G A *et al.* 2010 *Eur. Phys. J.* C **68** 619
[9] Agapov I, Blair G A, Malton S and Deacon L 2009 *Nucl. Instrum. Meth.* A **606** 708–712
[10] Babusci D *et al.* 2010 *Nucl. Instrum. Meth.* A **617** 81–84
[11] Babusci D *et al.* 2011 *arXiv:1109.2461 (Submitted to EPJC)* (*Preprint* 1109.2461)
[12] Low F E 1960 *Phys. Rev.* **120**(2) 582–583 URL http://link.aps.org/doi/10.1103/PhysRev.120.582
[13] Williams D A a 1988 *Phys. Rev. D* **38**(5) 1365–1376 URL http://link.aps.org/doi/10.1103/PhysRevD.38.1365
[14] Parisi J and Kessler P 1972 *Phys.Rev.* D **5** 2229–2237
[15] Terazawa H 1973 *Rev. Mod. Phys.* **45**(4) 615–662 URL http://link.aps.org/doi/10.1103/RevModPhys.45.615
[16] Czyz H and Ivashyn S 2011 *Computer Physics Communications* **182** 1338 – 1349 ISSN 0010-4655 URL http://www.sciencedirect.com/science/article/pii/S0010465511000671
[17] Ambrosino F *et al.* 2004 *Nuclear Instruments and Methods in Physics Research Section A: Accelerators, Spectrometers, Detectors and Associated Equipment* **534** 403 – 433 ISSN 0168-9002 URL http://www.sciencedirect.com/science/article/pii/S0168900204015633
[18] Calame C C 2001 *Physics Letters B* **520** 16 – 24 ISSN 0370-2693 URL http://www.sciencedirect.com/science/article/pii/S037026930101108X
[19] Calame C C, Montagna G, Nicrosini O and Piccinini F 2004 *Nuclear Physics B - Proceedings Supplements* **131** 48 – 55 ISSN 0920-5632 sIGHADO3 Proceedings of the Workshop on Hadronic Cross Section at Low Energy URL http://www.sciencedirect.com/science/article/pii/S0920563204000106
[20] Knecht M and Nyffeler A 2001 *The European Physical Journal C - Particles and Fields* **21**(4) 659–678 ISSN 1434-6044 10.1007/s100520100755 URL http://dx.doi.org/10.1007/s100520100755
[21] Behrend H J, Criegee L, Field *et al.* 1991 *Zeitschrift fur Physik C Particles and Fields* **49**(3) 401–409 ISSN 0170-9739 10.1007/BF01549692 URL http://dx.doi.org/10.1007/BF01549692
[22] Gronberg J, Hill *et al.* ((CLEO Collaboration)) 1998 *Phys. Rev. D* **57**(1) 33–54 URL http://link.aps.org/doi/10.1103/PhysRevD.57.33
[23] Jegerlehner F and Nyffeler A 2009 *Phys. Rept.* **477** 1–110
[24] Aubert B *et al.* (BABAR Collaboration) 2009 *Phys. Rev. D* **80**(5) 052002 URL http://link.aps.org/doi/10.1103/PhysRevD.80.052002

International Workshop: Meson Production at Intermediate and High Energies IOP Publishing
Journal of Physics: Conference Series **349** (2012) 012015 doi:10.1088/1742-6596/349/1/012015

Light meson physics with Crystal Ball/MAMI and at BES-III

Marc Unverzagt

Institut für Kernphysik, Johannes Gutenberg–Universität Mainz, Germany

E-mail: unvemarc@kph.uni-mainz.de

Abstract.
Rare decays of light mesons like η, η', and ω offer excellent possibilities to study fundamental symmetries and symmetry-breaking patterns. Furthermore, electromagentic transitions with vertices $P\gamma^*\gamma$ as in decays of light pseudoscalar mesons (P) or meson production in e^+e^- collisions give insight into hadronic structure through transition form factors and produce additional constraints for the determination of the anomalous magnetic moment of the muon $(g-2)_\mu$. In this article, recent results on η and ω decays from the Crystal Ball at MAMI experiment as well as feasibility studies on transition form factors for the Crystal Ball and the BES–III at BEPC–II facility are presented.

1. Introduction

The decays of the light mesons provide unique information on the understanding of low-energy Quantum Chromo Dynamics (QCD). Since perturbative QCD cannot be applied in the low-energy region, other methods like lattice QCD, chiral perturbation theory (χPT) or model-dependent approaches are used. These can be tested by studying the different decay modes of the η and η' mesons. Moreover, many models and theories of hadronic interaction can be tested. One can also search for the violation of lepton–family number and place limits on the masses and couplings of many proposed lepto–quark families, see refs [1, 2, 3, 4, 5]. Decays of light mesons are also suitable to search for violations of C, CP, and CPT invariance [6].

Other studies that can be addressed with light meson decays are electromagnetic transition form factors. These form factors shed light on the intrinsic structure of hadrons. Furthermore, measurements of electromagnetic transition form factors of mesons are indirectly related to the Light–by–Light contribution, which is of increasing relevance for the Standard Model prediction of the anomalous magnetic moment of the muon. The high meson production rates achievable with the Crystal Ball/TAPS setup at MAMI (Mainz Microtron) provide a good opportunity for a valuable contribution to the Light–by–Light calculations.

Transition form factors of light mesons can also be studied in e^+e^- scattering experiments in so–called $\gamma\gamma$–physics. This is done using the BES–III (Beijing Spectrometer) detector at the BEPC–II (Beijing Electron Positron Collider). These two different approaches are complementary, covering different momentum transfer regions. At MAMI this is $2m_l^2 < q^2 < m_P^2$ with m_l a lepton mass, while at BEPC–II momentum transfers of $-10\,\mathrm{GeV}^2 < q^2$ can be measured.

2. Experiments

The groups at the Johannes Gutenberg–University at Mainz, Germany, carry out measurements with two different experiments. At Mainz the accelerator MAMI together with the combined Crystal Ball/TAPS setup provides a unique environment to study symmetry–breaking patterns and transition form factors in meson decays. Furthermore, this experiment is ideal to study the lowest nucleon resonances and investigate the properties of light mesons. The goal of the BES–III detector at BEPC–II is to improve the knowledge of charmonium states and investigate D–meson decays. The mass of the τ lepton will also be measured very precisely. Another important topic at BEPC–II is $\gamma\gamma$–physics. Since the two facilities work at different q^2 regions, many studies carried out at both experimental setups complement each other.

2.1. Crystal Ball at MAMI

The photoproduction facility at MAMI, with a maximum beam energy of 1604 MeV, together with the combined Crystal Ball/TAPS setup provides a perfect environment for studies of light meson decays. The high statistics available together with the good signal–to–background conditions makes investigations of η, η', and ω decays especially competitive at MAMI.

The electron accelerator MAMI consists of a cascade of three Race-Track-Microtrons (RTM) [7], and a fourth stage, a Harmonic-Double-Sided Microtron (HDSM) [8]. MAMI provides a very stable electron beam (energy drift $\delta E < 100$ keV) with a maximum energy of $E_0 = 1604$ MeV and an energy width of $\sigma_E/E < 3 \cdot 10^{-5}$. High currents (110 μA) with electron polarisations up to 80 % can be achieved.

Within the A2 collaboration at MAMI, experiments are performed with a real photon beam. The photon beam for the production of light mesons is derived from the production of Bremsstrahlung photons during the passage of the MAMI electron beam through a thin radiator. The photon tagging spectrometer [9, 10] at Mainz (fig. 1) provides energy tagging of the photons through the relation $E_\gamma = E_0 - E_{e^-}$, by detecting the post–radiating electron energy, E_{e^-}. It can determine the photon energy with a resolution of 2 to 4 MeV depending on the incident beam energy, with a single–counter time resolution $\sigma_t = 0.17$ ns [11]. The maximum photon flux is $2.5 \cdot 10^5\,(\mathrm{s\,MeV})^{-1}$. Photons can be tagged in the energy range from 5 to 93 % of the incoming electron energy E_0.

Since the η' photoproduction threshold is at $E_{thr} \approx 1447$ MeV, most of the full photon energy range accessible at MAMI for η' production is not covered by the photon tagging spectrometer. Therefore, a new tagging device (End–Point Tagger) is currently being constructed. At the maximum electron beam energy, $E_0 = 1604$ MeV, photon energies between $E_\gamma \approx 1450$ MeV and $E_\gamma \approx 1590$ MeV can be tagged with this new device. To study the decays of the η' meson, the End–Point Tagger is essential for a clean trigger. The assembly of the End–Point Tagger is close to completion.

The central detector, surrounding the liquid hydrogen target (see fig. 2), is the Crystal Ball calorimeter (CB) [12, 13]. The CB is a highly segmented 672-element NaI(Tl), self triggering photon spectrometer. Each element is a 41 cm (15.7 radiation lengths) long truncated triangular pyramid. The CB has an energy resolution of $\Delta E/E = 0.020(E[\mathrm{GeV}])^{0.36}$, angular resolutions σ_θ of $2 - 3°$ and σ_ϕ of $\sigma_\theta/\sin\theta$ for electromagnetic showers. The Crystal Ball is equipped with a barrel detector of 24 scintillator strips (50 cm length, 4 mm thickness at a radius of ≈ 6 cm around the photon beam line), which allows discrimination between charged and neutral particles, and offers the possibility to distinguish between different charged particles (PID).

At forward polar angles $\theta < 21°$ the TAPS detector [14, 15] provides acceptance for particle detection. Hence, the full detector system, as shown in fig. 2, is almost hermetic. The TAPS forward wall is composed of 384 BaF_2 elements, each 25 cm in length (12 radiation lengths) and hexagonal in cross section, with a diameter of 59 mm. Every TAPS element is covered by a 5 mm thick plastic veto scintillator. The single counter time resolution is $\sigma_t = 0.2$ ns. The energy

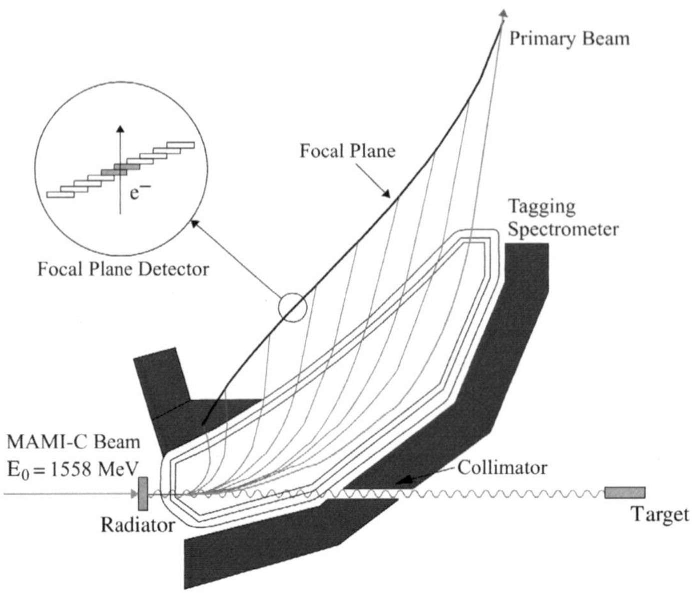

Figure 1. The photon tagging spectrometer at MAMI.

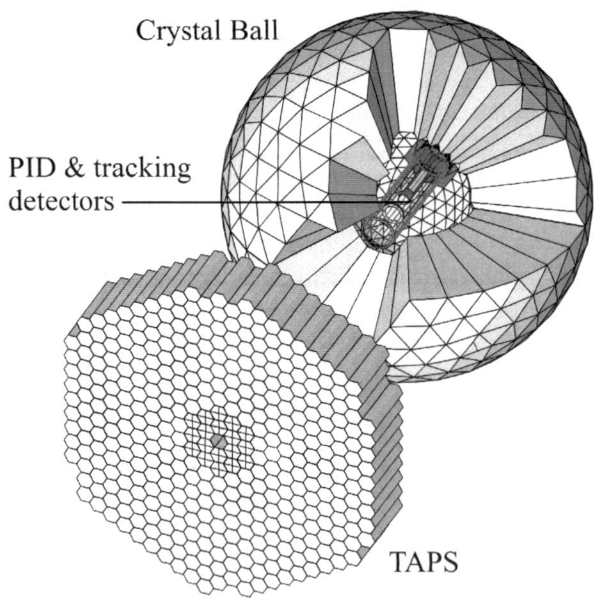

Figure 2. The Crystal Ball calorimeter with cut–away section showing the inner detectors and the TAPS forward wall.

Journal of Physics: Conference Series **349** (2012) 012015 doi:10.1088/1742-6596/349/1/012015

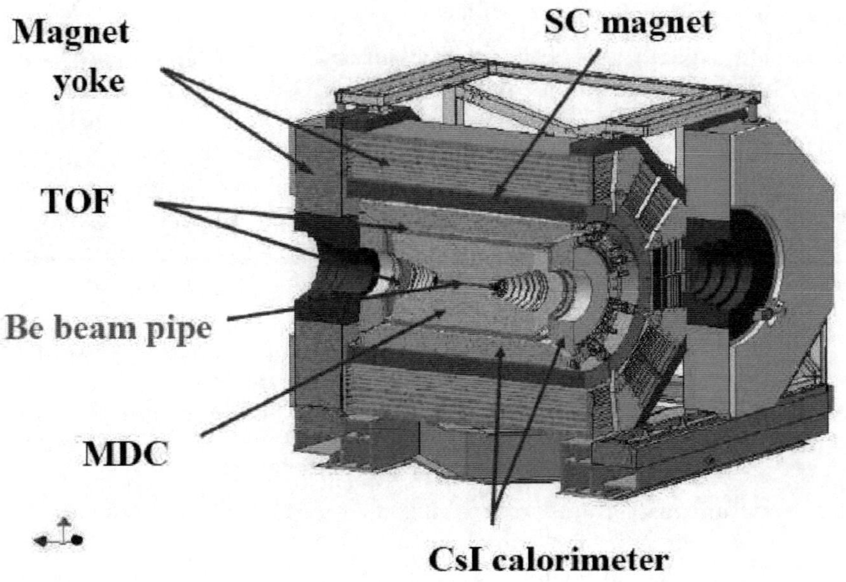

Figure 3. Cross section of the BES–III detector setup.

resolution can be described by $\Delta E/E = 0.018 + 0.008/(E[\mathrm{GeV}])^{0.5}$. The angular resolution in the polar angle is better than $1°$, and in the azimuthal angle it improves with increasing θ, being always better than 1/R radian, where R is the distance in centimetres from the central point of the TAPS wall surface to the point on the surface where the particle trajectory meets the detector. The 2 inner rings of 18 BaF_2 elements have recently been replaced by 72 $PbWO_4$ crystals, each 20 cm in length (22 radiation lengths), see fig. 2. The higher granularity improves the rate capability as well as the angular resolution.

2.2. BES-III at BEPC-II

The BEPC–II electron–positron collider features high–statistics measurements in the center–of–mass energy range between 2 GeV and 4.2 GeV (so–called τ–charm–factory). It came into operation in 2009, and has reached a peak luminosity of $6 \cdot 10^{32}\,\mathrm{cm}^{-2}\,\mathrm{s}^{-1}$ [16]. The world's largest data samples of J/ψ, ψ' and $\psi(3770)$ have already been produced.

The BES–III detector comprises the main drift–chamber (MDC), a time–of–flight system (TOF), an electromagnetic calorimeter (EMC), and muon detectors, see fig. 3. The cylindrical MDC and the TOF system both cover polar angles of $18.2° < \theta < 161.8°$ and sit inside a magnetic field to measure the momentum of the reaction products. The MDC also contributes to particle identification through specific energy deposition of the particles. The EMC consists of 6300 Cs(Tl)–crystals and covers polar angles of $21.6° < \theta < 158.4°$. It can measure precisely the position (5 mm at $E = 1$ GeV) and the energy of leptons and photons (2.5 % at $E = 1$ GeV) between 20 MeV and 2 GeV [16].

Recently a new detector was installed 3.49 m downstream from the interaction point. Its purpose is to measure photons close to $\theta = 0°$, Zero–Degree detector (ZDD), for measuring Initial–State–Radiation events. It consists of two lead bricks intermingled with scintillating fibers placed 5 mm above and below the beam axis. This detector will also improve the detection efficiency for $\gamma\gamma$–events.

Journal of Physics: Conference Series **349** (2012) 012015 doi:10.1088/1742-6596/349/1/012015

3. Symmetry laws

Light mesons are fundamental excitations of the QCD ground state. Investigating their spectrum and interactions gives insight into the complex nature of the QCD vacuum and several competing mechanisms of symmetry breaking. In the low–energy region it is sufficient to consider only the three light quarks, u, d, and s. The corresponding dynamical system exhibits gauge symmetry, and the discrete C, P, and T symmetries. Additionally it is subject to several approximate symmetries. Starting from the chiral symmetry for massless quarks, the interplay of three distinct symmetry–breaking patterns (explicit symmetry breaking due to finite quark masses, dynamical spontaneous symmetry breaking, and the U(1) axial anomaly) generates an extremely rich physics case.

The finite values of the quark masses lead to SU(3) flavour symmetry ($m_u = m_d = m_s \neq 0$). The fact that the s quark mass is much higher than the u and d quark masses ($m_u = m_d \neq m_s$) reduces the SU(3) flavour symmetry to SU(2) isospin symmetry. Isospin symmetry itself is also broken due to the different u and d quark masses ($m_u \neq m_d$), and also due to their coupling the electromagnetism.

For the dynamical spontaneous symmetry breaking no rigorous mathematical proof is yet available. But there are clear indications for a breakdown of the global $SU(3)_L \times SU(3)_R \times U(1)_V \times U(1)_A$ symmetry to $SU(3)_V$ in the chiral limit. This gives rise to eight massless pseudoscalar Goldstone bosons, which are assigned to the pions, kaons and the η_8 state.

The U(1) anomaly is generated through quantum fluctuations that destroy the singlet axial–vector current conservation. In the large–N_C limit the anomaly disappears, and the singlet axial–vector current is restored. But it is assumed to be spontaneously broken in the large–N_C limit, giving rise to a ninth Goldstone boson, which is addressed with η_0.

The physical states, η and η', are an admixture of the two Goldstone boson states. This can be expressed by a single mixing angle, or in a more evolved parametrisation two mixing angles. Recently the KLOE collaboration published data on the η–η' mixing angle and a possible gluonium content to the η' [17, 18] determined from $R = \Gamma(\phi \to \eta\gamma)/\Gamma(\phi \to \eta'\gamma)$. They found a non–zero gluonium content of the η', contrary to the theoretical calculation in [19].

In this context the η–η' system provides a unique tool for studying all three symmetry mechanisms simultaneously. By investigating decays like $\eta/\eta' \to 3\pi^0$, $\eta \to \pi^0\gamma\gamma$, , $\eta/\eta' \to 4\pi^0$, $\eta \to \pi^0\pi^0\gamma\gamma$, $\eta \to \pi^+\pi^-\gamma$, $\eta' \to \eta\pi^0\pi^0$, $\eta' \to \gamma\gamma$, and $\eta' \to 3\gamma$ it might be possible to disentangle the various sources of symmetry breaking. In the near future all the reactions mentioned above will be investigated with the Crystal Ball at MAMI. Also, further studies dedicated to ω decays (e.g. $\omega \to \pi^0\gamma$, $\omega \to \eta\gamma$, $\omega \to \pi^+\pi^-\pi^0$) will be carried out.

In the remainder of this section the measurements and results from the Crystal Ball at MAMI experiment of two symmetry breaking meson decays are described. First the isospin breaking decay $\eta \to 3\pi^0$, then the branching ratios of three C–violating ω decays are presented.

3.1. *Isospin breaking in the $\eta \to 3\pi^0$ decay*

The $\eta \to 3\pi^0$ decay violates isospin symmetry. Because electromagnetic contributions to the amplitude can be neglected [20, 21, 22], this decay occurs due to the isospin breaking part of the QCD Hamiltonian:

$$H = \frac{1}{2}(m_u - m_d)(\bar{u}u - \bar{d}d). \tag{1}$$

Hence, the amplitude is proportional to the mass difference $m_u - m_d$ of the two lightest quarks u and d. Calculations of the decay amplitude are usually based on the framework of χPT. Thus, this measurement also provides a very important test for χPT. The squared absolute value of the decay amplitude may be expanded around the centre of the Dalitz plot:

$$|A(\eta \to 3\pi^0)|^2 = |N|^2(1 + 2\alpha z + \ldots), \tag{2}$$

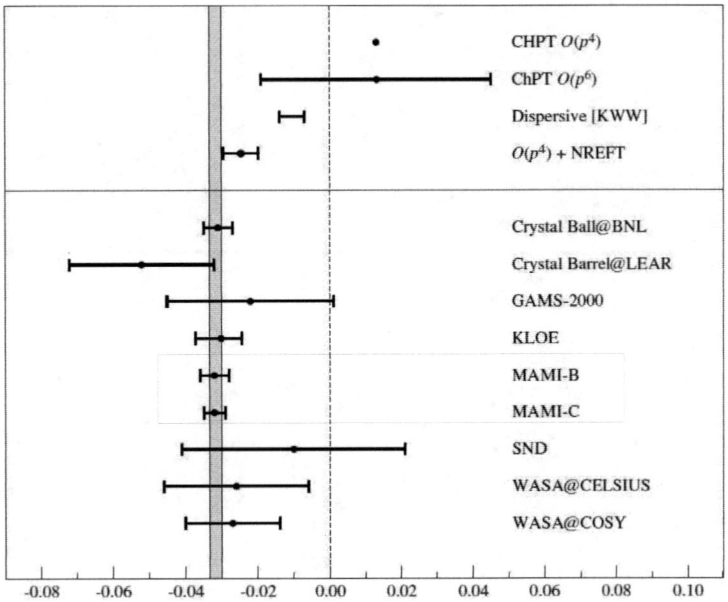

Figure 4. Summary of the world data on the Dalitz Plot parameter. Taken from [25]. The grey shaded area is the PDG average [26].

where N is a normalisation constant, which is equal to the amplitude that would apply, if the decay proceeded only according to the available phase space. The Dalitz plot parameter, α, describes the pion energy dependence of the squared absolute value of the decay amplitude up to first order of the expansion. The parameter z is given by

$$z = 6\sum_{i=1}^{3}\left(\frac{E_i - m_\eta/3}{m_\eta - 3m_{\pi^0}}\right)^2 . \tag{3}$$

Here E_i represents the pion energies in the η rest frame.

At MAMI two new determinations of the Dalitz plot parameter, α, from the $\eta \to 3\pi^0$ decay were performed using two different data sets with different electron beam energies, $E_0 = 883\,\text{MeV}$ [23] and $E_0 = 1508\,\text{MeV}$ [24], respectively. Thus, they can be considered as independent measurements. These results, with $1.8 \cdot 10^6$ and $3 \cdot 10^6$ events, represent the worlds two highest statistics measurements for the $\eta \to 3\pi^0$ decay. The two independent values for α from MAMI agree perfectly with each other and are consistent with other experiments (see fig. 4), [25] and references therein. Pure χPT calculations give the wrong sign for α, which is fixed by the experiments to be negative. Dispersion calculations give the correct sign, but differ in the absolute value by several standard deviations.

Recently, a new study of rescattering effects in $\eta \to 3\pi$ decays was presented [25]. Using the modified non–relativistic effective field–theory, and matching the coupling constants involved to χPT at $O(p^4)$, the Dalitz plot parameter $\alpha = -0.025 \pm 0.005$ could be calculated. Not only did this study give a value in good agreement with experimental results, but it could also partly explain why χPT calculations up to $O(p^6)$ are not sufficient to produce an agreement with experimental results.

The main goal of these studies is the u and d quark mass ratio, which can be extracted from the partial width. For a reliable determination a full understanding of all effects on the Dalitz

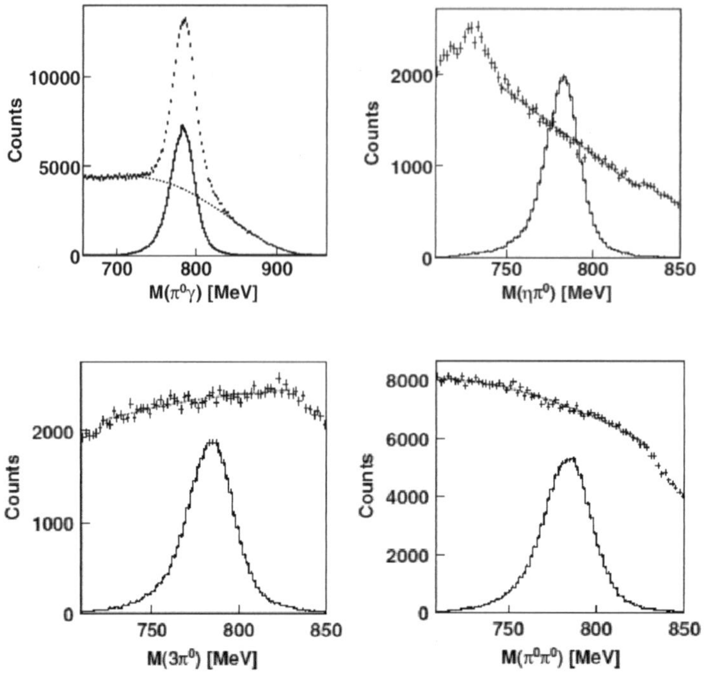

Figure 5. Invariant mass spectra for the $\omega \to \pi^0\gamma$, $\omega \to \eta\pi^0$, $\omega \to 3\pi^0$, and $\omega \to 2\pi^0$ decays (from upper left to lower right) measured with Crystal Ball at MAMI [28]. The crosses represent the data points. The solid histograms are the simulations. Dashed lines indicate the background contribution to the $\omega \to \pi^0\gamma$ invariant mass spectrum, and the combined fit functions for the others.

plot is necessary. Here, experiments are now at the precision border to the appearance of new effects. For instance, a cusp effect due to the $\pi^+\pi^- \to \pi^0\pi^0$ rescattering reaction, similar to the one observed by the NA48 collaboration in $K^+ \to \pi^+\pi^0\pi^0$ [27], should also appear in the $\eta \to 3\pi^0$ decay. This charge exchange reaction also has an impact on the Dalitz plot parameter. According to [22] a 5 % effect on α should be visible. The statistics extracted from the two recent MAMI measurements [23, 24] did not show a significant indication for such a cusp effect.

3.2. C–violating ω decays

With the Crystal Ball at MAMI three C–violating ω–decays were measured, and new upper limits were set [28]. The breaking of charge–conjugation symmetry is one of the three Sakharov criteria for explaining the excess of matter over antimatter in the known universe [29]. In strong and electromagnetic interactions C–symmetry should be conserved. Thus, any C–violation in such interactions would indicate physics beyond the Standard Model.

The three ω–decay channels investigated at MAMI are $\omega \to \eta\pi^0$, $\omega \to 3\pi^0$, and $\omega \to 2\pi^0$. The branching ratios $BR(\omega \to \eta\pi^0) < 1 \cdot 10^{-3}$ [30] and $BR(\omega \to 3\pi^0) < 3 \cdot 10^{-4}$ [31] were already determined by the GAMS2000–collaboration. A branching ratio for the $\omega \to 2\pi^0$ decay has never been set before.

From the Crystal Ball data the upper limits of the branching ratios for these three decays were determined by comparing the count rates of the decays with the number of $\omega \to \pi^0\gamma$ events. The 2008 PDG branching ratio $BR(\omega \to \pi^0\gamma) = 0.0892$ [32] was used for this. The numbers of events were determined from fits of a combined function for the background and a

Journal of Physics: Conference Series **349** (2012) 012015 doi:10.1088/1742-6596/349/1/012015

gaussian for the peak to the invariant mass spectra (see fig. 5). The measured branching ratios $BR(\omega \to \eta\pi^0) < 2 \cdot 10^{-4}$, $BR(\omega \to 3\pi^0) < 2 \cdot 10^{-4}$, and $BR(\omega \to 2\pi^0) < 2 \cdot 10^{-4}$ are all at the CL=90%. They are the world's most precise values today.

4. Transition form factors

The study of electromagnetic decays and hadron-photon interactions provide the possibility to study the intrinsic structure of hadrons through photon interactions with the electric charges of the quarks. The transition form factors describe the electromagnetic structure at $P\gamma^*\gamma$ vertices. Since only one hadron is involved in these processes the corresponding transition form factor defines the electromagnetic properties of the pseudoscalar meson P. Apart from the above mentioned importance for hadron structure physics, these transition form factors provide additional constraints on the Light–by–Light hadronic contribution to the anomalous magnetic moment of the muon $(g-2)_\mu$.

There are several methods available to determine the transition form factors of neutral pseudoscalar mesons. Two of these will be presented here. Vertices of the kind $P\gamma^*\gamma$ can be investigated in decays of pseudoscalar mesons: $P \to \gamma^*\gamma \to e^+e^-\gamma$. The Crystal Ball setup at MAMI is well suited to detect such decays with high statistics. The momentum transfer of the virtual photon $q^2 = m^2_{e^+e^-}$ is in this case limited to $(2m_{e^-})^2 < q^2 < m_P^2$, thus it lies in the time–like region. For the corresponding decay into muons $P \to \mu^+\mu^-\gamma$ the range starts only at $(2m_\mu)^2$.

Another method to determine transition factors at $P\gamma^*\gamma$ vertices is through e^+e^- collisions. The mesons are produced through the $e^+e^- \to e^+e^-\gamma^{(*)}\gamma^* \to e^+e^-P$ reaction. On the one hand, the momentum transfer of one of the two virtual photons in this reaction can be restricted to $q^2 \approx 0$ (see below), thus this photon can be considered as quasi–real. On the other hand, the momentum transfer of the second virtual photon lies in the negative region, the space–like region, limited by the accelerator energies. A program to extract space–like transition form factors from data measured with the BES-III detector at the BEPC–II collider in the range $-10\ \text{GeV}^2 < q^2$ is underway.

In the following the result from the Crystal Ball at MAMI experiment for the transition form factor of the $\eta \to e^+e^-\gamma$ decay is described. Then a feasibility study for the transition form factor in the $\eta' \to e^+e^-\gamma$ decay that will be measured with the Crystal Ball in the near future will be presented. The last part of this section contains feasibility studies for the BES–III experiment at BEPC–II aiming to determine the transition form factors of π^0, η, and η' in the space–like momentum transfer region.

4.1. Time–like transition form factors at MAMI

With the Crystal Ball/TAPS setup the Dalitz decay $\eta \to \gamma^*\gamma \to e^+e^-\gamma$ has been investigated [33]. The study of the electromagnetic transition form factors in such a decay not only gives insight into the intrinsic structure of hadrons, but is also important for validating hadronic models used in calculations of the hadronic Light–by–Light contribution to a_μ.

From the amplitude of conversion $P \to \gamma^*\gamma \to l^+l^-\gamma$ decays of pseudoscalar mesons (P) the effective mass spectrum of the leptonic pair (l^+l^-) normalised to the width of the corresponding radiative $P \to \gamma\gamma$ decay may be derived [34]:

$$\frac{d\Gamma(P \to l^+l^-\gamma)}{dq^2 \cdot \Gamma(P \to \gamma\gamma)} = [\text{QED}] \cdot \left|F(q^2)\right|^2 . \tag{4}$$

Here [QED] stands for a pure QED term containing constants such as the fine struture constant, α, the masses of the leptons and the pseudoscalar meson, and $F(q^2)$ is a normalised transition form factor.

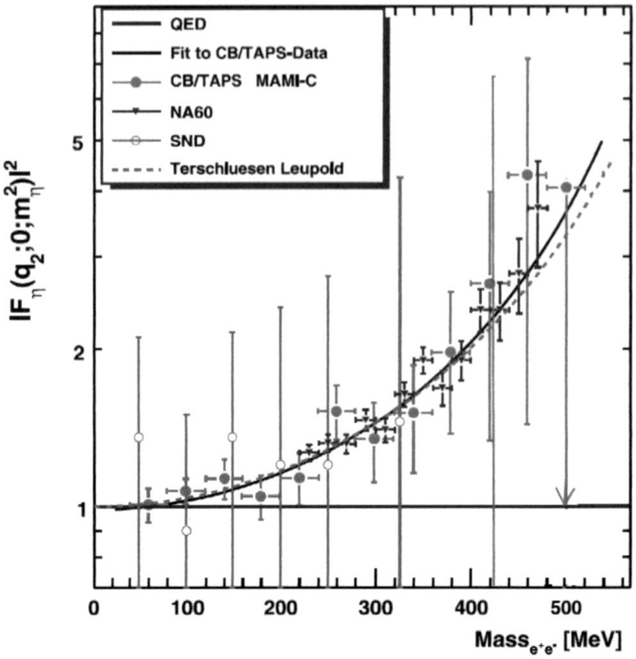

Figure 6. Transition form factor for η–Dalitz decays. Red circles are MAMI data [33]. The black curve is a fit to the MAMI data. The open green circles show the SND results [35]. For comparison the NA60 data points on the $\mu^+\mu^-$ channel [36] are also plotted. The dashed green line is a calculation performed by [37].

To describe the effect of the transition form factor, one may use the vector meson dominance model (VMD). According to this model photon coupling to hadrons is mediated by a virtual vector meson ($V = \rho, \omega, \phi$). The transition form factor is usually parametrised in the VMD assumption by a one–pole approximation [34]

$$F(q^2) = \frac{1}{1 - \frac{q^2}{\Lambda^2}}. \qquad (5)$$

The information from experiments on the η transition form factor is rather scarce. One result with 110 reconstructed $\eta \to e^+e^-\gamma$ events stems from the SND–collaboration [35]. With a value of $\Lambda^{-2} = (1.6 \pm 2.0)$ GeV^{-2} the accuracy was insufficient to establish a deviation from QED calculations. The η transition form factor may also be measured in $\eta \to \mu^+\mu^-\gamma$ decays. Recently, the NA60–collaboration performed such a measurement [36] and derived the very precise result ($\approx$ 9000 events): $\Lambda^{-2} = (1.95 \pm 0.17_{\text{stat}} \pm 0.05_{\text{syst}})$ GeV^{-2}. Though, the statistics are sufficient to show deviations from QED calculations, and are well suited to test different models, the NA60 result suffers from two disadvantages. Due to choosing the $\mu^+\mu^-$ channel, the NA60 data points only start at dilepton masses above 200 MeV. Furthermore, the photon was not detected, and the form factor was deduced by unfolding the $\mu^+\mu^-$ invariant mass spectrum [33].

At MAMI the $\eta \to e^+e^-\gamma$ decay was studied [33]. In a sophisticated analysis 1345 events could be reconstructed. The overall efficiency in this analysis was only 2%. An improved tracking device could help to increase this number greatly. The statistics achieved with the experiment at MAMI are a factor of 10 higher compared to the SND result. The NA60 statistics are clearly superior, but by choosing the e^+e^- channel the region below 200 MeV cold be reached

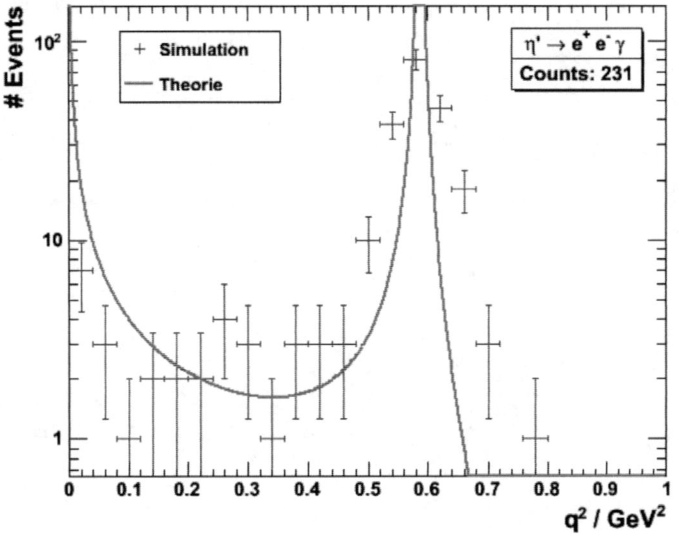

Figure 7. Distribution of simulated $\eta' \rightarrow e^+e^-\gamma$ events for the Crystal Ball experiment as a function of q^2. The red curve represents the shape of the probability distribution fed to the simulation program. It contains a one–pole approximation with $\Lambda^2 = 0.5883$ GeV^2 as a transition form factor.

at MAMI. The extracted η transition form factor is shown in fig. 6 together with the SND und NA60 results. All measurements agree within the uncertainties. From the fit to the MAMI data points the value $\Lambda^{-2} = (1.92 \pm 0.35_{\text{stat}} \pm 0.13_{\text{syst}})$ GeV^{-2} was deduced. In fig. 6 a theoretical calculation [37] is also shown which fits the MAMI data perfectly.

A feasibility study was carried out to probe a possible result from the Crystal Ball experiment for the transition form factor in the $\eta' \rightarrow e^+e^-\gamma$ decay [38]. This decay is not well measured, and only an upper limit for the branching ratio exists ($BR(\eta' \rightarrow e^+e^-\gamma) < 9 \cdot 10^{-4}$ at 90% CL) [26]. The simulation used realistic beam properties and the efficiency of 2% determined for the similar $\eta \rightarrow e^+e^-\gamma$ decay described above [33]. As a result a distribution of the simulated events as a function of q^2 is shown in fig. 7. This study, using 600 h of beamtime, which corresponds to the production of roughly 18 million η' mesons, means one might expect to extract 231 $\eta' \rightarrow e^+e^-\gamma$ events. Such a measurement with the Crystal Ball at MAMI would allow the first determination of the transition form factor from $\eta' \rightarrow e^+e^-\gamma$ decays.

4.2. Space–like transition form factors at BEPC–II

The experimental $\gamma\gamma$ physics program at BES–III for the measurement of pseudoscalar transition form factors will allow the region $Q^2 = -q^2 < 10$ GeV^2 to be covered. The investigations at BES-III will aim for an improved determination of the pseudoscalar transition form factors in the non–perturbative region below 3 GeV^2. Besides allowing for new testing grounds in this regime, such observables will serve as an important input to phenomenological calculations of the hadronic Light–by–Light contribution to $(g-2)_\mu$. For the low Q^2 region, BES-III data will have a statistical precision significantly better than the CLEO experiment [39]. Below $Q^2 < 1.5$ GeV^2 this will be the first ever precision measurement.

In the region between perturbative and non–perturbative QCD up to 10 GeV^2, BES–III data will reach a similar precision to the CLEO results [39]. In view of the unexpected Q^2 dependence

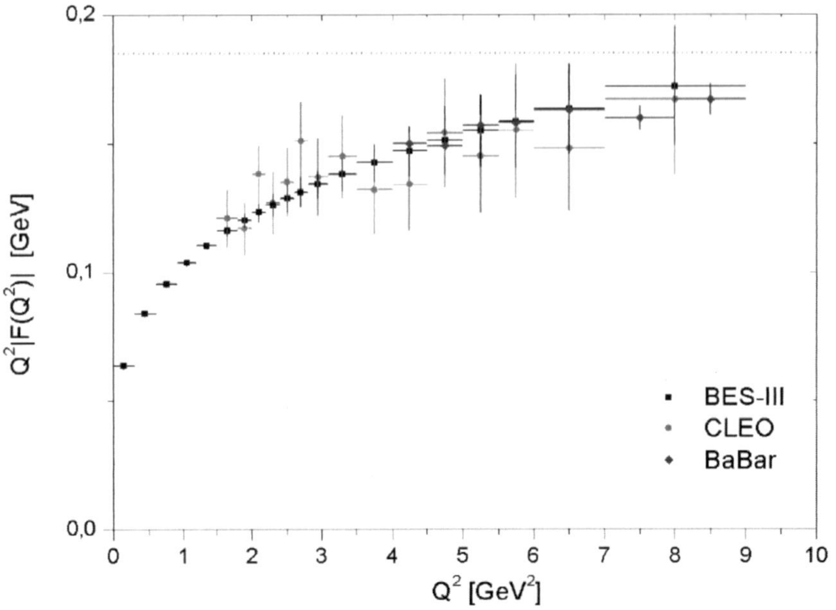

Figure 8. Measured $\gamma^*\gamma \rightarrow \pi^0$ transition form factors at CLEO [39] and BABAR [40] in comparison to the expected accuracy from a 10 fb^{-1} data sample at BES–III [42].

of the BABAR results [40, 41] in the high Q^2 regime, a cross check of the overlap region between CLEO and BABAR will be provided.

To demonstrate the feasibility of the meson transition form factor measurements, studies were carried out for $e^+e^- \rightarrow e^+e^-P$ reactions detected with BES–III. These studies were made, assuming an integrated luminosity of 10 fb^{-1}. Presently, a data sample of 3 fb^{-1} at $\sqrt{s} = 3.77$ GeV is already existing.

In fig. 8 the statistical accuracy for the $\gamma^*\gamma \rightarrow \pi^0$ transition form factor multiplied by Q^2 is shown, applying the so-called single-tag-method [42]. Taking into account the characteristics of the BES–III detector, this means that the beam electrons (or positrons) are simulated to be scattered in the fiducial polar angle area of the electromagnetic calorimeter, while the second beam particle is assumed to be scattered in forward direction. Since both the momentum of the scattered electron (positron) and the momentum of the neutral pion are fully reconstructed, specific cuts on the polar angle of the second (untagged) lepton can be applied. Through this one can assertain that the t-channel photon associated to this lepton is quasi-real. The form factor assumed in the simulation is a fit to the recent BABAR data [40]. The result of this feasibility study shows a similar accuracy to the CLEO data for Q^2 values above 5 GeV^2, while the accuracy will be unrivaled for lower Q^2 values.

Results from analogous studies [43] are presented for η production in fig. 9, and for η' production in fig. 10. At intermediate Q^2 values greater than 5 GeV^2, one observes a similar accuracy to previous measurements. Below 5 GeV^2, the η and η' transition form factors will be extracted with much higher precision at BES–III. Especially below 2 GeV^2 it will be the first precision measurement. The recently installed ZDD is ideally suited for detection of small angle leptons, and thus, will improve the extraction of the transition form factors significantly.

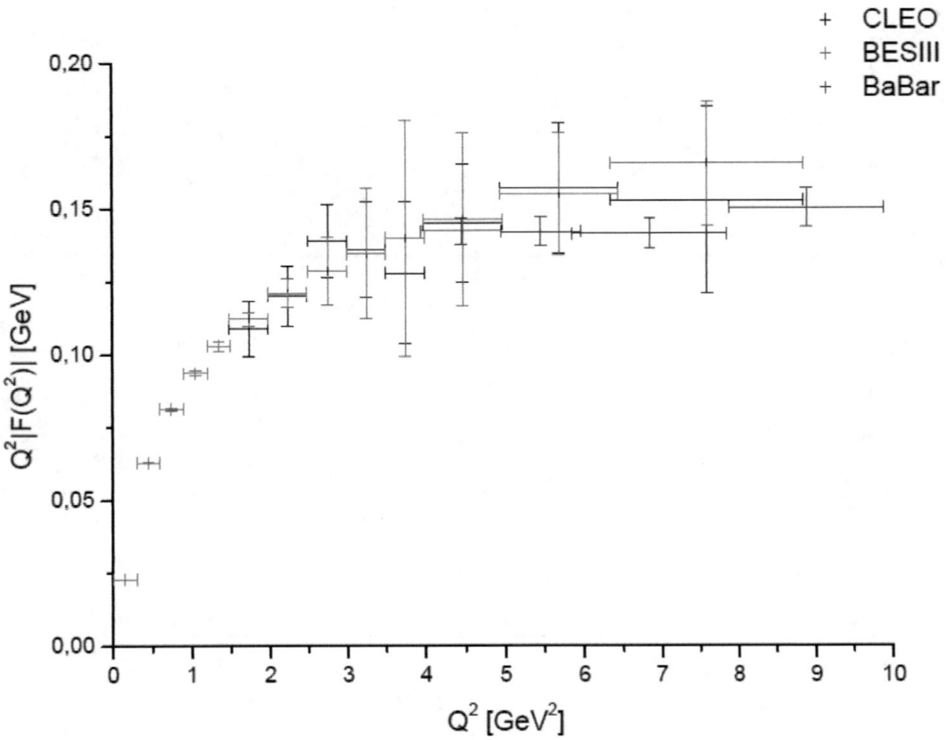

Figure 9. Measured $\gamma^*\gamma \to \eta$ transition form factors at CLEO [39] and BABAR [41] in comparison to the expected accuracy from a 10 fb^{-1} data sample at BES–III [43].

5. Summary

In this contribution results from the Crystal Ball at MAMI and BES–III at BEPC–II experiments were presented. At MAMI decays of the light mesons are studied to investigate the rich field of symmetries and symmetry-breaking patterns in low-energy QCD. As examples the measurements of the isospin-breaking $\eta \to 3\pi^0$, and three C–violating ω decays were described, which are currently the world's most precise results. Furthermore, the determination of the transition form factor in the $\eta \to e^+e^-\gamma$ decay and a feasibility study for the $\eta' \to e^+e^-\gamma$ for an upcoming beamtime with the Crystal Ball at MAMI were shown. Transition form factors of light pseudoscalar mesons can also be measured with the BES–III detector at the e^+e^- collider BEPC–II. With the results of feasibility studies the impact of this experiment on the π^0, η, and η' transition form factors were demonstrated.

Acknowledgments

This work was supported by the German state Rhineland-Palatinate (EMG "Elementarkräfte und mathematische Grundlagen"), the Deutsche Forschungsgemeinschaft (SFB 443, SFB/TR 16), the DFG-RFBR (Grant No. 05-02-04014), European Community-Research Infrastructure Activity under the FP6 "Structuring the European Research Area programme" (HadronPhysics, Contract No. RII3-CT-2004-506078).

References

[1] Goity J L, Bernstein A M, and Holstein B R 2002 *Phys. Rev.* D **66** 076014
[2] Ramsey-Musolf M J and Wise M B 2002 *Phys. Rev. Lett.* **89** 041601

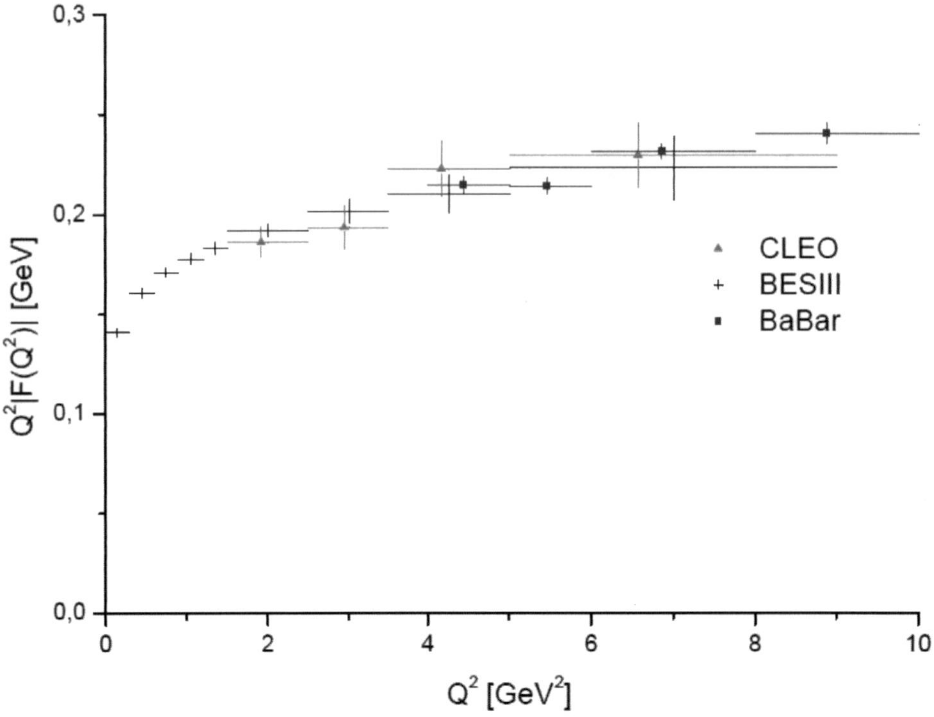

Figure 10. Measured $\gamma^*\gamma \rightarrow \eta'$ transition form factors at CLEO [39] and BABAR [41] in comparison to the expected accuracy from a 10 fb^{-1} data sample at BES–III [43].

[3] Martemyanov B V and Sopov V S 2005 *Phys. Rev.* D **71** 017501
[4] Nicola A G and Pelaez J R 2002 *Phys. Rev.* D **65** 054009
[5] Herczeg P 1995 *Phys. Rev.* D **51** 136
[6] Nefkens B M K 1995 Few Body Syst. Suppl. **9** 193
[7] Herminghaus H *et al.* 1983 *IEEE Trans Nucl. Sci.* **30** 3274
[8] Kaiser K H *et al.* 2008 *Nucl. Instrum. Meth.* A **593** 159
[9] Anthony I, Kellie J D, Hall S J, Miller G J, Ahrens J 1991 *Nucl. Instrum. Meth.* A **301** 230
[10] Hall S J, Miller G J, Beck R, Jennewein P 1996 *Nucl. Instrum. Meth.* A **368** 698
[11] McGeorge J C *et al.* 2008 *Eur. Phys. J.* A **37** 129
[12] Oreglia M *et al.* 1982 *Phys. Rev.* D **25** 2259
[13] Starostin A *et al.* 2001 *Phys. Rev.* C **64** 055205
[14] Novotny R 1991 *IEEE Trans. Nucl. Sci.* **38** 379
[15] Gabler A R *et al.* 1994 *Nucl. Instrum. Meth.* A **346** 168
[16] Asner D M *et al.* [BES–III collaboration] 2009 *Int. J. Mod. Phys.* **A24S1**
[17] Ambrosino F *et al.* 2007 *Phys. Lett.* B **648** 267
[18] DiMicco B *et al.* [KLOE collaboration] 2008 *Eur. Phys. J.* A **38** 129
[19] Escribano R, Nadal J 2007 *JHEP* **05** 006
[20] Sutherland D G 1966 *Phys. Lett.* **23** 384
[21] Baur R, Kambor J, Wyler D 1996 *Nucl. Phys.* B **460** 127
[22] Ditsche C, Kubis B, Meißner U G 2008 *Eur. Phys. J.* C **60** 83
[23] Unverzagt M *et al.* 2009 *Eur. Phys. J.* A **39** 169
[24] Prakhov S *et al.* 2009 *Phys. Rev.* C **79** 035204
[25] Schneider S, Kubis B, Ditsche C 2011 *JHEP* **1102** 028
[26] Nakamura K *et al.* [Particle Data Group] 2010 *J. Phys.* G **37** 075021
[27] Batley J R *et al.* 2006 *Phys. Lett.* B **633** 173
[28] Starostin A *et al.* 2009 *Phys. Rev.* C **79** 065201

[29] Sakharov A D 1967 *Sov. Phys. Jetp* **5** 24
[30] Alde D *et al.* 1994 *Phys. Lett.* B **340** 122
[31] Prokoshin Y D, Samoilenko V D 1995 *Phys. Dokl.* **40** 273
[32] C. Amsler *et al.*, Phys. Lett. B (Review of Particle Physics) **667**, 1 (2008).
[33] Bergäuser H *et al.* 2011 *Phys. Lett.* B **701** 562
[34] Landsberg L G *et al.* 1985 *Phys. Rep.* **128** 301
[35] Achasov M N *et al.* 2001 *Phys. Lett.* B **504** 275
[36] Arnaldi R *et al.* 2009 *Phys. Lett.* B **677** 260
[37] Terschlüsen C 2010 Diploma Thesis at the University of Gießen, Germany
[38] Wagner S 2011 Bachelor thesis at Johannes Gutenberg-University of Mainz, Germany, unpublished
[39] Gronberg J *et al.* [CLEO collaboration] 1998 *Phys. Rev.* D **57** 33
[40] Aubert B *et al.* []BABAR collaboration] 2009 *Phys. Rev.* D **80** 052002
[41] del Amo Sanchez P *et al.* [BABAR collaboration] 2011 arXiv:1101.1142 [hep-ex]
[42] Kloss B 2011 Bachelor thesis at Johannes Gutenberg-University of Mainz, Germany, unpublished
[43] Hahn A 2011 Bachelor thesis at Johannes Gutenberg-University of Mainz, Germany, unpublished

International Workshop: Meson Production at Intermediate and High Energies IOP Publishing
Journal of Physics: Conference Series **349** (2012) 012016 doi:10.1088/1742-6596/349/1/012016

Photoproduction of Vector Mesons

Valentina Vegna[1]

[1] Universität Bonn, Physikalisches Institut, Nußalle 12, 53115 Bonn

E-mail: vegna@physik.uni-bonn.de

Abstract. Recent experimental results from ϕ and ω photoproduction off the proton in the third resonance region will be shown. These results are not yet fully understood and leave open questions both from the experimental and from the theoretical point of view. Further experimental investigation could shed new light on the t-channel mechanisms responsible for vector meson production and on the role of resonances in case of ω photoproduction.

1. Introduction

The lightest vector mesons are $\rho(770)$, $\omega(782)$ and $\phi(1020)$. While the ρ has isospin I=1 and a large full width ($\Gamma_\rho = 149.1 \pm 0.8$ MeV), ω and ϕ can be observed in the only neutral charge state and show a much smaller width ($\Gamma_\omega = 8.49 \pm 0.08$ MeV and $\Gamma_\phi = 4.26 \pm 0.04$ MeV) [1]. In Fig.1 the cross sections of ρ_0, ω and ϕ photoproduction off the free proton are shown from the corresponding thresholds up to 6 GeV of incoming photon energy (data from Ref.[2]-[6]). The most probable process is the photoproduction of ρ_0 meson with a cross section $\sigma_{\rho_0} \simeq$ 20-30 μbarn in the whole energy range. The ω photoproduction cross section σ_ω is smaller but still comparable with σ_{ρ_0}, showing a maximum of $\simeq 8$ μbarn close to threshold. Finally it is observed that the ϕ cross section is much smaller with respect to σ_{ρ_0} and σ_ω: more than a factor 10 of difference is observed at ϕ threshold. This important difference could be explained in terms of SU(3) representation. In fact, while ρ_0 and ω mesons are described as combinations of up and down quarks and anti-quarks, ϕ meson is a pure $s\bar{s}$ state. Due to the Okubo-Zweig-Iizuka (OZI) rule [7], the direct interaction of ϕ mesons with particles "*built*" by up and down quarks (such as the proton) is strongly inhibited. At the same time $\omega - \phi$ mixing and strangeness content in the proton could be invoked to explain the OZI-violation resulting in a small but still not null cross section.

2. ϕ photoproduction

First measurements of cross sections and decay angular distributions for ϕ photoproduction off the free proton were performed at the end of the 60's and during the 70's ([3], [4], [8] and [9]). More accurate results have been published in the last decade. Examples of recently measured differential cross sections are shown in Fig. 2a and 2b (data are from Ref.[6] and [10], respectively). Results show the typical diffractive behavior, expected for vector meson photoproduction where the t-channel exchange mechanism dominates the production process. Data are well fitted by the exponential function $\frac{d\sigma}{d\Omega} = a \cdot e^{-b|t-t_{min}|}$ for t values corresponding to ϕ photoproduction angles up to 90°, confirming the dominant role of t-channel in this kinematics. From the parameters of the fit, the value of the differential cross section is extrapolated at

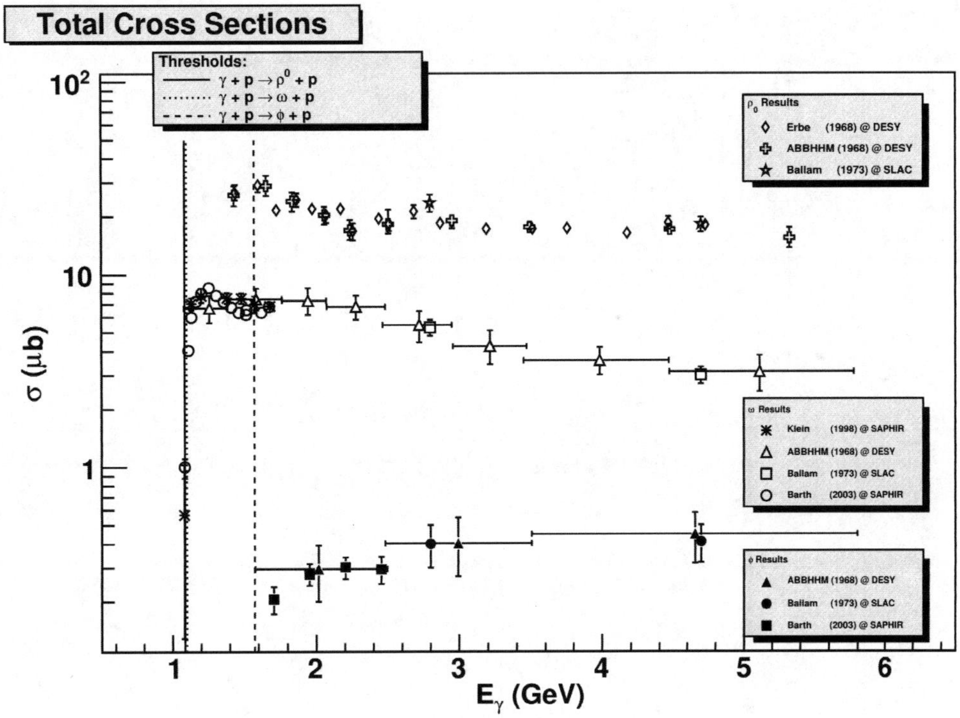

Figure 1: Total cross section for ρ_0, ω and ϕ photoproduction off free proton up to 6 GeV of incoming photon energy. Data are from references [2]-[6]. Solid, dotted and dashed lines: ρ_0, ω and ϕ threshold, respectively.

$|t|- = |t_{min}|$. Results from this procedure are shown in Fig.3. The accurate measurements from LEPS (black points) show an unexpected non-monotonic behavior already visible in data from Ref.[6], with the cross section showing a local maximum at 2.0 GeV. This effect is not yet explained and requires both experimental confirmation and theoretical interpretation.

Accurate measurements of the decay angular distributions were also performed for the decay $\phi \rightarrow K^+K^-$ both in the Gottfried-Jackson frame and in the Helicity frame to check t-channel and s-channel helicity conservation, respectively. In Fig. 4a and 4b angular decay distributions measured by the SAPHIR experiment are shown (Ref.[6]). The behavior expected if helicity is conserved is shown by solid line. Data are well described by the solid line in the Gottfried-Jackson frame indicating the dominance of helicity conserving reactions in the t-channel. It is the case of Pomeron exchange mechanism. On the contrary, in the Helicity frame there is evidence of helicity violating production mechanisms. It indicates that also contributions of unnatural parity exchange (e.g. π^0 exchange) take place in ϕ photoproduction off the free proton. A detailed investigation of the decay angular distributions was performed by LEPS in the Gottfried-Jackson frame, looking for a possible explanation of the non-monotonic behavior visible in Fig.3. In fact the angular distributions are measured in the energy range around to the local maximum of the cross section (E_γ=1.97-2.17 GeV) and above (E_γ=2.17-2.37 GeV) looking for any difference that could be a hint in the understanding of the non-monotonic behavior. Results are shown in Fig.5. The angular distributions show no difference in the two energy bins and it does not allow any conclusion on the origin of the local maximum in the cross section. The non-monotonic behavior still remains an open question.

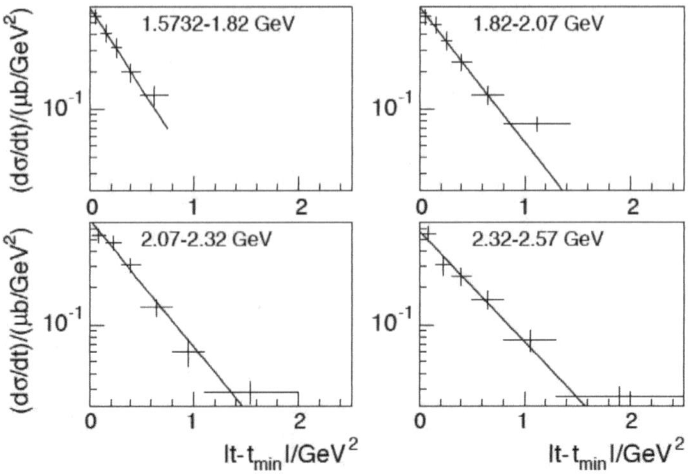

(a)

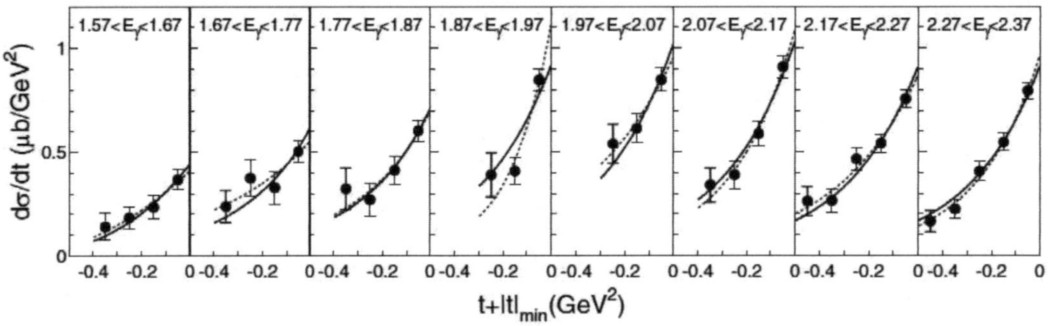

(b)

Figure 2: Differential cross section of ϕ photoproduction off free proton. 2a: results from Ref.[6]. 2b: results from Ref.[10].

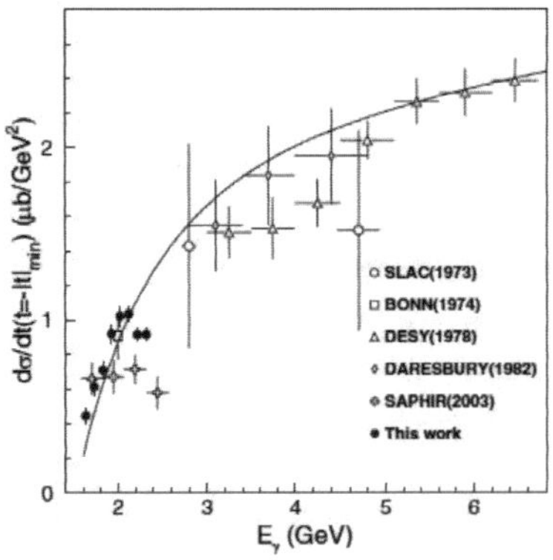

Figure 3: Values of the differential cross section extrapolated at the minimum possible value of transferred 4-momentum t (Fig.3 from Ref.[10]).

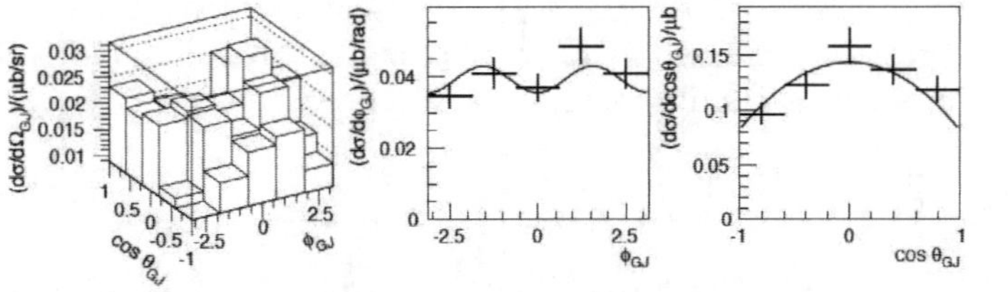

Fig. 12. Decay angular distributions in the Gottfried-Jackson frame.

(a)

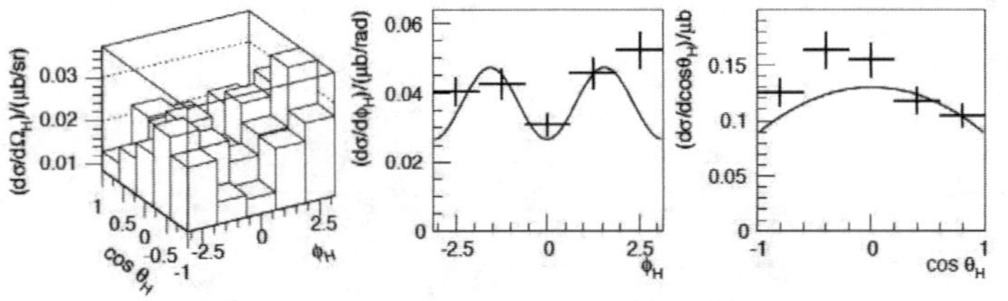

Fig. 13. Decay angular distributions in the helicity frame.

(b)

Figure 4: Decay angular distributions measured at SAPHIR in the Gottfried-Jackson frame and in the Helicity frame. Solid line: expected behavior if helicity is conserved.

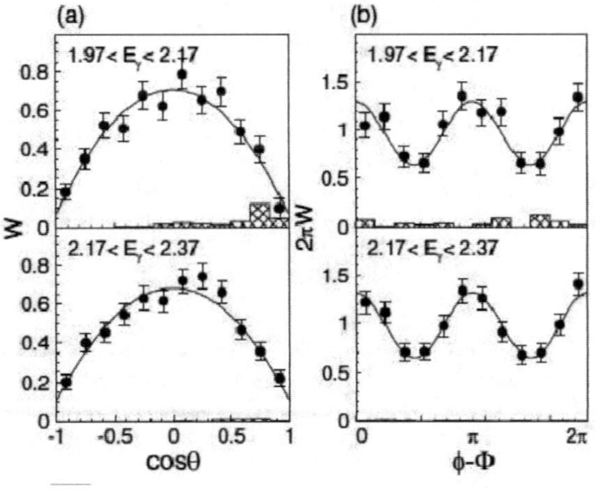

Figure 5: Decay angular distributions measured at LEPS in the Gottfried-Jackson frame.

3. ω photoproduction

The ω meson is described in the SU(3) representation as $\omega = \frac{1}{\sqrt{2}}(u\bar{u} + d\bar{d})$. Due to this, also contributions from s- and u-channels are supposed to play an important role in the ω photoproduction off the nucleon. As a consequence, while ϕ photoproduction is a powerful tool for the study of the t-channel reaction mechanisms, ω photoproduction is important in the study of the baryonic spectrum and in the investigation of the open issue of the *missing resonances*. Since it has isospin I = 0 and due to isospin conservation in strong interaction, the study of ω photoproduction allows to access only N^* nucleon resonances, which are the resonances with isospin I=$\frac{1}{2}$. The important role of the t-channel exchange contribution and the influence of resonant channels in ω photoproduction off the free proton is visible in the differential cross section, as it is shown in Fig. 6 (Fig 2 of Ref.[11]). t-channel exchange mechanisms dominate the differential cross section in the whole energy range for t up to 0.5 GeV^2. For higher values of the transfer momentum t, resonant contributions become visible.

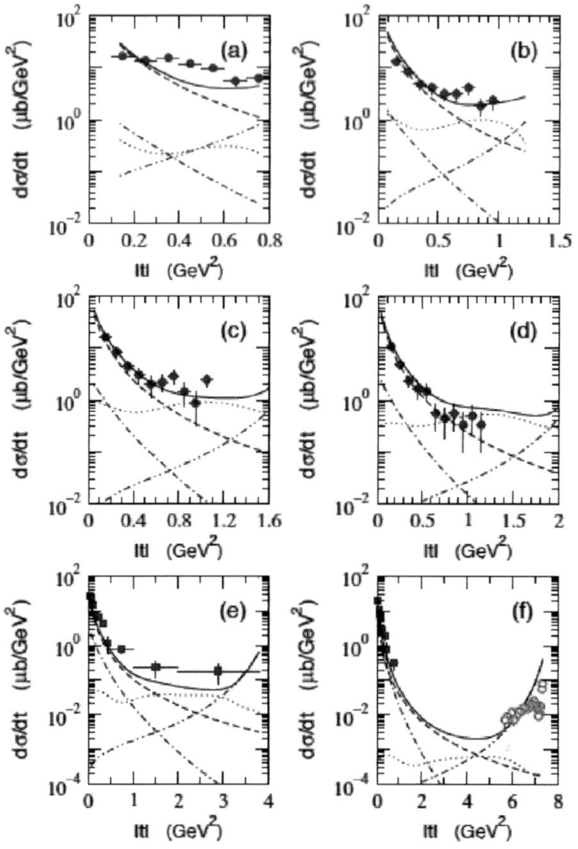

Figure 2. Differential cross sections for $\gamma p \to p\omega$ reaction as a function of $|t|$ at E_γ = (a) 1.23, (b) 1.45, (c) 1.68, (d) 1.92, (e) 2.8, and (f) 4.7 GeV. The results are from pseudoscalar-meson exchange (dashed), Pomeron exchange (dot-dashed), direct and crossed nucleon terms (dot-dot-dashed), N^* excitation (dotted), and the full amplitude (solid). Data are taken from Ref. [17] [filled circles in (a,b,c,d)], Ref. [18] [filled squares in (e,f)], and Ref. [19] [open circles in (f)].

Figure 6: See Ref.[11].

In order to better appreciate the resonances involved in the photoproduction of the ω meson, the beam asymmetry Σ should be measured. The asymmetry should be null if only contributions from t channel exchange are responsible of the production reaction. Any deviation is a strong indication of the role of the resonances. Figs. 7a and 7b show the already published results and theoretical descriptions for ω beam asymmetry. Experimental data suggested an important role of the resonance $P_{13}(1720)$ in ω photoproduction. The actual status is still confusing. New results from GRAAL and CB-ELSA collaborations should be published in the next future.

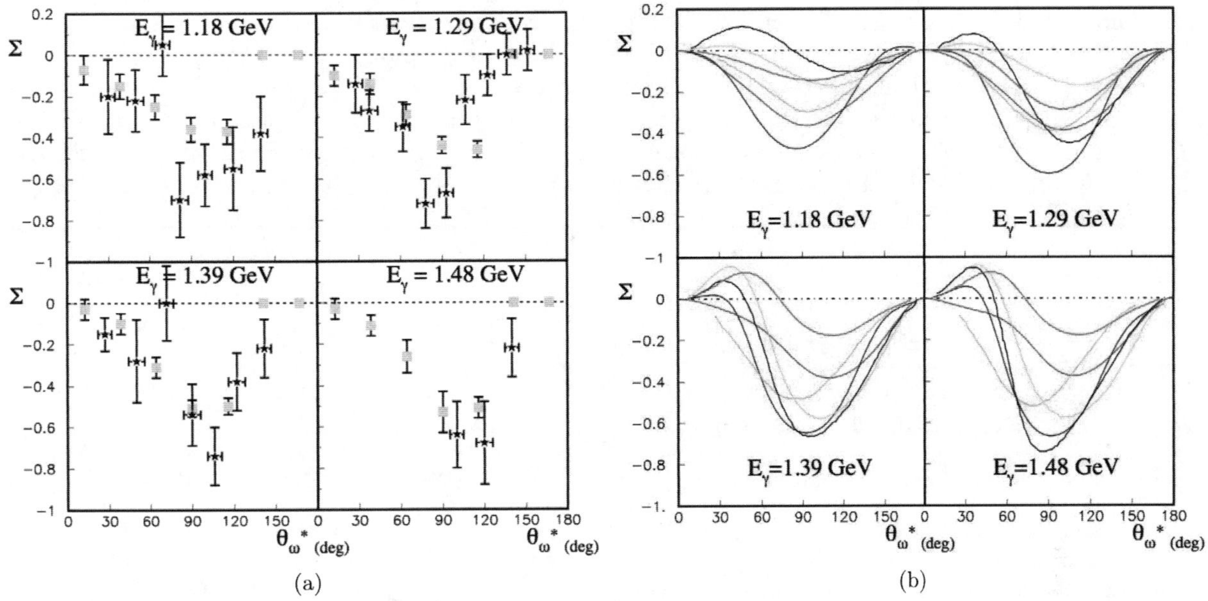

Figure 7: Experimental and theoretical results for the reaction $\vec{\gamma} + p \rightarrow \omega + p$.
Fig.7a Green squared and black stars are results from Ref.[12] and [13], respectively.
Fig.7b Black and light-blue lines from Ref.[14]. Purple line from Ref.[15]. Red line from Ref.[16]. Green line from Ref.[13]. Blue line from Ref.[17].

4. Perspectives

The new BGO-OD experiment installed at the ELSA facility in Bonn could give new results useful for a better understanding of ω and ϕ photoproduction off the nucleon. The experiment will use an unpolarized or linearly polarized photon beam produced by Bremsstrahlung from the ELSA electron beam. Photons will give rise to photonuclear reactions by impinging on an unpolarized target (e.g. Hydrogen, Deuterium or light nuclei). A large acceptance detector will allow photon detection and charged particle detection and identification. The central part of the detector consists of two cylindrical wire chambers for charged particle tracking, a barrel of plastic scintillating bars for charged/neutral particle discrimination and a BGO calorimeter optimized for photon detection. This set-up was already used in the GRAAL experiment (see Ref.[18] for more details on the central detector), giving high quality results in the study of several reactions (e.g. π^0 and η photoproduction off proton and neutron, $K\Lambda$ photoproduction off proton, etc.). The forward detector consists of detectors for charged particle tracking, an open dipole magnet and 8 drift chambers for charged particle identification (on the base of momentum reconstruction) and a time-of-flight detector for charged and neutral particle detection in the forward direction. The combination of the central detector - optimized for photon detection - and of the forward spectrometer will allow a detailed analysis of several photonuclear reactions by the study both of charged particle final states and multi-photons final states.
The detector could be used for a new measurement of the differential cross section of ϕ photoproduction at small values of the transfer 4-momentum t with ϕ identified by the decay $\phi \rightarrow K^+ + K^-$. In fact charged kaons can be detected and identified at small polar angle θ in the laboratory frame by the forward spectrometer and a new measurement of the differential cross section can be performed in the energy region corresponding to the local maximum of LEPS data. If the non-monotonic behavior will be confirmed, efforts will be needed to understand

the origin of this effect. For the first time, also the measurement of the cross section of ϕ photoproduction off the neutron could be performed using a Deuterium target.
If the results from LEPS will be confirmed, a new measurement of the cross section of ω photoproduction should also be performed to verify the presence of a local maximum in this case. First measurement of the cross section of ω photoproduction off the neutron could also be performed. With the polarized photon beam, new measurements of beam asymmetries and decay angular distributions could be performed both for ϕ photoproduction (with $\phi \rightarrow K^+ + K^-$) and for ω photoproduction (with $\omega \rightarrow \pi^+ + \pi^- + \pi^0$ and $\omega \rightarrow \pi^0\gamma$).

References

[1] K.Nakamura et al. (Particle Data Group) 2010 *J. Phys. G* **37**
[2] R.Erbe et al. (ABBHHM Collaboration) 1968 *Phys. Lett. B* **27** 54-57
[3] ABBHHM Collaboration 1968 *Phys. Rev.* **175** 1669-96
[4] J.Ballam et al. 1973 *Phys. Rev. D* **7** 3150-77
[5] F.J.Klein et al. (SAPHIR Collaboration) 1998 *πN Newslett* **14** 141
[6] J.Barth et al. 2003 *Eur. Phys. J. A* **18** 117-27
[7] S.Okubo 1963 *Phys. Lett.* **5** 165; G.Zweig 1964 (unpublished) *CERN Report No. 8182/TH* **401**; G. Zweig 1964 (unpublished) *CERN Report No. 8182/TH* **412**; J. Iizuka 1966 *Prog. Theor. Phys. Suppl.* **37/38** 21
[8] H.J.Besch at al. 1973 *Nucl. Phys.* **70** 257-71
[9] H.-J.Behrend et al. 1975, *Phys. Lett. B* **56** 408-12
[10] Mibe et al. (2005), *Phys. Rev. Lett* **95** 182001/1-5
[11] Y.Oh, A.I.Titov and T.-S.H.Lee (2001), *Phys. Rev C* **63** 025201/1-9
[12] A.Ajaka et al. (2006), *Phys. Rev. Lett.* bf 96 1132003/1-4
[13] F.Klein et al. (2008), *Phys. Rev. D* **78** 117101/1-5
[14] G.Penner and U.Mosel (2002), *Phys. Rev. C* **66** 055212/1-32
[15] V.Shklyar et al (2005), *Phys. Rev. C* **71** 055206/1-18
[16] Q.Zhao et al. (2005), *Phys. Rev. C* **71** 054004/1-10
[17] M.Paris (2009), *Phys. Rev. C* **79** 025208/1/14
[18] O.Bartalini et al. (2006), *Eur. Phys. J. A* **26** 399-419

AUTHOR INDEX